Advances in Plant Breeding Strategies

Volume 10

Series Editors

Jameel M. Al-Khayri, Department of Agricultural Biotechnology, King Faisal University, Al-Hassa, Saudi Arabia

Shri Mohan Jain, Department Agricultural Sciences, University of Helsinki, Helsinki, Finland

This book series is intended to provide a platform for plant scientists to present the latest advances in both innovative and traditional methodologies applied in plant breeding for the development of new crop varieties under different environmental conditions leading to sustainable food production, food and nutrition security, food supply chain, germplasm conservation and utilization. This series is an excellent reference source for plant breeders, geneticists, biotechnologists, and molecular biologists using molecular tools, bioinformatics, artificial intelligence, and other emerging technologies. The subject matter of all volumes will relate to postgraduate students specializing in agriculture, biotechnology, and molecular breeding as well as policymakers, seed companies, researchers, and professionals interested in plant genetic improvement.

Jameel M. Al-Khayri · Shri Mohan Jain ·
Suprasanna Penna
Editors

Biodiversity and Genetic Improvement of Medicinal and Aromatic Plants II

Springer

Editors
Jameel M. Al-Khayri
Department of Agricultural Biotechnology
King Faisal University
Al-Ahsa, Saudi Arabia

Shri Mohan Jain
Department of Agricultural Sciences
University of Helsinki
Helsinki, Finland

Suprasanna Penna
Amity Center for Nuclear Biotechnology,
Amity Institute of Biotechnology
Amity University of Maharashtra
Mumbai, Maharashtra, India

ISSN 3004-8737 ISSN 3004-8745 (electronic)
Advances in Plant Breeding Strategies
ISBN 978-3-031-81856-1 ISBN 978-3-031-81857-8 (eBook)
https://doi.org/10.1007/978-3-031-81857-8

This Springer imprint is published by the registered company Springer Nature Switzerland AG
The registered company address is: Gewerbestrasse 11, 6330 Cham, Switzerland

Preface of the Series

Contemporary plant breeders rely on conventional and emerging methodologies to provide sustainable food production worldwide. However, the human population is growing at an alarming rate, especially in developing countries, and food availability to feed new mouths could gradually become a serious problem. Agriculture production is negatively impacted by environmental pollution, rapid industrialization, water scarcity, soil erosion of fertile topsoil, limited areas of potentially arable land, lack of improvement of local plant types, erosion of genetic diversity, and dependence on just a few major crop species for the world's food supplies. Breeding programs of these crops have been very much dependent on the ready availability of genetic variation, whether spontaneous or induced. Plant breeders and geneticists are under constant pressure to sustain food production by using innovative breeding strategies and enhancing minor crops that are well adapted to the marginal land and source of nutrition as well as abiotic and biotic stresses. An inherent limitation in traditional breeding is the introgression of one or few genes in a cultivar is carried out via backcrossing for several generations requiring extended time.

With the development of new molecular tools, molecular marker-assisted breeding helped rapid introgression of transgene and reduction of linkage drag. The continued development and adaptation of plant biotechnology approaches including genetic engineering, genome editing, molecular markers, genomics, proteomics, bioinformatics, artificial intelligence, and nanotechnology have established new tools for the creation, analysis, and manipulation of genetic variation for the development of improved cultivars to achieve agricultural sustainability and food security. A multidisciplinary approach combining traditional plant breeding, mutation breeding, plant biotechnology, and molecular biology would be strategically ideal for developing improved crop varieties to feed the world, provide natural pharmaceutical sources, and supply fiber and ornamental crops for industry, esthetic and environmental suppose. However, breeders are challenged by the constraints of environmental stress, the limitation of arable land expansion, and erosion of genetic resources particularly under the impact of global climate change.

This book series is intended to provide a platform for plant scientists to present the latest advances in both innovative and traditional methodologies applied in plant breeding for the development of new crop varieties under different environmental conditions leading to sustainable food production, food and nutrition security, food supply chain, germplasm conservation, and plant sustainable utilization. This series is an excellent reference source for plant breeders, geneticists, biotechnologists, and molecular biologists using molecular tools, bioinformatics, artificial intelligence, and other emerging technologies. The subject matter relates to postgraduate students specializing in agriculture, biotechnology, and molecular breeding as well as policymakers, seed companies, researchers, and professionals interested in plant genetic improvement.

Al-Ahsa, Saudi Arabia Jameel M. Al-Khayri
Helsinki, Finland Shri Mohan Jain

Preface

Plant breeding approaches have always been the foundations of crop improvement to achieve higher plant productivity. Several climate, agronomic and genetic constraints are limitations, and hence, newer techniques are necessarily required to understand and breed for novel plant varieties towards meeting the challenges of food, nutritional and health security. Biotechnological and genomics advances have revolutionized opportunities for more efficient breeding of crop plants.

Medicinal and aromatic plants (MAPs) play an important role in the health care and industrial sectors. Since ancient time, MAPs have been cultivated for their application the therapeutic interventions of several diseases in traditional and modern health care. However, information on the breeding aspects of these important plant species is not much available, and there is a great need to develop comprehensive knowledge resource. This book has 11 chapters covering research advances in conventional and modern breeding technologies of various medicinal and important aromatic plants individually, including Indian ginseng, Licorice, Marjoram, Moringa, Neem, Opium poppy, Patchouli, Plumbago, St John's-wort, Stevia, and Upastree resin. Chapters comprehensively review the contemporary literature on the subject and reflect the experiences of the authors. Each chapter emphasizes introduction covering related backgrounds and provides in-depth discussion of the subject supported with high-quality color illustrations and relevant data. Chapters conclude with future research perspective and a comprehensive list of pertinent references.

The book provides an excellent reference source for plant biologists, breeders, biotechnologists and geneticists engaged in plant improvement using plant breeding programs. It is suitable for both advanced undergraduate postgraduate students and researchers specializing in agriculture, biotechnology, and molecular breeding as well as entrepreneurs and policymakers concerned with analysis improvement of plant varieties.

The editors extend their appreciation and gratitude to the contributing authors for their efforts and diligence and to Springer team for the opportunity and constant support to publish this book.

Al-Ahsa, Saudi Arabia — Jameel M. Al-Khayri
Helsinki, Finland — Shri Mohan Jain
Mumbai, India — Suprasanna Penna

Acknowledgments The editors extend their appreciation to the support of the Deanship of Scientific Research, Vice Presidency for Graduate Studies and Scientific Research, King Faisal University, Saudi Arabia (Project No. KFU251325).

Contents

Editors and Contributors

About the Editors

Prof. Jameel M. Al-Khayri is affiliated with the Department of Agricultural Biotechnology, King Faisal University, Saudi Arabia. He received B.S. in Biology in 1984 from the University of Toledo, M.S. in Agronomy in 1988, and Ph.D. in Plant Science in 1991 from the University of Arkansas. He is National Correspondent of the International Association of Plant Tissue Culture and Biotechnology. Al-Khayri has authored 130 research articles and 70 chapters and edited several journal special issues. In addition, he edited 30 reference books on plant biotechnology, genetic resources, breeding, genomics and nanotechnology. In addition to teaching, advising and research, he held administrative responsibilities as Assistant Director of Date Palm Research Center, Head of Department of Plant Biotechnology and Vice Dean for Development and Quality Assurance. Al-Khayri served as Member of Majlis Ash Shura (Saudi Legislative Council) for the 2009–2012 term. He is interested in agricultural innovations and sustainability.

Prof. Shri Mohan Jain, (Adjunct) is currently associated with the Department of Agricultural Sciences, University of Helsinki, Helsinki, Finland. He received M.Phil. in 1973 and Ph.D., 1978, Jawaharlal Nehru University, New Delhi, India. He served as Technical Officer, Plant Breeding and Genetics, International Atomic Energy Agency (IAEA), Vienna, Austria, 1999–2005, when the agency was awarded Nobel Peace Prize, 2005. He is on the Editorial Board of *Euphytica*, *In Vitro*, *Propagation of Ornamental Plants*; he is Reviewer in plant cell and organ culture, *Plant Cell Reports*, and *Plants*. His publications are more than 225 peer-reviewed journals, book chapters, and conference proceedings and edited 72 books; he is Invited Speaker and acted as Chairperson in several international conferences worldwide. He is Former Consultant to the IAEA, the European Union, The Government of Grenada, the Iranian Private Company, and the Egyptian Government.

Prof. Suprasanna Penna, Ph.D. (Genetics) is Director of Amity Institute of Nuclear Biotechnology, Amity University of Maharashtra, Mumbai, India. He was formerly with Bhabha Atomic Research Centre, Mumbai, as Head of Nuclear Agriculture and Biotechnology Division. He is also IAEA Expert Consultant. His significant contributions include plant biotechnology, mutation breeding, stress tolerance, and biopolymers for enhancing plant productivity in plants. His research on plant abiotic stress tolerance in crop plants and halophytes led to elucidating mechanisms of salinity tolerance. He is on the editorial board of several international journals and has published over 400 publications. He has edited several books on plant mutation breeding, salinity tolerance, plant genetic diversity, plant biotechnology and plant-metal interactions and was Guest Editor of special issues of International journals.

Contributors

Sawsan Abd-Ellatif Bioprocess Development Department, Genetic Engineering and Biotechnology Research Institute (GEBRI), City of Scientific Research and Technology Applications (SRTA-City), Borg EL-Arab, Alexandria, Egypt

B. Akshita Department of Life Sciences, Christ University, Bengaluru, Karnataka, India

Jameel M. Al-Khayri Department of Agricultural Biotechnology, College of Agriculture and Food Sciences, King Faisal University, Al-Ahsa, Saudi Arabia

Geoffrey Obinna Anoliefo Department of Plant Biology and Biotechnology, University of Benin, Benin, Nigeria

Chole Pranjali Bajrang Department of Life Sciences, School of Sciences, CHRIST (Deemed to be University), Bengaluru, Karnataka, India

Akshatha Banadka Department of Biotechnology and Genetics, School of Sciences, JAIN (Deemed to be University), Bengaluru, India

Versha Dixit Department of Life Sciences, School of Sciences, CHRIST (Deemed to be University), Bengaluru, Karnataka, India

Jozef Fejér Department of Ecology, Faculty of Humanities and Natural Sciences, University of Prešov, Prešov, Slovakia

Nikhil Gaikwad Laboratory of Cytogenetics and Plant Breeding, Department of Botany, Shivaji University, Kolhapur, Maharashtra, India

Amira A. Ibrahim Department of Botany and Microbiology, Faculty of Science, Arish University, Al-Arish, Egypt

Beckley Ikhajiagbe Department of Plant Biology and Biotechnology, University of Benin, Benin, Nigeria

Akbar Karami Department of Horticultural Science, School of Agriculture, Shiraz University, Shiraz, Iran

Kavyashree Kashyap Department of Life Sciences, Christ University, Bengaluru, Karnataka, India

N. Manikanda Boopathi Department of Plant Biotechnology, Centre for Plant Molecular Biology and Biotechnology, Tamil Nadu Agricultural University, Coimbatore, India

B. T. Manjunath Department of Life Sciences, School of Sciences, CHRIST (Deemed to be University), Bengaluru, Karnataka, India

Amit Mirgal Laboratory of Cytogenetics and Plant Breeding, Department of Botany, Shivaji University, Kolhapur, Maharashtra, India

Seyyed Sasan Mousavi Department of Horticultural Science, School of Agriculture, Shiraz University, Shiraz, Iran

Tahereh Movahhed Haghighi Institute of Agriculture, Iranian Research Organization for Science and Technology (IROST), Tehran, Iran

Barka Peter Mshelmbula Department of Plant Science and Biotechnology, Federal University of Lafia, Lafia, Nigeria

Praveen Nagella Department of Life Sciences, School of Sciences, CHRIST (Deemed to be University), Bengaluru, Karnataka, India

M. Nandhini Department of Microbiology and Botany, School of Sciences, JAIN (Deemed to be University), Bengaluru, India

Efeota Bright Odozi Department of Medical Laboratory Science, College of Medical Sciences, University of Benin, Benin City, Nigeria

Rashmi Ramakrishnan Department of Life Sciences, Christ University, Bengaluru, Karnataka, India

El-Sayed S. Abdel Razik Livestock Research Department, Arid Lands Cultivation Research Institute (ALCRI), City of Scientific Research and Technology Applications (SRTA-City), Borg EL-Arab, Alexandria, Egypt

Khaled F. M. Salem Department of Biology, College of Science and Humanitarian Studies, Shaqra University, Qwaieah, Saudi Arabia; Department of Plant Biotechnology, Genetic Engineering and Biotechnology Research Institute (GEBRI), University of Sadat City, Sadat City, Egypt

Wudali Narasimha Sudheer Department of Life Sciences, School of Sciences, CHRIST (Deemed to be University), Bengaluru, Karnataka, India

Yogesh Chandrakant Suryawanshi Department of Yoga and Naturopathy, Vishwakarma University, Pune, India

Varsha Toppo Department of Life Sciences, Christ University, Bengaluru, Karnataka, India

Suraj Umdale Department of Botany, Jaysingpur College Jaysingpur, (Affiliated to Shivaji University), Kolhapur, Maharashtra, India

V. L. Vasantha Department of Life Sciences, School of Sciences, CHRIST (Deemed to be University), Bengaluru, Karnataka, India

Jasini Alexander Wahedi Department of Biological Sciences, Federal University of Health Sciences, Benue State, Otukpo, Nigeria

Prachi Saurabha Zimare Department of Botany, Comrade Godavari Shamrao Parulekar College of Arts, Commerce and Science, Talasari, Palghar, Maharashtra, India

Saurabha Bhimrao Zimare Department of Botany, Sir Parashurambhau College (Autonomous), Pune, India

Abbreviations

CE	Crude Extract
CRICPR/Cas9	Clustered Regularly Interspaced Short Palindromic Repeats/ CRISPR-Associated Protein 9
DNA	Deoxyribose Nucleic Acids
EMS	Ethyl Methane Sulfonate
InDels	Insertion Deletions
LD 50	Lethal Dose 50
MAS	Marker-Assisted Selection
MP	Moringa Plant
PKM	Periyakulam
QTL	Quantitative Trait Loci
RAPDs	Randomly Amplified Polymorphic DNAs
RNA	Ribo Nucleic Acid
RNAi	RNA interference
RNA-Seq	RNA Sequencing
SMRT	Single-Molecule Real-Time
SNPs	Single Nucleotide Polymorphisms
SSR	Simple Sequence Repeat
WBC	White Blood Cell

Chapter 1
Revolutionizing Ashwagandha Breeding: From Traditional to Modern Breeding Techniques for Indian Ginseng, *Withania somnifera* (L.) Dunal

Akshatha Banadka, Varsha Toppo, Kavyashree Kashyap, M. Nandhini, B. Akshita, W. N. Sudheer, and Praveen Nagella

Abstract *Withania somnifera*, known as Ashwagandha, holds great medical significance due to its excellent therapeutic properties. Rich in bioactive compounds with potential health benefits, the plant has garnered substantial market demand. However, overutilization due to increased demand poses a challenge to the sustainable supply of this medicinal herb. To meet the market demands for *W. somnifera*, various breeding strategies are being employed. Traditional breeding methodologies, which include selective and hybrid breeding, are the most commonly utilized methods for the cultivation of *W. somnifera*. The lengthy breeding cycle and low genetic variation of these approaches, however, may restrict the effectiveness of trait development. Modern

A. Banadka
Department of Biotechnology and Genetics, School of Sciences, JAIN (Deemed to be University), Bengaluru 560027, India
e-mail: b.akshatha@jainuniversity.ac.in; akshatha.rlt95@gmail.com

V. Toppo · K. Kashyap · B. Akshita · W. N. Sudheer · P. Nagella (✉)
Department of Life Sciences, Christ University, Bengaluru 560029, Karnataka, India
e-mail: praveen.n@christuniversity.in

V. Toppo
e-mail: varsha.toppo@res.christuniversity.in

K. Kashyap
e-mail: kavyashree.a@mbot.christuniversity.in

B. Akshita
e-mail: akshita.barwad@science.christuniversity.in

W. N. Sudheer
e-mail: wudali.sudheer@res.christuniversity.in

M. Nandhini
Department of Microbiology and Botany, School of Sciences, JAIN (Deemed to be University), Bengaluru 560027, India
e-mail: m.nandhini@jainuniversity.ac.in

J. M. Al-Khayri et al. (eds.), *Biodiversity and Genetic Improvement of Medicinal and Aromatic Plants II*, Advances in Plant Breeding Strategies 10,
https://doi.org/10.1007/978-3-031-81857-8_1

breeding strategies serve as an alternative approach for enhanced production of *W. somnifera*. Tissue culture strategies enable large-scale propagation and production of elite germplasm. Simultaneously safeguarding the plant's genetic diversity through conservation approaches of genetic resources, such as in vitro, in situ, cryopreservation, and gene bank preservation, which preserve the plant's genetic diversity and ensure its long-term availability for breeding and cultivation. Mutation breeding, including polyploid breeding, offers opportunities for generating novel genetic variations. Furthermore, molecular marker-assisted breeding and transgenic breeding through gene editing techniques and genetic engineering facilitate the identification, selection, and improvement of desired traits, thereby serving as efficient breeding programs. Taking into account the existing literature, this chapter focuses on the promotion of ashwagandha cultivation and withanolide production using various plant breeding strategies as well as modern biotechnological tools.

Keywords Ashwagandha · Germplasm · Genetic improvement · Traditional breeding · Modern breeding

1.1 Introduction

Withania somnifera is revered for its medicinal properties and has a long history deeply rooted in traditional Ayurvedic medicine (Gaurav et al. 2023). The withanolides present in the plant contribute to its adaptogenic properties that helps to boost the body's immune system and cope with stress. This medicinal herb aids in lowering blood sugar and fat levels, fights cancer, boosts fertility and promotes overall well-being. Due to its wide range of potential health benefits, companies have launched various ashwagandha-based herbal products. The preference for natural, plant-based solutions to enhance health and well-being has increased the demand for Ashwagandha-based herbal formulations. Thus, breeding has been spurred to meet the growing market requirements and optimize cultivation. Breeding focuses on enhancing the plant's medicinal properties by increasing the concentration of bioactive compounds, such as withanolides. Furthermore, breeding *W. somnifera* aims to develop improved varieties with enhanced quality, disease resistance, adaptability to diverse environmental conditions, and, most importantly, improved yield, thereby playing a crucial role in agricultural improvement. In this context, the current chapter discusses the conventional and modern breeding strategies employed to produce new cultivars of Ashwagandha, along with large-scale propagation techniques to meet the increasing global demand for this herb.

1.1.1 Botanical Classification and Distribution

The plant *W. somnifera* is sometimes referred to as poison gooseberry, winter cherry, Indian ginseng, and ashwagandha (Srivastava et al. 2019). It is a member of the Solanaceae family, which has about 3000 species spread over 84 genera (Afewerky et al. 2021). The ashwagandha bush may reach a height of three to four feet. It is usually upright and tomentose. The leaves of the ashwagandha plant are oval, simple, petiolate, exstipulate, and hairless. The base of the full-bordered leaves is cuneate or oblique, and the tip is acute to thick. Large leaves are arranged in pairs above the vegetative stem. On flower branches, nevertheless, they are placed in the reverse order. Cymose inflorescences that are gamosepalous and light green in hue are the blossoms of ashwagandha.

The flowers have slender filaments and intrinsic epipetalous stamens with oval anthers that protrude from the base of the petals. A gynoecium that is syncarpous features a small, protruding ovary and a long, thin style. As the fruit matures, the calyx, which encloses it, gets bigger and more inflated (Poojari et al. 2019).

Kingdom	Plantae
Division	Angiosperma
Class	Dicotyledons
Order	Tubiflorae
Family	Solanaceae
Subfamily	Solanoideae
Tribe	Physaleae
Subtribe	Withaninae
Genus	*Withania*
Species	*somnifera*

Botanical classification of *Withania somnifera* L. Dunal

According to Srivastava et al. (2018), climatic zones with a dry, subtropical environment are ideal for the cultivation of this crop. Its roots may be found in the Mediterranean area of North Africa and in North Western and Central India (Srivastava et al. 2018). In desert and subtropical regions, ashwagandha is frequently grown, respectively. It is found in Africa, Asia, and Europe. It is found in African countries including South Africa, Morocco, Congo, Egypt, and Jordan in the Middle East, as well as Asian nations like Afghanistan, India, and Pakistan (Kumar et al. 2019) (Fig. 1.1).

Two species of the *Withania* genus are known in India: *Withania coagulans* (L.) Dunal (Panir) and *Withania somnifera* (L.) Dunal (Ashwagandha). The plant *Withania coagulans* (L.) grows wild. The states of Madhya Pradesh, Gujarat, Rajasthan, Maharashtra, Uttar Pradesh, and Punjab are the main producers of ashwagandha. Because rainfed farming requires little irrigation, this late kharif crop thrives in desert environments. The districts of Neemuch and Mandsaur in Madhya Pradesh,

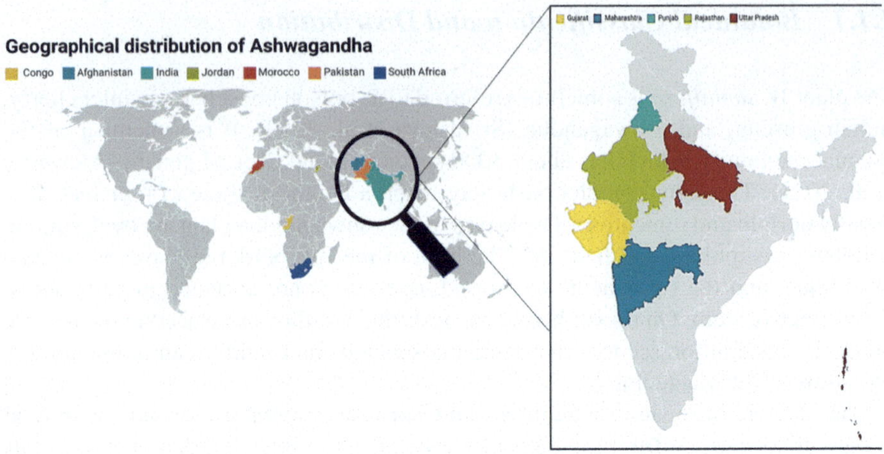

Fig. 1.1 Distribution of *Withania somnifera* in various geographical locations

as well as Kota, Jhalawar, Pratapgarh, Chittorgarh, and Barand in Rajasthan, are where it is mostly grown in marginal regions (Joshi et al. 2015).

1.1.2 Importance of Ashwagandha (Economic and Health Benefits)

Numerous metabolites, such as steroids, fatty acids, alkaloids, sugars, saponins, glycosides, terpenoids, tannins, and coumarin, are abundant in ashwagandha (Poojari et al. 2019). Withanolides (including 2-D withastramonolide and withanolides A, B, D, F, and G) are mostly derived from ashwagandha. Furthermore, due to their pharmacological potential, Withanamides (A to F, I to R), Withanosides (I–VI and IX), and alkaloids (withasomine, withanine, somniferine, anhygrine, cuscohygrine, tropine, pseudotrophine, tropanol, anaferine, isopelletriene, choline, pseudotropanol, etc.) have been used in traditional medicine for 300 years (Arpan et al. 2014). Because of its anticancer properties, withaferin A, one of the well-known withanolides, became an important medicinal component of ashwagandha. Some herbalists refer to ashwagandha as Indian ginseng because the plant is used in Ayurveda in a way that is similar to how ginseng is used in traditional Chinese medicine (Ekka et al. 2021).

Because ashwagandha root extract contains a wide range of minerals and phytochemicals, it is used as a dietary supplement and for health restoration. The roots are used to extract and identify a variety of withanolides, around 12 alkaloids, sitoindosides, and withanamides (Trivedi et al. 2017). According to Abdelwahed et al. (2023), the plant's leaves also include sitoindosides, reducing sugars, withanamides, glycosides, starch, withanicil, peroxidases, dulcitol, 2, 3-dihydroxycinnamic acid, benzyl alcohol, phenyl acetic acid, and benzoic acid.

Withasomnilide, somniferawithanolide, somniferanolide, withasomniferanolide, and somniwithanolide are among the metabolites found in stem bark. According to Saleem et al. (2020), metabolites found in fruit include linoleic acid, tetracosanoic acid, palmitic acid, oleic acid, and elaidic acid. Owing to the different metabolites, it is widely used in many different medicinal systems, such as Siddha, Ayurveda, and Unani.

India's ancient medicine, Ayurveda, dates back to 6000 BC, according to the 1949 Charaka Samhita. Throughout these 6000 years, ashwagandha has been used as a Rasayana treatment. The anthelmintic, astringent, diuretic, narcotic, stimulant, tonic, and thermogenic qualities of ashwagandha root make it valuable. Ashwagandha gets its name from the root's horse-like ("ashwa") smell and its potential to give people equine vigour. It is widely used to treat a variety of illnesses, including goitre, rheumatism, leucoderma, constipation, insomnia, neurological breakdown, and childhood emaciation (Singh et al. 2011). To cure joint inflammation, a paste produced by crushing roots with water is given (Bhandari 2021). The paste is also useful for treating inflammation, ulcers, and carbuncles locally. The root is used in conjunction with other drugs for treating snake venom and scorpion stings. It is also used to treat leucorrhoea, boils, piles, worms, and flatulent colic (Singh et al. 2011). The greatest ashwagandha type on the market is Nagori Ashwagandha. The effects are most when using fresh powdered ashwagandha (Singh et al. 2011). Additionally, the herb possesses strong immune-modulatory, antioxidant, anti-cancer, anti-stress, hepatoprotective, and rejuvenating properties (Manivel et al. 2017). Apart from its medicinal application, it finds application in phytoextractions as a bioremediation technique (Abhilash et al. 2008).

Plant-based raw materials are the main source of preparations and formulations utilized in the Unani, Siddha, and homeopathic systems. Nearly two thousand kinds of therapeutic plants have been identified by the Indian medical system, or Ayurveda. The traditional raw material sources—the forest, in particular—for Ayurvedic treatments are under a lot of stress, due to the need for and widespread revival of herbal medicine. It is obvious that there is a greater demand for raw materials than there are available natural sources, which pushes the business to look for alternate sources like farmed fields. This has led to the commercial cultivation of a number of well-known Ayurvedic medicinal plants, including Safed Musli, Chitrak, and ashwagandha.

Nevertheless, the genetic resources of the species are threatened by unethical and unscrupulous gathering practices. As a result, it is now illegal to gather several therapeutic plants, which are now regarded as critically endangered. Putting into practice the plan for medicinal plant use and sustainable plant conservation in the vicinity of endangered species (Shinde et al. 2015).

1.1.3 Domestication, Selection, and Early Improvements

It has been grown in India for many centuries in order to meet the demands of both home and foreign markets because of its broad application in the food, medicinal, cosmetic, and agricultural industries. Although 7000 tonnes of ashwagandha are needed annually, only 1500 tonnes are produced. The annual plants that mature in 4–6 months are presently available since this species has been domesticated and cultivated over time (Rama Reddy Nagaraja Reddy and Deore 2017).

With so many people interested in this plant and so much demand for its roots, there's no shortage of space to grow it commercially. The crop is currently inexpensive to farm in rainfed environments, and it may be made compatible with traditional crops by crop sequencing. Other options for cultivation include selling opportunities for both leaf and seed, profitable by-product utilisation, and the possibility of adding value to increase profits. Research into large-scale exports of roots and value-added commodities is necessary, despite these advantages, the amount of roots that are now exported is little. In order to generate high yielding varieties of this vital medicinal crop, minimal breeding work has been done. According to Rajeswara Rao et al. (2012), there is a considerable likelihood that genotypes with high dry root production and withanolide content would emerge. Therefore, one long-term breeding objective has been to measure the feature of genetic variation. Evaluating the variety of ashwagandha accessions is crucial for improving this crop (Joshi et al. 2015).

1.2 Current Cultivation Practices and Challenges

1.2.1 Current Cultivation Practices

Ashwagandha is a drought-tolerant annual plant that preferably grows in dry, semi-arid tropical, and sub-tropical regions. The plant grows in slightly basic to neutral soil with a pH of 7.5–8. Sandy, loam soil, and red or black soil with a light texture and adequate drainage are beneficial for the plant's growth. This particular crop needs a dry season for the majority of its growth stage, with the added benefit of one or two late winter rainfalls that promote root growth. It is commonly sown as a Kharif crop during the latter phase of the monsoon season, especially in semi-tropical zones with precipitation levels ranging from 500 to 750 mm, making it ideal for rain-dependent agricultural practices. The crop tolerates temperatures as low as 10 °C and thrives in reasonably dry conditions throughout its growth period. It is temperature tolerant from 20 to 38 °C. It displays versatility at altitudes ranging from sea level to as high as 1500 m (Poojari et al. 2019) (TNAU 2013). The best season to grow ashwagandha is between April and July.

W. somnifera is propagated through seeds. The preferred methods for sowing the seeds include the line and the broadcasting method. However, the line method is preferred over other methods due to its ability to facilitate proper intercultural practices and enhance root production. The seeds are typically sown during the months of late June-Early August at a depth of 1–3 cm and with rows spaced 5 cm apart. If there is heavy rainfall, the sowing can be done up to September. Thus, it is vital to remember that as "Ashwagandha" being a rainy season crop, the date of planting depends on when the local monsoon season begins. After being sown, the seeds have to be covered with light soil. The seeds are pretreated with Dithane, Indofil, or Thiram at a dosage of 3 g/kg of seed. It is then sown at the rate of 500–750 g of seeds/hectare of field. Germination typically initiates within 6–7 days of sowing, and the germination completes within ten days after sowing. Light showering after sowing ensures better germination. Thinning of plants sown through broadcasting or line furrows should be carried out to maintain a plant population of approximately (30–60 plants/m^2). The sufficiently tall seedlings are transplanted into the field around 6 weeks after sowing and are transplanted in 60X60 cm spaced well-prepared land. The choice of plant density varies depending on the soil's characteristics and fertility. Post transplantation, the successful establishment of seedlings can be aided by light showers (TNAU 2013; Shinde et al. 2015). Figure 1.2 represents various growth stages of *W. somnifera.*

For cultivation of ashwagandha, it is recommended to utilize organic fertilizers like Farm Yard Manure (FYM), Green Manure, and Vermi-Compost instead of spraying chemical fertilizers and pesticides. Frequent rainfall eliminates the need for irrigation, while excessive rainfall or water might harm the crop. However, in dry lands with less rainfall, irrigation may be applied. Around 150–180 days after seeding, the crop typically begins to flower and bear fruit in December and is suitable for harvest between January and March. The drying of the leaves and the emergence of yellow–red berries are signs that the plant is mature and is now ready to harvest. During harvest, the whole plant is uprooted to get the roots (TNAU 2013; Shinde et al. 2015).

Fig. 1.2 **A** Seeds of *Withania somnifera*, **B** and **C** growth stages of *Withania somnifera*, **D** Habit of *Withania somnifera*

The roots are then detached from the aerial portions by cutting the stem approximately 1–2 cm above the crown. Subsequently, the roots can then be dried in the sun directly or sliced into smaller pieces approximately 7–10 cm. Berries are manually plucked, dried, and then crushed to extract the seeds. The dried roots (whole/small pieces), require cleaning, trimming, and grading. The thin, fragile lateral rootlets and the soil adhering to the roots are cut off using a club. The remaining stem, lateral branches, and root crowns are all carefully pruned away using a knife. In commercial farming, one hectare of land can produce between 50 and 75 kg of seeds and 3–5 quintals of dried roots on average (Kothari et al. 2003; Inagaki 2017).

1.2.2 Current Agricultural Challenges Faced by Ashwgandha

While being able to grow in a variety of climates, ashwagandha is still prone to face some agricultural challenges associated with its cultivation and production. Ashwagandha prefers to grow well in warm temperatures and is frost sensitive thereby making its cultivation challenging in colder climates. The plants are susceptible to root rot when grown in soil that retains moisture or waterlogged soil. Further, the plants are prone to be attacked by leaf-eating insects, shoot borers, beetles, aphids, and mites. Aphids and other pests are controlled with foliar sprays of 0.3% Kelthane and 0.5% Malathion applied every 15 days. Shoot borers can be controlled by spraying Sumicidin 10 ml/L and mite can be controlled by spraying 10 ml/L ethion. Like any other crop, ashwagandha is susceptible to diseases such as seedling rot and blight that kills seeds and seedlings and is caused by insects or nematodes. Seedling rot can be treated organically using neem. Leaf spot is another bacterial, fungal or a viral plant disease. To prevent leaf spot Dithane M-45 can be sprayed at the rate of 3 g/L of water. Similar to this, spraying 0.3% Mancozeb or Copper oxychloride or treating seeds with Thiram/Mancozeb can suppress fungi. As a preventative precaution to safeguard the crop from pests and illnesses, plants grown organically are treated with natural insecticides including garlic and chilli extract, custard apple leaf decoction, cow urine, and neem products (Rajeswara Rao et al. 2012; Apnikheti 2023).

1.2.3 Genetic Improvement

Genetic improvement is one of the most effective methods of overcoming agricultural constraints. Plant breeders can develop crop varieties with desired traits that improve productivity, quality, and resilience to various constraints through genetic improvement. Here are some of the traits that can be considered for genetic improvement:

- Higher yield (root, stem, leaf, and fruits (seeds)
- Improved withanolide content

- Disease resistance
- Stress tolerant with climatic adaptability
- Size and morphology for facilitating cultivation.
- Improve nutritional content.

1.3 Germplasm Biodiversity and Conservation

1.3.1 Germplasm Biodiversity

The states of Madhya Pradesh, Rajasthan, Gujarat, Maharashtra, Punjab, and Uttar Pradesh are the main locations for ashwagandha cultivation, while *Withania coagulans* (L.), the wild equivalent, are found in natural settings. Rainfed agriculture is an option for this crop, which is normally planted as a late kharif crop. It grows well in arid regions and requires little irrigation. In areas like Neemuch and Mandsaur in Madhya Pradesh and Kota, Jhalawar, Pratapgarh, Chittorgarh, and Baran in Rajasthan, the majority of farming takes place on marginal lands. The most widely grown medicinal herbs is ashwagandha, which is grown on 15,000 hectares and produces 60,000 quintals at a productivity rate of 4.00 quintals a hectare (Joshi et al. 2015). Although this species grows naturally as a perennial, via domestication and cultivation, beneficial features have been selected, resulting in the formation of annual plants that mature in 4–6 months. Although they have been grown for more than 200 years in India, serious breeding attempts didn't start until a few decades ago. Consequently, certain varieties have been approved for commercial cultivation, including JA20, JA134, and JA100 from Rajmata Vijayaraje Scindia Krishi Vishwa Vidyalaya (RSVKVV) in Mandsaur, India, and Rakshita and Poshita from the CSIR-Central Institute for Medicinal and Aromatic Plants (CIMAP) in Lucknow, India (Rama Reddy Nagaraja Reddy and Deore 2017). Throughout history, herbal plants have been crucial in providing medicinal remedies for human health. Even today, nearly 80% of the global population traditionally use the basic healthcare strategy in underdeveloped nations that has been centered on plant-based medicinal systems. The global market for herbal products is presently valued at $ 6.2 billion, with projections indicating a significant growth to reach $5 trillion by the year 2050 as reported by the World Health Organization (WHO) (Manivel et al. 2017). Medicinal plants possess significant potential as renewable natural resources. Consequently, the sustainable utilization and preservation of these plants require comprehensive, long-term, and scientifically driven action plans. Such strategies should encompass essential elements such as preservation, protection, exploitation, sustainable utilization and maintenance.

1.3.2 Genetic Resource Conservation Approaches

Superior pure lines can be utilized as cultivars, as parents in hybridization, and as valuable research tools for mutation studies and various biological aspects such as medicine, biochemistry, immunology, and physiology. Additionally, the genetic diversity among pure lines significantly influences the selection of parents for hybridization purposes (Manivel et al. 2017). Despite the availability of various molecular marker techniques that have proven effective in studying diversity and establishing relationships between cultivars, there is scarcity of scientific knowledge on the genetic variance of *W. somnifera*. PCR-based markers have been widely utilized to collect data on evolutionary history, genetic variation, and phylogenetic relationships among plant species. However, in the case of *W. somnifera*, limited studies have been conducted in this area (Parita et al. 2018).

1.3.3 Germplasm Diversity of Ashwagandha

Ashwagandha, a versatile medicinal herb, was cultivated in India centuries ago to cater to the growing demand. Although ashwagandha has been grown for over two centuries in India, focused breeding attempts didn't start until the last few decades. As a result, only a few varieties, such as JA134, JA20, and JA100 from Rajmata Vijayaraje Scindia Krishi Vishwa Vidyalaya (RSVKVV) in Mandsaur, India, and Rakshita and Poshita from the CSIR-Central Institute for Medicinal and Aromatic Plants (CIMAP) in Lucknow, India, have been released for commercial cultivation (Shinde et al. 2015). Ashwagandha cultivars have been established by repeatedly self-pollinating a small number of local cultivars or wild-sourced plants from the fields of local farmers (Fig. 1.3). The vital objective of ashwagandha breeding is to enhance the production of desirable secondary metabolites in its roots. Efforts are focused on developing varieties with increased levels of these beneficial compounds to maximize their medicinal properties (Manivel et al. 2017) (Table 1.1).

Accurately determining the amount and distribution of genetic variation can lead to improved crop performance. The examination of genetic variation within cultivars, the discovery of varied parent combinations to yield offspring with high genetic variability for subsequent selection, and the introgression of desirable genes from a variety of germplasm into the existing genetic pool are three of these methods. These are only a few instances of how accurate assessments of genetic diversity can be applied to efforts to enhance crops (Mohammadi and Prasanna 2003). Pure lines are most important in self-pollinated species, whereas inbreds are essential as the main genetic stocks for improvement in crop-pollinated species. It can be quite advantageous in many ways to have a good understanding of the genetic links between these pure or inbred lineages. It aids in the strategic planning of crosses, facilitating the assignment of lines to specific heterotic groups, and ensuring accurate identification for plant varieties protection purposes. Such knowledge of genetic

Fig. 1.3 CIMAP-Pratap and CIMAP-Chetak varieties of ashwagandha. *Source* Venugopal et al. (2022). https://doi.org/10.1016/j.indcrop.2022.115696. Accessed on 2 July 2023; reproduced with permission from the publisher; License No: 5580691109049

relationships among inbreds or pure lines is valuable in enhancing crop improvement efforts (Hallauer et al. 2010). Selecting parents with different genetic backgrounds and exceptional genotypes is necessary for effective hybridization programs. This selection mechanism guarantees the existence of strong heterosis (hybrid vigor) and a large variety of transgressive segregants, or people who display features beyond the range seen in their parents (Fig. 1.4). These qualities are essential for promoting cultivar growth. Hybridization programs can maximize the potential for desirable features and speed up the breeding process to produce new cultivars by selecting genetically diverse parents and superior genotypes (Chauhan et al. 2018).

Simple sequence repeats/microsatellites, which consist of tandem repeats of DNA motifs ranging from 1 to 6 nucleotides in length have garnered substantial significance in the fields of plant breeding and genetics. This is primarily due to their multi-allelic nature, hypervariability, co-dominant inheritance, widespread occurrence, reproducibility, chromosome-specific localization, comprehensive genome coverage (including organellar genomes), compatibility with automation, and high-throughput genotyping capabilities. These attributes collectively contribute to the popularity and utility of simple sequence repeats/microsatellites as valuable tools in various applications related to plant genetics and breeding (Mondini et al. 2009).

The utilization of Expressed Sequence Tag (EST) databases enables the generation of markers that are particularly associated with a specific trait. These markers can exhibit transferability between closely related genera or species, highlighting their

Table 1.1 Pure lines with distinct physical traits derived from the JA134 population (Manivel et al. 2017)

No.	Distinct morphological character	Name of pure line
1.	Dwarf plant (< 30 cm)	DWS327
2.	Procumbent type plant	DWS6
3.	Male sterile plant	DWS10
4.	Straight margin/non-wavy margin	DWS12
5.	Yellowish color leaf	DWS16
6.	Compact plant with less fibrous root	DWS23
7.	Downward curling leaves	DWS37
8.	Triangular shape leaves	DWS61
9.	Creamy colour berry	DWS65
10.	Non-wavy and elongated leaves	DWS69
11.	Broad leaves	DWS110
12.	Big size flowers	DWS112
13.	More hairs	DWS114
14.	Open calyx/berries visible	DWS117
15.	Yellow colour young leaves	DWS127
16.	Red berry, non-wavy leaves, tall and leaf margin entire	DWS131
17.	Yellow green leaves	DWS144
18.	Branches on every node	DWS139
19.	More wavy leaves	DWS195
20.	Profuse branches	DWS207
21.	Spreading type (Procumbent)	DWS217
22.	More hairs on stem and leaves	DWS260
23.	Long leaves	DWS268
24.	Orange colour berries	DWS135

significance as a valuable resource for identifying candidate genes. In the study by Varshney et al. (2005), the genetic diversity among 36 ashwagandha genotypes was assessed using EST-SSR markers. By employing these markers, the study aimed to assess the variations and relationships within the ashwagandha genotypes at the genetic level.

Each marker's Polymorphic Information Content (PIC), ranged from 0.21 to 0.86 with an average PIC value of 0.732. The resolving power was calculated as 3.23 and the percent polymorphism was found to be 94.71. By employing Jaccard's dissimilarity coefficients, the ashwagandha genotypes could be categorized into three distinct groups. The cophenetic correlation coefficient, which measures the consistency of the clustering, was determined to be 0.933. The genotypes utilized in the recent investigation performed by Parita et al. (2018) exhibited significant genetic divergence, with their genetic similarity ranging from 0.069 to 0.846. The ashwagandha

Fig. 1.4 Phenotypic variations across different *W. somnifera* accessions: **A** berries, **B** branching, pod and berry color, **C** harvested roots. *Source* Anuj Kumar et al. 2023: https://doi.org/10.1007/s13237-023-00423-9. Accessed on 2 July 2023; reproduced with permission from the publisher; License No: 5580690375693)

genotypes MWS-205-3-2 and MWS-322-1-2 shared the most similarities among these genotypes, whereas MPAS-15-3-1 and RAS-23-2-1 shared the least similarities. According to a NJ (neighbor-joining) analysis, the RAS-11-3-2 genotype may be the oldest of all the genotypes under investigation. Due to their propensity to traverse species borders, EST-SSRs from the Solanaceae family can be used to explore the genetic diversity of *W. somnifera.*

1.3.4 Genetic Resource Conservation Approaches (Including in Vitro in Situ, Cryopreservation, and Gene Bank)

W. somnifera is an important medicinal plant that is used extensively in the global pharmaceutical sector. In its native environment, Withania mostly spreads through vegetative means. But because of its comparatively slow rate of multiplication, it is difficult to supply the high-quality planting material needed for commercial cultivation. The *W. somnifera* plant is an important resource for which conservation and management plans must be thoughtful, effective, and long-lasting. Therefore, to prevent the plant from going extinct, urgent conservation action is required. The preservation of extremely medicinal plants must be given top priority in conservation plans (Aslam et al. 2017).

According to the Global Forest Resources Assessment 2020, there has been a net loss of approximately 7 million hectares of forests annually worldwide between 2010 and 2020. Alarmingly, several hundred species are disappearing from the Earth each day, posing a grave threat to the survival of humanity. These distressing trends highlight the urgent need for conservation and sustainable practices to protect our planet's biodiversity and secure a viable future for all. The IUCN's 2020 Red List of Threatened Plants reveals a troubling fact: over 32,000 plant species, including 20,339 tree species, face extinction risk. Among these endangered species, more than 900 tree species are now listed in the CITES appendices, resulting in restricted commercial activities related to them (Food and Agriculture Organization 2020).

1.3.4.1 In Vitro Conservation

In the mid-1970s, in vitro conservation emerged as a recommended approach. However, alternative methods are necessary for species that vegetatively propagate (e.g., potato), lack viable seeds (e.g., banana), or produce recalcitrant seeds. In these cases, in vitro plant conservation is considered the most effective option. This procedure entails the aseptic removal of cells or tissues from the donor plant and their growth on an artificial medium in a suitable container under regulated climatic conditions. In vitro conservation provides a means to maintain and preserve plant species that face reproduction and seed viability challenges, offering a valuable tool for long-term preservation (Ding et al. 2008).

The preservation of *W. somnifera* (also known as ashwagandha) through tissue culture methods using in vitro-derived germplasm has been successfully achieved. In order to facilitate this investigation, an improved technique for rapid in vitro multiplication of the species was developed as a necessary prerequisite. The study focused on developing a quick, efficient, and time-saving strategy for regeneration of *W. somnifera.* The MS (Murashige and Skoog) medium showed to be the most effective in stimulating shoot multiplication when used with particular concentration and combinations of cytokinin and auxin. Shoot buds were directly induced from cotyledon explants, and within a period of six weeks in culture, numerous propagules developed in clusters, displaying a high sprouting ability. This optimized approach offers a valuable means to propagate *W. somnifera* rapidly and lays the foundation for further investigations and applications in the field of plant tissue culture (Aniel Kumar et al. 2013).

1.3.4.2 Cryopreservation

Cells and tissue preservation at very-low temperatures is achieved through cryopreservation methods, utilizing substances like deep freezers with the lowest temperature (− 80 °C), solid carbon dioxide (− 79 °C), liquid nitrogen (− 196 °C) or vapor nitrogen (− 150 °C). Subjecting the cells to such extremely low temperatures can be effectively inactivated and maintained in a viable state for an extended period (Kaviani 2011). Two advanced cryopreservation techniques are employed to reduce damage to cells caused by the formation of ice crystals. One approach involves vitrification, which converts cellular water into a non-crystalline vitreous state using cryoprotective chemicals (Katkov 2012). *W. somnifera* is preserved genetically using a process called cryopreservation, which is done at extremely low temperatures—typically below − 150 °C. Using this method, plant tissues—such as embryos or shoot tips—are meticulously collected and preserved in a cryogenic setting. The first step in the procedure is to choose healthy plant material. Next, little tissue samples are removed. Cryoprotectants are applied to these samples to reduce cellular damage during freezing and thawing. The tissues are then carefully chilled using regulated freezing rates and placed in liquid nitrogen to be stored for an extended period of time. For the preservation of *W. somnifera*, cryopreservation provides a number of benefits. It guarantees the preservation of important genetic features by enabling the upkeep of a varied gene pool. It also acts as a safety net against losses brought on by illnesses, natural catastrophes, or other unanticipated circumstances. Additionally, material that has been cryopreserved can easily be resurrected and utilized for research or propagation as necessary. As there are limited or no studies on ashwagandha cryopreservation, there is scope for application of one of the above mentioned cryopreservation methods. However, it is important to optimize these cryopreservation methods as per the specific requirements of the plant. The optimization includes parameters such as cryoprotectant concentrations, exposure times, dehydration procedures, and the choice of suitable method (encapsulation or vitrification techniques).

1.3.4.3 Gene Banks

Out of the 5.43 million accessions stored in gene banks worldwide by the end of 2019, only 5.8% were stored as living field collections, and the majority of them were cryopreserved as DNA. This indicates that a small proportion of the stored genetic resources are actively grown and maintained in field conditions. However, it is worth noting that as of December 2019, a total of 290 gene banks across the globe were able to safeguard 96,000 accessions belonging to 1700 critically endangered species identified by the IUCN. Important crop plants and their wild relatives are included in these collections, which play a vital role in ensuring food security at both local and global levels. The efforts of gene banks in protecting and conserving these endangered species and their genetic diversity are crucial for the long-term stability and sustainability of agriculture and food production systems (Byrne et al. 2018).

The National Bureau of Plant Genetic Resources (NBPGR) has played a vital role in enhancing and diversifying multiple crop plants through its germplasm transfer and collection efforts (Dempewolf et al. 2023). This has been accomplished by the bureau through the importation of germplasm from various research facilities based abroad, as well as the acquisition of germplasm from India and other nations. The NBPGR's activities have led to the acquisition of 452,212 accessions, a large quantity. Of these accessions, 2194 are kept in the DNA gene bank, 11,932 are cryopreserved, and 1916 are maintained using in vitro methods. These accessions represent 1762 plant species, a broad spectrum of plant species. An important resource for crop development, agricultural diversification, and conservation, the NBPGR's enormous collection of genetic resources is the cornerstone of India's resilient and sustainable agricultural practices (Walters et al. 2013). A gene bank for various cultivars of *Withania somnifera* is maintained under the ICAR National Fellow Project by the ICAR National Bureau of Plant Genetic Resources in Delhi, India. The ICAR-NBPGR National Gene Bank maintains approximately 413 accession numbers (unique identifiers assigned to records in a sequence) for different cultivars of ashwagandha collected from various states in India. Cultivar names such as Ashwagandha, Asgandh, Gado, Asavagandha, Kala Aak, Kuvarpathu, Amukkiran Kizhangu, AGB002, NPIT and LI002, along with specific identity cultivars like DWS-327, Gujarat Anand Ashwagandha 1 (GAA1/AWS1), Withania mandsaur-29, 30, 39, 31, 34, 35, 6, 16, 25, 17, 9, Withania nauni-9, 13, Withania Rishikesh-1, Withania Gopeshwar-1, Withania Bhopal-1,3, Withania Jaipur-7a, 1d, 1b, 7b, Raj Vijay Ashwagandha-100, SKA 11, HWS 08-18, and HWS 12-12 were collected and preserved from November 1982 to January 2022.

1.4 Traditional Breeding

Nowadays, the creation of a superior variety of rare medicinal plants and herbs like *Withania somnifera* is the primary goal of any plant scientist. In everyday life, both traditional and unconventional ways are used to modify or change the qualities or traits of these essential Ayurvedic plants. Conventional breeding approaches include germplasm evolution, mutagenesis, hybridization, interspecific hybridization, etc. (Srivastava and Sangwan 2020). Crop improvement efforts rely on genetic variation, providing diverse genotypes for breeding high-yielding cultivars. Nevertheless, due to the limited success of traditional breeding reliant on observable traits for low metabolite production, contemporary breeding methods, such as those employing DNA-based molecular markers like RAPD (random amplified polymorphic DNA), are increasingly employed to assess genetic diversity within plant species (Mirjalili et al. 2009).

1.4.1 Traditional Breeding Methodologies (Selective and Hybrid Breeding; Drawbacks of Traditional Breeding Strategies)

Ashwagandha is an extremely popular crop both worldwide and domestically, but its supply is limited because of the insufficient productivity of the Ashwagandha wild germplasm. Maximizing its productivity by employing a breeding strategy focused on harnessing hybrid vigor and targeted selection is essential to surpass these yield constraints (Ahmed et al. 2023).

1.4.1.1 Selective Breeding

When a population possesses sufficient inherent or deliberately induced variation, breeders can initiate the process of enhancing the population or engaging in selective breeding. In a population, selection accumulates the genes that regulate the desirable trait manifestations. It maintains fitness while promoting uniformity via inbreeding (De 2017). Because numerous species remain in their natural state and demonstrate significant genetic diversity, opting for selection within wild populations or landraces stands out as the preferred method in medicinal plant breeding. Only a few selection stages produce already satisfactory results. Common selection strategies in medicinal and aromatic plant breeding include mass selection, recurrent selection, pedigree, and clonal breeding (Saidaiah 2018).

Certain improved varieties of Ashwagandha were also cultivated through half-sib selection. Rather than solely relying on the physical characteristics of the parent plants, the selections predominantly hinge on the results of offspring assessments. In unreplicated progeny rows, seeds from selected half-siblings are grown, enabling selection based on chance pollination from the population (this means that only the female parent is known and selected, hence the term "half-sib"). A portion of the seed is planted to evaluate each plant's yielding ability, breeding value ability, or breeding value (Saidaiah 2018).

1.4.1.2 Hybrid Breeding

Hybrid breeding is frequently utilized in crucial agricultural and horticultural crops, whereas its adoption in Medicinal and Aromatic Plants (MAPs) such as *W. somnifera* is still nascent (De 2017). Because ashwagandha is a self-pollinated crop, natural variability tends to diminish over time owing to selection. As a result, hybridization can be used to increase genetic diversity (Vekariya et al. 2022). The remarkable performance of hybrid varieties is attributable to hybrid vigor and homogeneity, and they also come with built-in protection for the rights of plant breeders by preventing unauthorized seed propagation, which can result in the growth of unwelcome segregating populations. In order to produce hybrid seeds, carefully chosen parents with high combining ability are crossed, producing the F1 generation. Good combining ability suggests that the F1 children of these outstanding parents function remarkably well. Certain parents' F1 seed production demands regulated pollination (De 2017). To lower the cost of emasculation, lines with genetic-cytoplasmic pollen sterility should be used for hybrid breeding or outbreeding. Hybrid varieties require the development of distinct lines: (a) male fertile with maintenance properties to sustain the male sterile line; (b) male sterile; (c) male fertile pollinator with strong combining ability for F1 seed production; and (d) if generative organs are harvested and the pollinator parent lacks restorer properties, a male fertile restorer must be included in the F1-seeds to ensure adequate seed set in the production field. There has been advancement in implementing a hybrid cultivar system (De 2017).

Crop hybrids must additionally exhibit heterosis, sometimes referred to as hybrid vigour, in order to be taken into consideration for production. The extent of heterosis significantly influences the success of a hybrid, and those exhibiting substantial heterosis are consistently favored (Ahmed et al. 2023). Commercially, hybrid vigor exploitation could aid in breaking through the yield barrier. Ahmed et al. (2023) studied the 45 F1 crosses of ashwagandha using a line x tester mating arrangement consisting of three testers and fifteen lines, during Kharif 2018–19 and observed the 11 different yield-attributing factors to calculate the amount of heterosis. The findings indicated a considerable amount of diversity within the breeding material and highlighted three promising crosses: UWS-305 UWS-10, UWS-301 RVA-100, and UWS-305 RVA-100, which showed superior heterotic characteristics for root yield at harvest. Additionally, three excellent hybrids, UWS-305 RVA-100, UWS-314 RVA-100, and UWS-309 RVA-100, were identified for total alkaloid content. The combination of RVA-100 and UWS-305 demonstrated that combining heterosis in these crosses leads to highly positive and valuable outcomes, suggesting their potential advantage in ashwagandha breeding efforts (Ahmed et al. 2023).

Many other studies have discovered that heterosis breeding can aid in the genetic improvement of Ashwagandha (Vekariya et al. 2022). In one of their investigations, four lines and seven testers were crossed in a line tester mating design to evaluate crossability success, level of heterosis, nature of gene activity, and the ability of parents and cross to combine. Therefore, their findings indicated a significant genetic variation among the parents and their crosses. *W. somnifera* uses mixed mating, with 66% self and 34% cross-pollination, and a thorough grasp of its reproductive biology and the breeding system would be advantageous in the creation of both its recombinant and autogamous seeds (Mehta and Raina 2018).

1.4.2 *Drawbacks of Traditional Breeding and Potential Advancements in Plant Breeding*

With conventional breeding techniques, *W. somnifera* encounters specific obstacles including reduced seed viability, poor germination rates, ineffective propagation procedures, and decreased withanolide accumulation in wild plants. Additionally, the genetic and genomic data of *W. somnifera* is meager. Conventional breeding which relies on phenotypic selection makes it difficult for the breeders to recognize and choose out individuals with desired genetic features without having access to full genetic and genomic data (Lamichhane and Thapa 2022). Therefore, the assessment of genetic diversity or heterosis potential for root yield, withanolides, biomass and, total alkaloids is currently conducted using advanced breeding techniques such as RAPD, RNA sequencing, and similar methodologies (Arpan et al. 2014). Several biotechnological and genetic engineering approaches like plant tissue culture, modern breeding strategies, elicitation methods, and artificial stressed conditions are nowadays being explored to boost biomass and bioactive chemicals. Given that traditional breeding methods fall short in meeting the requirements for large-scale biomass production (Sharma et al. 2023).

1.5 Tissue Culture Studies for *Withania somnifera*

W. somnifera, is a renowned botanical species within the ancient medicinal practices and has been employed in various formulations for human well-being. It is rich in a wide range of phytochemical compounds, and extensive research has been conducted to investigate its pharmacological properties (Dar et al. 2015). To meet the rising demand for ashwagandha as a pharmaceutical ingredient, tissue culture techniques have been implemented to enhance biomass production. Furthermore, employing tissue culture strategies facilitates the extraction of both abundant and high-quality bioactive compounds from ashwagandha, which possess significant medicinal value. The limited viability and inefficient seed germination of ashwagandha further emphasize the efficacy of tissue culture techniques as a viable alternative for plant cultivation. *Withania* is well suited for regeneration using both direct and indirect modes of approaches (Shasmita et al. 2018).

1.5.1 Direct Organogenesis

The process of direct organogenesis facilitates the direct formation of specific organ cultures from cultured explants. This approach is particularly useful in tissue culture studies as it provides a streamlined and reliable method for obtaining organ cultures with desired characteristics. Multiple shoots were successfully induced on MS (Murashige and Skoog) medium fortified with 2.0 mg/L of BAP and 2.0 mg/L of IAA when shoot tips were employed as explants. Additionally, the GA3-supplemented medium helped in shoot elongation at 0.3 mg/L concentration (Sivanesan 2007). Apart from the MS medium, the revised tobacco medium (RT medium) also exhibited a positive response in inducing shoots from shoot tip explants. Supplementation of 1.0 mg/L 2,4-Dichlorophenoxyacetic acid (2,4-D) to the RT medium facilitated the shoot induction and subsequent rooting on the same medium (Kanungo and Sahoo 2011). Studies indicate that ashwagandha seeds were inoculated to an MS medium augmented with 1.5 mg/L of BAP (6-benzylaminopurine) and 0.5 mg/L of IAA (indole-3-acetic acid) to stimulate the emergence of multiple shoots. Subsequently, shoot propagation was conducted on MS media enriched with 0.3 mg/L of gibberellic acid (GA3) and 3.0 mg/L of indole-3-butyric acid (IBA) (Tata et al. 2019).

Leaf explants obtained from in vitro cultured ashwagandha plantlets were grown on an MS medium supplemented with a combination of indole-3-acetic acid (IAA) and BAP. This culture method effectively induced the growth of multiple shoots. Notably, an average of 13.37 ± 1.66 adventitious shoots were observed when the medium was enriched with 7.99 μM IAA and 13.2 μM BAP, representing a statistically significant increase (Kulkarni et al. 1996). Likewise, the combination of 2.0 mg/L of BAP and 0.1 mg/L of naphthalene acetic acid (NAA) facilitated the generation of multiple shoots from the leaf explants (Ghimire et al. 2010). The most significant shoot induction was observed in the epicotyl explants when cultured on MS media supplemented with 2.0 mg/L BAP and 0.2 mg/L IAA. Subsequently, these stimulated shoots were proliferated on MS media supplemented with 1.0 mg/L GA3 (Udayakumar et al. 2013). In a similar fashion, the highest rates of shoot initiation and proliferation were observed in cotyledonary nodes cultured on MS media enriched with 1.0 mg/L BAP. In this scenario, an average of 16.93 shoots per explant was recorded (Nayak et al. 2013).

The highest regeneration efficiency for shooting in nodal explants was observed when they were cultured on MS medium supplemented with 2.5 μM BAP in combination with 0.5 μM NAA (Fatima and Anis 2012). Callus formation and shoot development happened when nodal explants were cultured in MS media supplemented with 1.0 mg/L NAA and 0.5 mg/L BAP. Further accelerating shoot growth was subsequent subculturing of the callus on MS medium fortified with 1.0 mg/L BAP and 0.5 mg/L KN (Mishra 2014). When nodal explants were cultivated in MS medium supplemented with 2.0 mg/L BAP, they showed the maximum shoot induction, with the greatest number of shoots (210.057) and the highest length seen (Saema et al. 2015). Using MS medium supplemented with 1.5 mg/L BAP and 1.5 mg/L IAA, it was successfully possible to induce numerous shoots in nodal explants. A dosage of 0.15 mg/L of GA3 was additionally added to promote shoot proliferation (Kumar et al. 2011). Furthermore, the addition of 3.0 mg/L BAP to MS medium promoted the initiation of multiple shoots, while the combination of 3.0 mg/L BAP and 0.5 mg/L NAA in MS medium facilitated shoot multiplication and proliferation (Jhankare et al. 2013). The greatest number of shoots was observed when nodal explants were cultured on MS media supplemented with 2.0 mg/L BAP and 0.5 mg/L NAA (Goswami et al. 2022). Compared to traditional plant growth regulators (PGRs), the inclusion of meta-Topolin at a concentration of 2.5 μM in the growth medium demonstrated the highest efficacy in inducing shoot formation (Kaur et al. 2021).

1.5.2 Indirect Organogenesis

The production of desirable organ cultures is the culmination of a gradual regeneration process that begins with the beginning of callus formation during indirect organogenesis. Given the right circumstances and inputs, the callus will eventually go through organogenesis and differentiation, resulting in the growth of the appropriate organ cultures. When leaf explants were grown on MS medium enriched with 2.0 mg/L BAP and 0.5 mg/L IAA, the highest efficiency in callus formation, followed by shoot induction, was observed (Dewir et al. 2010). In another trial, leaf explants obtained from in vitro cultivated plants were cultured on MS medium supplemented with IBA and BAP, leading to callus formation. Subsequent subculture of the callus from the leaf explants on MS media supplemented with 2.0 mg/L BAP resulted in simultaneous shooting (Chakraborty et al. 2013). Additionally, MS media supplemented with 0.5 mg/L BAP and 1.0 mg/L NAA made it easier to induce callus from leaf explants. Shoot induction and regeneration were aided by later subculturing the callus on conditions containing 0.5 mg/L NAA and 2.0 mg/L BAP (Rani et al. 2014).

Callus formation in ashwagandha was effectively initiated by culturing epicotyl explants on MS medium with 2.0 mg/L of 2,4-D and 0.6 mg/L of KN. Subsequent shoot induction was accomplished by transferring the resultant callus to media augmented with 1.0 mg/L of BAP and 20 mg/L of adenine sulphate (Udayakumar et al. 2013). According to Goswami et al. (2022), the highest shoot production from callus derived from nodal segments was observed when cultivated on MS medium supplemented with 1.0 mg/L of BAP and 0.5 mg/L of NAA. For hypocotyl explants, the most effective callus induction was achieved on MS medium supplemented with 2.0 mg/L of 2,4-D and 0.2 mg/L of KN. The same medium was found suitable for inducing shoots from the resulting callus. Furthermore, the maximum proliferation efficiency was observed when the callus was cultured on MS medium supplemented with 2.0 mg/L of BAP (Rani et al. 2003). The findings emphasize the significance of choosing particular hormone combinations and their proper doses for effective callus formation and in vitro shoot regeneration from leaf explants. These elements are critical in determining the effectiveness and success of the tissue culture method of plant regeneration.

1.5.3 Rooting of Plantlets

Overall, various auxins have proven to be efficient plant growth regulators (PGRs) for root induction in in vitro regenerated shoots. Studies indicate that auxin-rich media are beneficial for root induction in ashwagandha. In previous studies, it was found that in vitro-grown shoots exhibited effective root induction and multiplication when cultured on a 1/2-strength MS medium supplemented with 2.0 mg/L IBA (Sivanesan 2007; Rani et al. 2003). Likewise, other research showed root formation in in-vitro-grown shoots when cultured on media enriched with 1.0 mg/L IBA (Nayak et al. 2013; Ghimire et al. 2010). Mishra (2014) reported that the optimal efficiency for root induction was achieved by using a 1/2-strength MS medium supplemented with 1.4 mg/L IBA. Additionally, Fatima and Anis (2012) observed 100% rooting efficiency when a 1/2-strength MS medium was supplemented with 0.5 μM NAA. Higher IBA concentrations, like 5 mg/L, demonstrated improved rooting and increased effectiveness (Kumar et al. 2011). Along with auxins, BAP also showed an important role in rooting. The in vitro-induced shoots exhibited the highest rooting efficiency when cultured on MS medium supplemented with 0.44 M BAP, as reported by Kulkarni et al. (1996). Rooting success was also observed using either full-strength or 1/2-strength MS medium supplemented with 0.01 mg/L BAP (Kulkarni et al. 2000).

1.5.4 Acclimatization of In-Vitro-Grown Plantlets

Acclimatization plays a significant role in the regeneration of plantlets, allowing them to adapt to a new environment. During acclimatization, the plantlets are carefully transferred to specific potting mixtures to facilitate their adjustment in controlled climatic conditions like greenhouses and polyhouses. Once the plantlets have acclimated to the potting mixture, they are subsequently transferred to the external environment. Table 1.2 provides an overview of the various potting mixtures employed for the acclimatization of the in vitro raised plantlets.

1.6 Mutation Breeding

The breeding objectives of ashwagandha are enhancing root yield, uniform crop cover, High root and withanolide yield, and development of dwarf high root yielding varieties (Manivel et al. 2017). Mutagenesis or mutation breeding is an essential tool supporting traditional crop improvement breeding practices. Utilizing physical and chemical mutagens, the practice of mutation breeding is used to purposefully cause mutations in crops. Crops can have their genetic makeup altered using both chemical and physical mutagens, such as sodium azide (SA) and ethyl methanesulfonate (EMS) (Kolakar et al. 2018). It also contributes to the broadening of the gene pool as a

Table 1.2 Various potting mixtures for the acclimatization of *Withania somnifera* plantlets

Potting mixture	Ratio	References
Sand: soil	1:1	Dewir et al. (2010)
Sand: soil: FYM (manure)	1:1:1	Jhankare et al. (2013)
Soil: vermicompost	1:3	Chakraborty et al. (2013)
Soil: sand: vermiculite	1:2:1	Udayakumar et al. (2013)
Vermicompost: soil: sand	1:2:1	Mishra (2014)
Garden soil: vermicompost: sand	1:1:1	Kanungo and Sahoo (2011)
Soil: perlite along with MS salts in the mixture	3:1	Bimal et al. (2010)
Garden soil: vermicompost along with half strength MS salts without vitamins	3:1	Fatima and Anis (2012)

result of allelic variation at 1 or several gene loci of interest. EMS and gamma rays' treatment of *W. somnifera* has led to recombinational changes and induced genetic variations in the further generations. It was also reported that the hydroxylamine (HA) and ethyl methane sulphonate (EMS) were effective in inducing mutations compared to gamma rays in viable mutation frequency (Mohsina and Datta 2006).

Most breeding procedures for ashwagandha, which concentrates on root output, are carried out through self-pollination of plants from local or wild cultivars. The popular varieties cultivated are Chetak, Nmitli 118, Pratap, Jawahar 20, Poshita, Rakshita, WSR and CIM-Pushti (Kumar et al. 2023). Application of advanced breeding practices has been reported which can aid in traditional practices. While mutational breeding has been effectively applied to various crops, there is limited information prevailing in regard to mutational breeding in ashwagandha. After EMS treatments were applied to eight Poshita and Jawahar mutants, it was discovered that two of the mutants had improved root and shoot characteristics, seed yield, secondary metabolite concentration, and root fiber content (Das et al. 2010). Bharathi et al. (2013) had used gamma radiation in different concentrations in ashwagandha to induce physical mutagenesis and found that the plant height, number of leaves, seed yield and root yield increased in lower doses of gamma radiation (Bharathi et al. 2013). Low level laser irradiation has also been found to be beneficial in improving the germination of seeds, growth characters and withanolide content in *W. somnifera* (Thorat et al. 2021).

Polyploidy breeding is a boon to improve the crop where they exhibit increased vigor and outpace their diploid plants. The main purpose of employing polyploidy breeding is that they increase heterozygosity and hybrid vigor (Koli et al. 2021). Basu (2019) successfully induced polyploidy, examined the morphometric specifics of the induced autotetraploid, and evaluated the genetic stability over generations (Basu 2019). The cell size of stomata, leaves, pollen and seed size were found to be larger than the diploid plant. The seed fertility was also found to be high in the autotetraploid. Colchicine treatment resulted in induction of polyploids in varieties JA-20, Poshita, and KRC-11 and found that Poshita variety was most suitable for polyploidy induction (Chinapolaiah 2008). Another report by Vidya et al. (2013) supports the suitability of polyploid generation in Poshita variety where the root yield and alkaloid content was higher compared to JA-20 variety (Vidya et al. 2013). Sharma (2013) have developed HPLC–PDA for quantification of withaferin-A and withanolide-A in diploid ($n = 12$) and tetraploid ($n = 24$) cytotypes and found that tetraploid cytotype had higher metabolite content than diploid cytotypes.

1.7 Molecular and Molecular Marker-Assisted Breeding

Molecular breeding is the process of enhancing crop plants by changing their genes through the use of various molecular tools and procedures. The emphasis is on locating and altering genes linked to desirable qualities. On the other hand, molecular marker-assisted breeding is a particular form of molecular breeding that makes use of molecular markers to help in the selection of plants with desirable features (De 2017). In each of these techniques, DNA fragments are used as markers to look for desired features including high productivity, higher quality, and robust resistance. Through focused breeding efforts, they are essential to the creation of new plant varieties.

To enhance the breeding practices, information about genetic diversity is needed. To determine the genetic diversity of plants and choose varieties with higher yields, a number of molecular markers including RAPD, inter simple sequence repeat (ISSR), EST-SSR, random fragment length polymorphism (RFLP), SAMPL, and AFLP were utilized. RAPD and ISSR markers were used to evaluate the genetic diversity of 25 Indian ginseng germplasms. The study found that the various genotypes under investigation have a high genetic variety. In addition to a marker-assisted selection of genotypes, the genotypes were also screened for metabolite production through HPLC and one genotype was found to harbor high withanolide content and also high root yield (Chauhan et al. 2022). Parmar et al. (2015) have successfully developed 11 EST-SSR markers using a pool of 742 available ESTs derived from *W. somnifera* EST sequences. Cross-species transferability is another criterion for assessing the genetic analysis of ashwagandha (Parita et al. 2018; Radhika et al. 2019).

Sequence-based molecular markers for molecular breeding have not been developed so far. Preliminary screening of different varieties of molecular markers to assess the genetic diversity has been carried out to date. Different genotypes of ashwagandha have been assessed through RAPD markers (Sahu et al. 2018; Khanna et al. 2014), ISSR markers (Bamhania et al. 2013; Niraj and Awasthi 2018; Hiremath et al. 2021), AFLP (Mir et al. 2011). Sunita et al. (2014) have also assessed the inter and intra-specific diversity in genotypes by protein profiling and found that genotypes of different locations did not cluster together, suggesting genetic variability. Mehmood et al. (2020) have identified 10 mutational hotspots by plastome sequencing, which could be developed as molecular marker development and species identification. These DNA and protein markers serve as a reference base and guide for the breeding of new varieties.

1.8 Transgenic Breeding Through Genetic Engineering and Gene Editing

Transgenic breeding through genetic engineering and gene editing techniques can be employed to introduce specific genes or modify existing genes in ashwagandha. These techniques offer the potential to develop plants with enhanced withanolide production, disease and pest resistance, abiotic stress tolerance etc. Large-scale production of withanolide was achieved by *Agrobacterium rhizogenes-mediated* genetic transformation, which induced hairy root culture (Thilip et al. 2015; Dehdashti et al. 2017; Singh et al. 2022). Reports on transgenic breeding through genetic engineering and gene editing in ashwagandha are meager.

1.9 Conclusion and Prospects

Ashwagandha holds economic value and is highly regarded in traditional medicine systems due to its recognized potential therapeutic properties, including anti-inflammatory effects, stress reduction and immune enhancement. Also, the demand for it as a herbal supplement and ingredient in the pharmaceutical and nutraceutical sectors of the economy is consistently rising. Traditional breeding strategies, such as selective and hybrid breeding, have been employed to develop cultivars with desired traits. However, these approaches have limitations, such as lengthy breeding cycles and limited genetic variations. Therefore, the implementation of modern breeding strategies offers promising prospects for the improvement of ashwagandha. Tissue culture strategies have been employed to propagate disease-free plantlets rapidly. Other modern breeding strategies such as mutation breeding, including polyploid breeding, have been explored to induce genetic variation and develop plants with novel traits. Molecular and molecular marker-assisted breeding techniques have accelerated breeding progress and aided in the selection of desirable traits. There is limited research in transgenic breeding of ashwagandha through genetic engineering and gene editing despite being an emerging field.

In the future, there is room for further exploration of transgenic breeding through gene editing technologies to develop high yielding disease resistant varieties. As there are a limited number of institutions in conservation of germplasm diversity, there is a need for improving the preservation of germplasm diversity through in situ and in vitro methods, cryopreservation, and gene banks. The conservation of the germplasm thereby ensures the availability of genetic material for breeding programs and further scientific research. In conclusion, consistent research, development of breeding strategies and germplasm conservation efforts, will contribute to the development of cultivars with desired traits and increased resilience. Further the modern breeding strategies of ashwagandha hold aids in meeting the increasing demand for this medicinal herb. These developments will not only help farmers and the herbal sector, but also open doors for other industries.

1.9.1 Appendices

Appendix 1: Withania Germplasm Accessions and Cultivar Details from ICAR-NBPGR National GeneBank Database

Accession number	Cultivar name	Location	Accession number	Cultivar name	Location
IC421053	Ashwagandha	Andhra Pradesh	IC553933	Withania Mandsaur-29	Madhya Pradesh
IC310620	Gado	Gujarat	IC553934	Withania Mandsaur-30	Madhya Pradesh
IC282715 IC315274 IC418292	Ashwagandha	Rajasthan	IC553935	Withania Mandsaur-31	Madhya Pradesh
IC361042-043	Asgandh	Uttarakhand	IC553936	Withania Mandsaur-34	Madhya Pradesh
IC588697	DWS-327	Gujarat	IC553937	Withania Mandsaur-35	Madhya Pradesh
IC614019 IC614010	Asgandh	Haryana	IC553938	Withania Mandsaur-39	Madhya Pradesh
IC527324-325 IC527336-357	Asgandh	Rajasthan	IC553939	Withania Nauni-9	Himachal Pradesh
IC614772	Gujarat Anand Ashwagandha 1 (GAA 1/AWS 1)	Gujarat	IC553940	Withania Nauni-13	Himachal Pradesh
IC272705	Ashwagandha	Andhra Pradesh	IC553930	Withania Mandsaur-6	Madhya Pradesh
IC281141	Kala Aak	Rajasthan	IC553931	Withania Mandsaur-16	Madhya Pradesh
IC281159	Asgandha	Rajasthan	IC553932	Withania Mandsaur-25	Madhya Pradesh
IC310595	Kuvarpathu	Gujarat	IC553941	Withania Nauni-17	Himachal Pradesh
IC273842	Amukkiran kizhangu	Tamil Nadu	IC553942	Withania Rishikesh-1	Uttarakhand

(continued)

(continued)

Accession number	Cultivar name	Location	Accession number	Cultivar name	Location
IC281139	Kala Aak	Rajasthan	IC553943	Withania Gopeshwar-1	Uttarakhand
IC400317	Indian ginseng	Karnataka	IC553944	Withania Bhopal-1	Madhya Pradesh
IC430867	Asgard	Delhi	IC553945	Withania Bhopal-3	Madhya Pradesh
IC553946	Withania Jaipur-7a	Rajasthan	IC553948	Withania Jaipur-1b	Rajasthan
IC553947	Withania Jaipur-1d	Rajasthan	IC566524	AGB 002,NPIT LI002	Jammu and Kashmir
IC587064	Raj Vijay Aswagandha-100 (RVA100)	Madhya Pradesh	**IC639095**	SKA 11	Gujarat
IC623186 IC623187 IC623188 IC623189	Amukkira Kilangu	Tamil Nadu	**IC642106**	HWS 08-18	Haryana
IC646383	Ashwagandha	Madhya Pradesh	**IC642107**	HWS 12-12	Haryana

Appendix 2: List of *Withania somnifera* Cultivars and Their Distinctive Properties

Cultivars	Institute/place	Characteristic and seed yield	References
Poshitha Ashwagandha	CSIR-CIMAPS, Lucknow	Improved high yielding variety released Dry root yield 5–6 q per hectare	Sai et al. (2022)
Rakshitha	CSIR-CIMAPS, Lucknow	Improved high yielding variety released. Dry root yield 6–8 q per hectare	TNAU (2021)
Nagori	Local variety from Nagori district of Madhya Pradesh	Starchy roots	TNAU (2021)
Jawahar Asgand-20 Jawahar Asgand-134	Jawaharlal Nehru Krishi Vishwavidyalaya, Madhya Pradesh	High alkaloid variety., 0.3% ashwagandha, short plant with higher planting density	Apnikheti (2023)

(continued)

(continued)

Cultivars	Institute/place	Characteristic and seed yield	References
Raj Vijay Ashwagandha-100	Jawaharlal Nehru Krishi Vishwavidyalaya, Madhya Pradesh	Dry root yield 6–7 q per hectare	Jat et al. (2015)
WSR	CSIR-Regional Research Laboratory, Jammu	Adaptability and root yield is high	Sharada et al. (2008)
NIMTLI-118	CSIR-NMITLI	Higher root quality and leaf yield at maturity	Anjanidevi et al. (2022)
Arka Ashwagandha	ICAR-IIHR, Bengaluru Karnataka	High metabolite	ICAR-IIHR (2023)
CIMAP-Chetak	CSIR-CIMAP	Highly stable for dry root yield and root starch (%)	Kumar et al. (2022)
CIMAP-Pratap			
CIMAP Pushti	CSIR-CIMAP	High withaferin A, leaf blight tolerant	Jnanesha et al. (2023)

Appendix 3: Research Institutes Working on Ashwagandha Germplasm Conservation and Management

Institution	Specialization research activities	Contact information and website
ICAR National Bureau of Plant Genetic Resources	Germplasm collection and management	Agriculture Knowledge Management Unit, ICAR National Bureau of Plant Genetic Resources, New Delhi-110012, India, akmu.nbpgr@icar.gov.in
Central Institute of Medicinal and Aromatic Plants—CIMAP	Releasing new varieties with high alkaloid and withaferin content	Director, Central Institute of Medicinal and Aromatic Plants PO- CIMAP, Lucknow-226,015, India Ph: + 91-522-2,718,639, 2,718,641, 2,718,505 E-mail: director@cimap.res.in, Website: www.cimap.res.in

(continued)

(continued)

Institution	Specialization research activities	Contact information and website
Regional Plant Resource Centre (RPRC)	Collection and conservation of germplasm	Regional Plant Resource Centre Forest, Environment and Climate Change Department, Government of Odisha Nayapalli, Bhubaneswar 751,015,Odisha, India, Phone: (0674) 2,552,002, 2,557,925 Fax: 91-674-2,550,274 E-mail: rprcbbsr@gmail.com
Leibniz Institute of Plant Genetics and Crop Plant Research (IPK)—Germany	Collection, conservation characterization, and documentation	Corrensstr. 3, 06,466 Gatersleben, Germany. https://www.gbif.org/dataset/863 4d730-f762-11e1-a439-00145e b45e9a
ICAR National Research Centre for Medicinal and Aromatic Plants (NRCMAP)	Characterisation, preservation and utilization of local cultivars	National Research Centre for Medicinal and Aromatic Plants Boriavi, Anand—387310, Gujarat, India
ICAR Indian Institute of Horticultural Research (IIHR)	Collection, conservation and development of new varieties	ICAR-IIHR, Hessaraghatta Lake Post, Bengaluru-560089. E-mail director.iihr@icar.gov. in sanjaydr2@gmail.com website: https://www.iihr.res.in Contact details: 080-28,466,471 080-28,466,353

References

Abdelwahed MT, Hegazy MA, Mohamed EH (2023) Major biochemical constituents of *Withania somnifera* (ashwagandha) extract: a review of chemical analysis. Rev Anal Chem 42:1–15

Abhilash PC, Jamil S, Singh V et al (2008) Occurrence and distribution of hexachlorocyclohexane isomers in vegetation samples from a contaminated area. Chemosphere 72:79–86

Afewerky HK, Ayodeji AE, Tiamiyu BB et al (2021) Critical review of the *Withania somnifera* (L.) Dunal: ethnobotany, pharmacological efficacy, and commercialization significance in Africa. Bull Natl Salmon Resour Cent 45:176

Ahmed I, Dubey RB, Jain D et al (2023) Heterosis studies for root-yield-attributing characters and total alkaloid content over different environments in *Withania somnifera* L. Collect FAO Agric 13:1025

Aniel Kumar O, Jyothirmayee G, Subba Tata S (2013) In Vitro conservation of *Withania somnifera* (L) dunal (Ashwagandha)—a multipurpose medicinal plant. Asian J Sci Res 3:852–861

Anjanidevi KS, Padma M, Kumar MR et al (2022) Performance of ashwagandha (*Withania somnifera* L.) promising lines on growth, yield, quality and productivity. The Pharma Innovation J 11:2547–2551

Apnikheti (2023) Ashwagandha farming in Punjab. In: Apnikheti. https://www.apnikheti.com/en/pn/agriculture/horticulture/medicinal-plants/ashwagandha. Accessed 28 Jun 2023
Arpan A, Shanker K et al (2014) RNA-Seq analysis of two elite lines of *Withania somnifera* to evaluate potential of heterosis. Indian J Genet Plant Breed 74:612–619
Bamhania K, Khatakar S, Punia A, Yadav OP (2013) Genetic variability analysis using ISSR markers in *Withania somnifera* L. Dunal genotypes from different regions. J Herbs Spices Med Plants 19:22–32
Basu S (2019) Meiotic correlations and bound arm associations between the diploid and autotetraploid of *Withania somnifera* (L.) Dunal, and implications on genetic stability. Nucleus 62:15–20
Bhandari CVC (2021) An encyclopedia of Indian botanies and herbs (Part I). Chaukhamba Sanskrit sansthan.
Bharathi T, Gnanamurthy S, Dhanavel D, Murugan S (2013) Induced physical mutagenesis on seed germination, lethal dosage and morphological mutants of Ashwagandha (*Withania somnifera* (L.) Dunal). Int J Adv Res 1:136–141
Bimal KG, Eun SS, Eun HK et al (2010) Direct shoot organogenesis from petiole and leaf discs of *Withania somnifera* (L.) Dunal. Afr J Biotechnol 9:7453–7461
Byrne PF, Volk GM, Gardner C et al (2018) Sustaining the future of plant breeding: the critical role of the USDA-ARS national plant germplasm system. Crop Sci 58:451–468
Chakraborty N, Banerjee D, Ghosh M et al (2013) Influence of plant growth regulators on callus mediated regeneration and secondary metabolites synthesis in *Withania somnifera* (L.) Dunal. Physiol Mol Biol Plants 19:117–125
Chauhan S, Joshi A, Rajamani G, Jain D (2018) Genetic diversity analysis in ashwagandha [*Withania somnifera* (L.) Dunal] genotypes. Int J Curr Microbiol App Sci 7:1574–1583
Chauhan S, Mandliya T, Jain D et al (2022) Early selective strategies for higher yielding bio-economic Indian ginseng based on genotypic study through metabolic and molecular markers. Saudi J Biol Sci 29:3051–3061
Chinapolaiah A (2008) Studies on germplasm evaluation and induction of autotetraploidy in Ashwagandha (*Withania somnifera* (L.) Dunal). Ph.D., University of Agricultural Sciences, Dharwad, Karnataka
Dar NJ, Hamid A, Ahmad M (2015) Pharmacologic overview of *Withania somnifera*, the Indian ginseng. Cell Mol Life Sci 72:4445–4460
Das A, Datta AK, Bhattacharya A et al (2010) EMS induced mutagenesis in Poshita and Jawahar 22 of *Withania somnifera* (L.) Dunal (Solanaceae). Cytologia 75:305–311
Dehdashti SM, Acharjee S, Kianamiri S, Deka M (2017) An efficient *Agrobacterium rhizogenes*-mediated transformation protocol of *Withania somnifera*. Plant Cell Tissue Organ Cult 128:55–65
De LC (2017) Breeding of medicinal and aromatic plants-an overview. Int J Bot Res 7:25–34
Dempewolf H, Krishnan S, Guarino L (2023) Our shared global responsibility: safeguarding crop diversity for future generations. Proc Natl Acad Sci USA 120:e2205768119
Dewir YH, Chakrabarty D, Lee S-H et al (2010) Indirect regeneration of *Withania somnifera* and comparative analysis of withanolides in in vitro and greenhouse grown plants. Biol Plant 54:357–360
Ding F, Jin S, Hong N et al (2008) Vitrification-cryopreservation, an efficient method for eliminating *Candidatus Liberobacter asiaticus*, the citrus Huanglongbing pathogen, from in vitro adult shoot tips. Plant Cell Rep 27:241–250
Ekka A, Tirkey A, Kujur N (2021) Cluster and principal component analysis (PCA) in ashwagandha [*Withania somnifera* (L.) Dunal] for root traits. Int J Chem Stud 9:3012–3016
Fatima N, Anis M (2012) Role of growth regulators on in vitro regeneration and histological analysis in Indian ginseng (*Withania somnifera* L.) Dunal. Physiol Mol Biol Plants 18:59–67
Food and Agriculture Organization (2020) The state of the world's forests 2020 The state of the world's forests 2020: forestry, biodiversity and people. Food & Agriculture Organization of the United Nations (FAO), Rome, Italy

Gaurav H, Yadav D, Maurya A, Yadav H, Yadav R, Shukla AC, Sharma M, Gupta VK, Palazon J (2023) Biodiversity, biochemical profiling, and pharmaco-commercial applications of *Withania somnifera*: a review. Molecules 28(3):1208. https://doi.org/10.3390/molecules28031208

Ghimire BK, Seong ES, Kim EH et al (2010) Direct shoot organogenesis from petiole and leaf discs of *Withania somnifera* (L.) Dunal. Afr J Biotechnol 9:7453–7461

Goswami B, Khan S, Banu TA et al (2022) In vitro mass propagation of Withania somnifera (L.) Dunal, an important medicinal plant of Bangladesh. Bangladesh J Bot 51:191–197

Hallauer AR, Carena MJ, Miranda Filho JB (2010) Quantitative genetics in maize breeding. Springer

Hiremath C, Philip R, Sundaresan V (2021) Molecular characterization of India ginseng *Withania somnifera* (L) using ISSR markers. Mol Biol Rep 48:3971–3977

ICAR-IIHR (2023) Arka ashwagandha. In: ICAR-IIHR. https://iihr.res.in/variety/arka-ashwagandha. Accessed 28 Jun 2023

Inagaki K (2017) *Withania somnifera*; Drug-induced liver injury: case report. React Wkly 1671:382–382

Jhankare A, Tripathi MK, Tiwari G et al (2013) Efficient plantlet regeneration from cultured nodal segment of *Withania somnifera* (L.) Dunal. Plant Cell Biotech Mol Biol 14:99–110

Jnanesha AC, Venugopal S, Kumar RS et al (2023) Transvaluate of genotype × environment × season interaction of agronomic traits and bioactive compounds of *Withania somnifera* (L.) Dunal in the Deccan plateau region. J Appl Res Med Aromat Plants 100501

Joshi NR, Patel MA, Prajapati KN et al (2015) Genetic diversity in ashwagandha (*Withania somnifera* (L.) Dunal). Electron J Plant Breeding 6:870–874

Kanungo S, Sahoo SL (2011) Direct organogenesis of *Withania somnifera* L. from apical bud. Int Res J Biotechnol 2:58–61

Katkov I (2012) Current frontiers in cryopreservation. BoD—Books on Demand

Kaur K, Kaur K, Bhandawat A, Pati PK (2021) In vitro shoot multiplication using meta-Topolin and leaf-based regeneration of a withaferin a rich accession of *Withania somnifera* (L.) Dunal. Ind Crops Prod 171:113872

Kaviani B (2011) Conservation of plant genetic resources by cryopreservation. Aust J Crop Sci 5:778–800

Khanna PK, Chandra R, Kumar A et al (2014) Correlation between morphological, chemical and RAPD markers for assessing genetic diversity in *Withania somnifera* (L.) Dunal. J Crop Sci Biotechnol 17:27–34

Kolakar SS, Nadukeri S, Jakkeral SA et al (2018) Role of mutation breeding in improvement of medicinal and aromatic crops: review. J Pharmacogn Phytochem 7:425–429

Koli GK, Arya R, Nimbal S, Kumar D (2021) Genetic improvement for immunity boosting traits in medicinal plants. In: Anil Kumar A (ed) Advances in medicinal plants, pp 61–73

Kothari SK, Singh CP, Vijay Kumar Y, Singh K (2003) Morphology, yield and quality of ashwagandha (*Withania somnifera* L. Dunal) roots and its cultivation economics as influenced by tillage depth and plant population density. J Hortic Sci Biotechnol 78:422–425

Kulkarni AA, Thengane SR, Krishnamurthy KV (1996) Direct in vitro regeneration of leaf explants of *Withania somnifera* (L.) Dunal. Plant Sci 119:163–168

Kulkarni AA, Thengane SR, Krishnamurthy KV (2000) Direct shoot regeneration from node, internode, hypocotyl and embryo explants of *Withania somnifera*. Plant Cell Tissue Organ Cult 62:203–209

Kumar A, Lal RK, Gupta AK et al (2022) Exegesis of cultivars-multi-year/environment interactions for agro morphological traits in Ashwagandha (*Withania somnifera* (L.) Dunal). S Afr J Bot 151:523–531

Kumar A, Venugopal S, Jnanesha AC, Lal RK (2023) Agricultural-based challenges, genetic enhancement, and obstacles to an industrially important medicinal plant, ashwagandha (*Withania somnifera* (L.) Dunal): a review. Ecol Genet Genomics 28:100183

Kumar M, Muthusamy A, Kumar V, Bhalla-Sarin N (2019) In vitro plant breeding towards novel agronomic traits: biotic and abiotic stress tolerance, 1st edn. Springer, Singapore

Kumar OA, Jyothirmayee G, Tata SS (2011) Multiple shoot regeneration from nodal explants of Ashwagandha *Withania somnifera* (L.) Dunal. Asian J Exp Biol Sci 2:636–640

Lamichhane S, Thapa S (2022) Advances from conventional to modern plant breeding methodologies. Plant Breed Biotechnol 10:1–14

Manivel P, Nagaraja RR, Deore HB (2017) Genetic diversity for root yield and its component traits in ashwagandha (*Withania somnifera* (L) Dunal) pure lines derived from JA134 population. Science 6:1694–1710

Mehmood F, Abdullah UZ et al (2020) Comparative plastomics of ashwagandha (Withania, Solanaceae) and identification of mutational hotspots for barcoding medicinal plants. Plants 9:752

Mehta A, Raina R (2018) Breeding system studies in *Withania somnifera*-an important commercial medicinal plant. Med Plants Int J Phytomed Relat Ind 10:236–242

Mir BA, Koul S, Kumar A et al (2011) Assessment and characterization of genetic diversity in *Withania somnifera* (L.) Dunal using RAPD and AFLP markers. Afr J Biotechnol 10:14746–14756

Mirjalili MH, Fakhr-Tabatabaei SM, Alizadeh H et al (2009) Genetic and withaferin a analysis of Iranian natural populations of *Withania somnifera* and W. coagulans by RAPD and HPTLC. Nat Prod Commun 4:337–346

Mishra T (2014) An efficient regeneration of *Withania somnifera* (L.) Dunal through direct organogenesis. Med Plants-Int J Phytomedicines Relat Ind 6:143–146

Mohammadi SA, Prasanna BM (2003) Analysis of genetic diversity in crop plants—salient statistical tools and considerations. Crop Sci 43:1235–1248

Mohsina I, Datta AK (2006) Mutagenic effectiveness and efficiency of gamma-rays, hydroxylamine and EMS in *Withania somnifera* (L.) Dun. (Ashwagandha). Plant Archives 6:533–535

Mondini L, Noorani A, Pagnotta MA (2009) Assessing plant genetic diversity by molecular tools. Diversity 1:19–35

Nayak SA, Kumar S, Satapathy K et al (2013) In vitro plant regeneration from cotyledonary nodes of *Withania somnifera* (L.) Dunal and assessment of clonal fidelity using RAPD and ISSR markers. Acta Physiol Plant 35:195–203

Niraj T, Awasthi A (2018) Molecular diversity analysis among *Withania somnifera* genotypes collected from Madhya Pradesh using RAPD markers. Ann Plant Soil Res 20:416–421

Parita B, Kumar SN, Darshan D, Karen P (2018) Elucidation of genetic diversity among ashwagandha [*Withania somnifera* (L.) Dunal] genotypes using EST-SSR markers. Res J Biotechnol 13:52–59

Parmar EK, Fougat RS, Patel CB et al (2015) Validation of dbEST-SSRs and transferability of some other solanaceous species SSR in ashwagandha [*Withania somnifera* (L.) Dunal]. 3 Biotech 5:933–938

Poojari P, Kiran KR, Swathy PS, Muthusamy A (2019) *Withania somnifera* (L.) Dunal: an overview of bioactive molecules, medicinal properties and enhancement of bioactive molecules through breeding strategies. In: Kumar M, Muthusamy A, Kumar V, Bhalla-Sarin N (eds) In vitro plant breeding towards novel agronomic traits: biotic and abiotic stress tolerance. Springer, Singapore, pp 1–25

Radhika RM, Sukanya DH, Himabindu K (2019) Study on cross-species transferability and DNA fingerprinting of Ashwagandha genotypes using SSR markers. Med Plants Int J Phytomed Relat Ind 11:444–454

Rajeswara Rao BR, Rajput DK, Nagaraju G, Adinarayana G (2012) Opportunities and challenges in the cultivation of Ashwagandha *Withania somnifera* (L.) Dunal. J Pharmacogn 3:88–91

Rama Reddy Nagaraja Reddy PM, Deore HB (2017) Genetic diversity for root yield and its component traits in ashwagandha (*Withania somnifera*(L) dunal) pure lines derived from JA134 population. Int J Curr Microbiol Appl Sci 6:1694–1710

Rani A, Kumar M, Kumar S (2014) In vitro callus induction and shoot regeneration from leaf explant of *Withania somnifera*. Adv Appl Res 6:62

Rani G, Virk GS, Nagpal A (2003) Callus induction and plantlet regeneration in *Withania somnifera* (L.) dunal. In Vitr Cell & Dev Biol-Plant 39:468–474

R, Jat, R Nagaraja R, Ruchi B, P, Manivel (2015) Good agricultural practices for Ashwagandha. Directorate of Medicinal and Aromatic Plants Research, Boriavi, Anand—387310, Gujarat, India

Saema S, Ahmad IZ, Misra P (2015) Rapid in vitro plant regeneration from nodal explants of *Withania somnifera* (L.) Dunal: a valuable medicinal plant. Int J Sci Res 4:1649–1652

Sahu V, Dodiya NS, Joshi A et al (2018) Elucidation of genetic diversity among ashwagandha [*Withania somnifera* (L.) Dunal] genotypes using EST-SSR markers. Res J Biotechnol 13:52–59

Saidaiah P (2018) Advances in genetics and plant breeding, 1st edn. AkiNik Publications

Sai K, Thanuja Sivaram G, Yuvaraj KM et al (2022) Effect of different dates of sowing, organic manures on protein, fiber and alkaloid content in leaves and roots of ashwagandha (*Withania somnifera*). Pharma Innov 11:522–524

Saleem S, Muhammad G, Hussain MA et al (2020) *Withania somnifera* L.: insights into the phytochemical profile, therapeutic potential, clinical trials, and future prospective. Iran J Basic Med Sci 23:1501–1526

Sharada M, Ahuja A, Vij SP (2008) Applications of biotechnology in Indian ginseng (ashwagandha): progress and prospects. In: Recent advances in plant tissue culture and its applications. K International Pvt. Ltd, New Delhi, India, pp 45–667

Sharma L, Maurya B, Rai SP (2023) An overview of biotechnological interventions and abiotic elicitors on biomass and withanolide biosynthesis in *Withania somnifera* (L.) Dunal. Ind Crops Prod 193:116238

Sharma V (2013) HPLC-PDA method for quantification of Withaferin-A and Withanolide-A in diploid (n=12) and tetraploid (n=24) cytotypes of "Indian Ginseng" *Withania somnifera* (L.) Dunal from North India. Int J Indigenous Med Plants 46:1245–1250

Shasmita, Rai MK, Naik SK (2018) Exploring plant tissue culture in *Withania somnifera* (L.) Dunal: in vitro propagation and secondary metabolite production. Crit Rev Biotechnol 38:836–850

Shinde A, Gahunge P, Rath SK (2015) Conservation and sustainability of ashwagandha: a medicinal plant. J Biol Sci Opin 3:94–99

Singh M, Suyal DC, Dinkar N et al (2022) *Agrobacterium rhizogenes* as molecular tool for the production of hairy roots in *Withania somnifera*. J Appl Biol Biotechnol 11:1–10

Singh N, Bhalla M, de Jager P, Gilca M (2011) An overview on ashwagandha: a Rasayana (rejuvenator) of Ayurveda. Afr J Tradit Complement Altern Med 8:208–213

Sivanesan I (2007) Direct regeneration from apical bud explants of *Withania somnifera* Dunal. Indian J Biotechnol 6:125–127

Srivastava A, Gupta AK, Shanker K et al (2018) Genetic variability, associations, and path analysis of chemical and morphological traits in Indian ginseng [*Withania somnifera* (L.) Dunal] for selection of higher yielding genotypes. J Ginseng Res 42:158–164

Srivastava A, Srivastava P, Pandey A et al (2019) Phytomedicine: a potential alternative medicine in controlling neurological disorders. In: Ahmad Khan MS, Ahmad I, Chattopadhyay D (eds) New look to phytomedicine. Academic Press, pp 625–655

Srivastava Y, Sangwan NS (2020) Improving medicinal crops through phytochemical perspective: *Withania somnifera* (Ashwagandha). In: Tuteja N, Tuteja R, Passricha N, Saifi SK (eds) Advancement in crop improvement techniques. Woodhead Publishing, pp 275–295

Sunita K, Yadav OP, Ramender K (2014) Inter-and intra-specific genetic diversity analysis in thirty genotypes of *Withania somnifera* (Ashwagandha) using protein profile pattern. Int J Chem Sci 12:1655–1667

Tata SS, Jyothirmayee G, Kumar OA (2019) In vitro plant regeneration from mature seed explants of *Withania somnifera* (L.) Dunal, an important, rare and endangered medicinal plant. Not Sci Biol 11:387–391

Thilip C, Soundar Raju C, Varutharaju K et al (2015) Improved *Agrobacterium rhizogenes*-mediated hairy root culture system of *Withania somnifera* (L.) Dunal using sonication and heat treatment. 3 Biotech 5:949–956

Thorat SA, Poojari P, Kaniyassery A et al (2021) Red laser-mediated alterations in seed germination, growth, pigments and withanolide content of Ashwagandha [*Withania somnifera* (L.) Dunal]. J Photochem Photobiol B 216:112144

TNAU (2013) Horticulture. In: TNAU. https://agritech.tnau.ac.in/farm_enterprises/Farm%20enterprises_%20Ashwagantha.html. Accessed 21 Jun 2023

TNAU (2021) Aswagandha. In: TNAU. https://agritech.tnau.ac.in/horticulture/horti_medicinal%20crops_aswagandha.html. Accessed 28 Jun 2023

Trivedi MK, Panda P, Sethi KK, Jana S (2017) Metabolite profiling in *Withania somnifera* roots hydroalcoholic extract using LC/MS. GC/MS and NMR Spectroscopy. Chem Biodivers 14:e1600280

Udayakumar R, Choi CW, Kim KT et al (2013) In vitro plant regeneration from epicotyl explant of *Withania somnifera* (L.) Dunal. J Med Plants Res 7:43–52

Varshney RK, Graner A, Sorrells ME (2005) Genic microsatellite markers in plants: features and applications. Trends Biotechnol 23:48–55

Vekariya RG, Patel PC, Patel JS et al (2022) Genetic studies in *Withania somnifera* (L.) Dunal through intraspecific hybridization. The Pharma Innovation Journal 11:44–56

Venugopal S, Padma M, Rajkumar M, et al (2022) Comparative investigation of the genetic components of *Withania somnifera* L. yield and quality traits using multivariate and associations analysis in the Deccan plateau region. Ind Crops Prod 188:115696

Vidya KM, Hegde L, Saraswathi G (2013) Evaluation of diploids and autotetraploid genotypes of ashwagandha (*Withania somnifera* Dunal.) for growth, yield and quality characters. Asian J Hort 8:97–100

Walters C, Berjak P, Pammenter N et al (2013) Plant science. Preservation of recalcitrant seeds. Science 339:915–916

Chapter 2
Breeding of Licorice (*Glycyrrhiza glabra* L.)

Seyyed Sasan Mousavi, Tahereh Movahhed Haghighi, and Akbar Karami

Abstract Legumes are crucial components of sustainable agriculture and have the potential to improve both human and animal health. Through their capacity to fix atmospheric nitrogen, recycle nutrients, increase soil carbon content, and diversify cropping systems, legumes also offer other crucial benefits to agriculture. Licorice (*Glycyrrhiza glabra* L.), family; Fabaceae, the strategic and very valuable medicinal plant, which has many advantageous pharmaceutical features, makes a significant contribution to the export of herbs. Iran's biodiversity of licorice is currently threatened by rising demand and a sharp loss in its natural habitats. Thus, domestication of licorice would soon be required, and molecular breeding would aid in the introduction of genotypes suitable for agriculture. An essential first step in the agroecological transformation of the world's agri-food systems is the expansion and profitability of the cultivation of legumes. To enhance yield, resistance to stressors, grain quality, and nutritional and nutraceutical qualities, legume breeders and geneticists have used a variety of traditional and modern breeding approches. Conventional breeding techniques are time-consuming, costly, and ineffective for achieving the required results. However, cutting-edge breeding methods including omics, transgenic technology, genomics-assisted breeding, speed breeding, association and mapping research, and alien gene introgression will support sustainable agriculture and food security. This chapter focus on progress in the breeding background and methods for sustainable *Glycyrrhiza* genus cultivation and breeding.

Keywords Domestication · Germplasm diversity · Licorice · Molecular breeding · Traditional breeding

---

S. S. Mousavi · A. Karami (✉)
Department of Horticultural Science, School of Agriculture, Shiraz University, Shiraz, Iran
e-mail: akbarkarami@shirazu.ac.ir

T. Movahhed Haghighi
Institute of Agriculture, Iranian Research Organization for Science and Technology (IROST), Tehran, Iran

J. M. Al-Khayri et al. (eds.), *Biodiversity and Genetic Improvement of Medicinal and Aromatic Plants II*, Advances in Plant Breeding Strategies 10,
https://doi.org/10.1007/978-3-031-81857-8_2

2.1 Introduction

2.1.1 Background and Chapter Objectives and Content Description

In this chapter, an attempt has been made to provide complete and comprehensive information about the molecular and classical breeding studies of the very valuable medicinal and industrial plant, licorice. In addition, the information on licorice plants from the perspective of botany, phytochemistry, and cultivation is fully introduced. This chapter covers topics such as botanical classification and distribution, economic and health benefits, domestication, current cultivation practices, germplasm biodiversity, main objectives of licorice breeding, hybridization, mutation breeding, tissue culture, and molecular breeding related to licorice.

2.1.2 Botanical Classification and Distribution

German scientist Leonhard Fuchs (1501–1566) made the first attempt to codify a system of botanical nomenclature, precisely describing and characterizing this plant. During this time, the works of famous scientist Al Razi made significant contributions in this direction. Giuseppe Donzelli, a philosopher and chemist from Naples, first described licorice in the eighteenth century and used its current name (Öztürk et al. 2018). Carl von Linne, a Swedish naturalist, identified three distinct species after dividing plants into genera and species of *Glycyrrhiza*: *G. glabra*, *G. echinata*, and *G. hirsute* (Fiore et al. 2005). Although, the latest scientific classification of the genus *Glycyrrhiza* is given by Sharma et al (2017) is as in (Fig. 2.1).

The *Glycyrrhiza* genus has more than 30 species that are widely dispersed throughout the world (Shibata 2000). Since the sixteenth century, licorice has grown in Europe. Since the former Han dynasty, it has been the herb most frequently recommended in Ancient Egypt, Rome, Greece, East China, and the West (Rizzato et al. 2017). Different types of licorice are grown in various regions of the world. Additionally, licorice is grown for sale in Sicily, China, England, Spain, Iraq, and Turkey

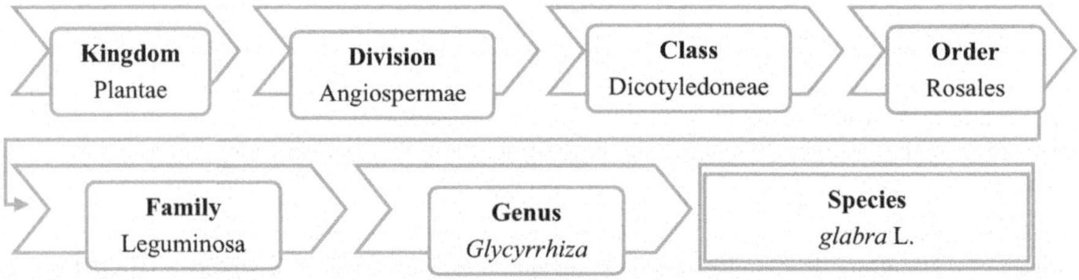

Fig. 2.1 Classification of the genus *Glycyrrhiza*

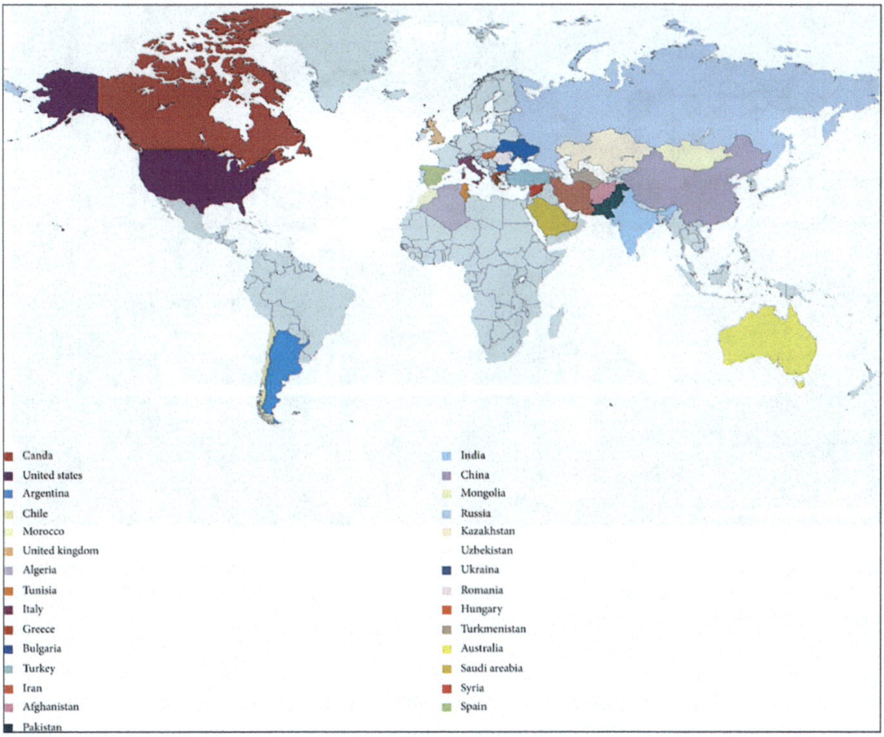

Fig. 2.2 World map showing the major countries producing *Glycyrrhiza* spp.

(Tohma and Gulçin 2010). Iran, Pakistan, Azerbaijan, Turkmenistan, and Uzbekistan are other countries that are licorice producers. Figure 2.2 displays a map of the world with the nations that cultivate licorice extensively (Sharifi-Rad et al. 2021).

2.1.3 Importance (Economic and Health Benefits)

Licorice has a substantial economic impact on global trade, whether it is used as a raw ingredient or processed into root extracts. *G. glabra*, *G. inflata*, and *G. uralensis* are among the known species, extensively researched for nutritive and therapeutic properties. In 2020, the countries that had the largest trade value in exports than in imports of licorice extract were Uzbekistan ($32.5M), Turkmenistan ($19.5M), and Iran ($17.9M). The nations with the highest trade value in licorice extract exports vs imports in 2020 were China ($15.4M), the United States ($12.9M), and Germany ($6.44M). (Fig. 2.3) (https://oec.world/en/profile/hs92/liquorice-extract).

G. glabra has been described in three different cultivars, each of which is produced in different countries. These cultivars include *G. glabra* var. *violacea* (Persian and

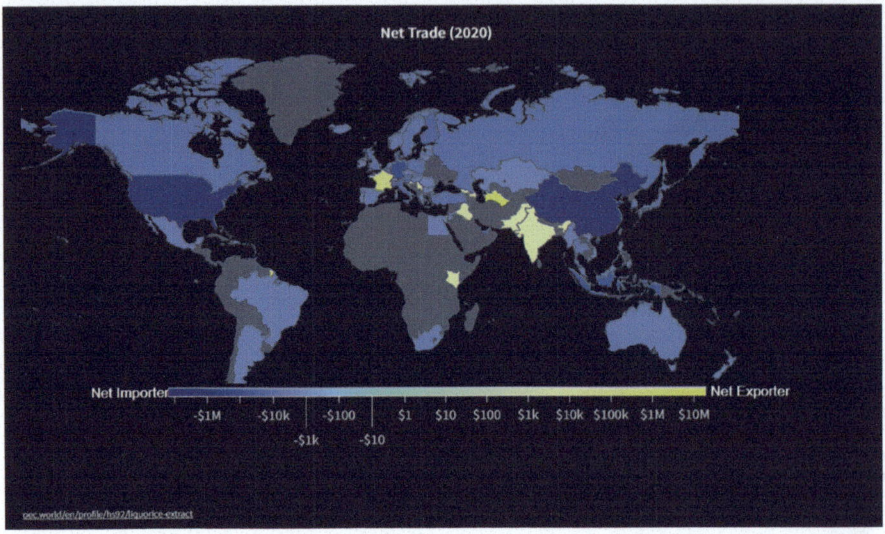

Fig. 2.3 This map shows which countries export or import more licorice extract. Each country is colored based on the difference in exports and imports of licorice extract during 2020. https://oec.world/en/profile/hs92/liquorice-extract

Turkish), *G. glabra* var. *gladulifera* (Russian), and *G. glabra* var. *typica* (Spanish and Italian) (Husain et al. 2021). Numerous licorice chemical components have been investigated for their potent pharmacological effects, including those against cancer, bacteria, inflammation, heart disease, hepatitis, respiratory infections, and numerous others (Wahab et al. 2021) (Fig. 2.4). It is also used commercially as a sweetener in candies, smoking mixtures, and cigarettes. Licorice is employed as a depigmenting agent in cosmetology. Glycyrrhizinic acid (GA) is present in several of health products, such as licorice tea, flavored gum, thrand oat and cough remedies. Licorice root extracts of all kinds are used in alcoholic beverages and chewing tobacco. Licorice cakes, bricks, sticks, toffee, balls, and bars are common in sweets. *Glycyrrhiza* is extensively used in the production of bioenergy and biomass. Additionally, it could be transformed into a widely used livestock feed (Wahab et al. 2021). The International Nomenclature of Cosmetic Ingredients recognizes “Glabridin-40” as a glabridin-rich extract of *G. glabra* that is sensibly used in cosmetic compositions (Husain et al. 2021). The leading producer of licorice products in the world, providing consumers with more than 70% of the world’s licorice tastes.

Licorice root is used by Mafco Worldwide, the world’s leading manufacturer of natural plant-based licorice products used in a wide range of applications, to create a variety of licorice products, which are then sold to confectioneries, cosmetics, food processors, and pharmaceuticals all over the world (Lim 2016).

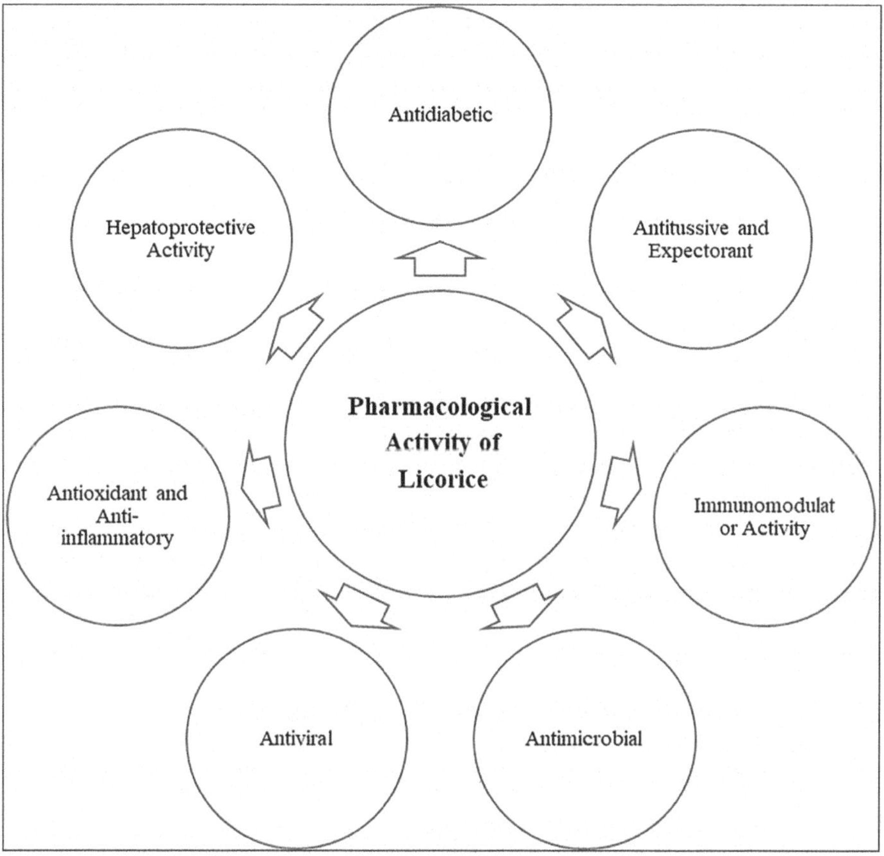

Fig. 2.4 Various pharmacological activities of licorice

2.1.4 Domestication and Selection

When humans cultivate wild plants, a co-evolutionary process known as domestication takes place that results in the emergence of new species that are essential to human survival (Purugganan 2019). Crop species that have undergone domestication as a result of exposure to the novel, selective settings brought on by human cultivation and use are the product of an evolutionary process (Purugganan and Fuller 2009). When one species (the domesticator) starts to regulate the reproduction and dispersal of another species (the domesticated) to suit the demands of the former, most notably (but not entirely) for food, it is a process of speciation and/or species change. 120–160 taxonomic families comprise the estimated 1000–2500 semi- and completely domesticated plant species. All of these species are relatively new; some have only been around for a few decades or have only been discovered since the

Neolithic (Purugganan 2019). The selection of good variations from mixed populations was the most crucial breeding strategy in the early phases of medicinal plant breeding. In this approach, good plant breeding should concentrate on the following important ideas. First, it's important to identify the features for which adjustments are wanted and to establish specific objectives for those changes. Among various *G. uralensis* Fisch. Strins, strains A-19 (high glycyrrhizin content) and G-6 (vigorous growth) were chosen for industrial production (Ozaki et al. 2010).

2.1.5 Selection

The earliest technique for breeding plants is selection, which is based on a rudimentary understanding of inheritance laws. The selection of plants within landraces is predicated on the expectation that the descendants of the greatest individuals will outperform those of a randomly chosen population sample (Breseghello and Coelho 2013). From the first crossing through cultivar registration, breeding a new cultivar takes 10–15 years. The goal of selection is to get a good selection answer with little effort and money spent (Koli et al. 2021).

2.2 Current Cultivation Practices, Challenges and Breeding Objects

2.2.1 Current Cultivation Practices

A perennial crop, licorice is usually harvested in the third or fourth year after planting. A licorice plantation is established in November or December. The process begins with soil preparation, then involves plowing at a depth of 30 cm, harrowing, leveling, and chiseling to a depth of 20–25 cm. Following the creation of the irrigation furrows, licorice is planted using either seed or, preferably, cuttings that have been prepared for planting. Typically, planting involves manually placing licorice cuttings vertically into the soil that has been loosened by a tractor. During the first years of growing, up until the developed roots reach the groundwater level, licorice requires irrigation water application farmers often irrigate at least twice every season when there is a sufficient quantity of water for irrigation. There may be some conflict with irrigation water availability because farmers grow licorice on abandoned fields or low-productive areas. Water supply is typically prioritized for more productive areas growing traditional crops like cotton and winter wheat. By using a (diesel) pump to pump groundwater for irrigation of their licorice crops, farmers can deal with water scarcity. Farmers must perform some agricultural tasks during the second year of licorice cultivation, including irrigation, fertilizer, weeding, cultivation, and harvesting of the green foliage. It must be noted that studies on the cultivation of licorice have shown that the plant requires little irrigation, which is crucial for carrying out irrigated agricultural production in areas with unpredictable water supplies. If a 4-year production cycle is used, the licorice field must be fertilized, irrigated, and have its green foliage picked in the third year. If a 3-year production cycle is used, the field may also require irrigation, fertilizer, and the harvesting of foliage and roots. Late in November, roots are harvested, and then they are quickly brought to the processor since incorrect storage, transportation, or drying might cause the quality of the roots to decline. To guarantee the regrowth of licorice seedlings naturally, at least 25% of the rhizomes must be left in the soil. On the same piece of land, the subsequent cycle of licorice production will likewise take three to four years, but won't involve planting operations. A few farming techniques must be used following the mechanical licorice root harvest. The first year of the second cycle calls for rough field leveling, irrigation furrow installation, fertilizing, and cultivation. Irrigation, fertilizer application, and harvesting of the green foliage are farming techniques that are used the next year. The grower only needs to harvest the roots in the last year of the second cycle. In the following cycles, it is anticipated that the output of licorice roots will rise by at least 30% (http://www.cacilm.org/en/technologies/section/licorice/for_farmers). The *Glycyrrhiza* genus contains taxa with highly developed rhizome and root structures. The root can descend up to 7 m (Öztürk et al. 2018). Additionally, *G. glabra* favors sandy soils, and its rhizomes are typically vertical. In loamy or slightly clayey soils, the rhizomes spread out horizontally. A bud on the rhizome in *G. glabra* can develop into a single plant. The rhizome typically has many buds in *G. echinate* (Öztürk et al. 2018). For optimal root development, sandy-loam-richh soils with a pH of 6–8.2 are excellent. The plant may inhabit areas with an annual rainfall of 50–100 cm. To raise a plantation, utilize stolons with two or three buds that are 15–25 cm long and cut the from newly dug rootstock. Old crowns can also be used as planting material. The cuttings of the underground stem/root are positioned in the field at a distance of 90 × 45 cm, 6–8 cm deep in the soil. For plantation, 250–300 kg of stem cuttings is needed per hectare. After planting, the cuttings start to grow 15–20 days later. During the spring season, irrigation must be done lightly and frequently until a new plant sprouts from the vegetative section that was cut. To encourage the formation of underground roots, farm yard manure is applied to the field at a rate of 10 tons per hectare. The crop receives irrigation every 30–45 days throughout the dry summer months. To prevent fungal disease infection, it is advisable to trim the plants 10 cm above the ground in the first few days of November. Avoid water logging in the field since it could result in soil-borne illnesses, which would then cause root rotting. The crop is harvested in the winter to provide a high output of glycyrrhizic acid. The maximum moisture content for dry roots while being stored in polythene-lined bags should be less than 10% (Pohtam et al. 2019). It has been noted that combining arbuscular mycorrhizal fungi (AMF) with the roots of certain plants might boost plant biomass and production (Amanifar et al. 2019; Haghighi and Saharkhiz 2022; Xie et al. 2019). Liu et al. (2007) have revealed that in the center and western parts of Inner Mongolia, China, if naturally grown *G. glabra* is infected with AM fungus, there is a 30% colonization rate (Liu et al. 2007). Liu et al. (2014) have examined the percentage of licorice roots colonized by several species of AM fungus (either alone or in combination), and the results ranged up to roughly 60%. Additionally, their findings demonstrated that, among all the examined clones, *Glomus intraradices* could reach the greatest level of mycorrhizal root colonization (Liu et al. 2014). It has been hypothesized that AM fungi start a communication cascade that leads to colonization, and that the host does not respond negatively in defense as a result (Dowarah et al. 2021). As a result, AM symbiosis is likely a complicated genetic mechanism that requires in-depth study to clarify the potential mechanisms influencing the differential colonization of different plant roots by different AMF species (Sangwan and Prasanna 2021). According to Yadav et al., by developing efficient mycorrhizal symbiosis, *G. glabra* micropropagated survival rate and growth enhancement. When *G. glabra* plants are inoculated with various AMF treatments at the start of the acclimatization period, their survival rate increases noticeably. *Glomus mosseae* inoculation increases plantlet survival by up to 100% compared to control, which increases plantlet survival by just 66.66% (Öztürk et al. 2018). It is discovered that the incorporation of AMF significantly affects the growth of micropropagated *G. glabra* plantlets. However, this enhancement differs in terms of the fungus strain that was employed. Mycorrhizal inoculation enhances the height of the herb in comparison to the control. After 120 days of inoculation, the dual combination of *Acaulospora laevis* + *G. mosseae* produced the best results. The number of leaves per plant has seen a similar development. Significant increases in root length have been seen with *G. mosseae*, then with *G. mosseae* and *A. laevis* together. The *A. laevis* treatment produced the largest leaf area (Öztürk et al. 2018). The infected plant exhibits a considerable increase in leaf area, which may be the result of higher phosphorus uptake that improves biosynthetic processes and causes a faster rate of growth and development (Mousavi et al. 2022).

2.2.2 Current Agricultural Challenges

Since the irrigated property is reserved for cotton and wheat, farmers could need official approval to produce licorice in the deserted areas. Roots expand and take control of a region as the licorice crop matures. Herbicides might be required to control the crop's "weediness". Farmers noted that because the seeds are so tiny, they can only grow around 20% of them into plants. They asked for and may need help to more effectively develop their crop by sowing or root cuttings. Farmers also said that because roots are now harvested by hand, mechanizing the procedure would enhance production (Chen et al. 2014).

2.2.3 Genetic Improvement Objectives

G. glabra develops seeds inside protective pods. These seeds are created when the plant is effective in moving pollen from one plant to another. This is accomplished by harnessing the energy of the wind. The germination rate of the seeds was calculated through studies and was found to be around 20%. The plant's commercial application has been restricted as a result of the seeds' poor rate of germination (Yu et al. 2011). So, the main aims of licorice breeding may be:

Wider adaptability: A variety's adaptability is its capability for widespread cultivation under a variety of environmental situations. An essential goal in plant breeding is adaptability since it aids in maintaining crop production across areas and seasons (Wei et al. 2012).

Abiotic resistance: Abiotic elements that affect crop plants include drought, salinity in the soil, high temperatures, wind, cold, and frost. Breeders must create varieties that are resistant to these environmental circumstances (Chapman et al. 2012).

Higher yield: Plant breeding's final goal is to increase the yield of'economic produce on economic part". Depending on the crop species, it could be grain, fodder, fiber, tuband cane yield, or oil yield (Lee et al. 2020). The main objectives of licorice breeding are presented in Fig. 2.5.

2.3 Germplasm Biodiversity and Conservation

2.3.1 Germplasm Diversity

In terms of distribution, Eurasia is home to a total of 21 licorice species (Li and Lu 2015). However, during the past ten years, the resources of *Glycyrrhiza* have dramatically declined and its habitat has been seriously harmed as a result of the overexploitation and excavation of wild licorice (Huang et al. 2010). As a result, soy and maize crops have displaced some native licorice populations, resulting in a distribution pattern of continuous patches (Huang et al. 2010). As a result of this drastic population decline as well as rising medical demand, *Glycyrrhiza* biodiversity faces an unprecedented threat. Despite the fact that cultivation eases part of this production pressure, cultivated species have lower levels of glycyrrhizic acid, the plant's primary therapeutic ingredient (Huang et al. 2010; Wang 2000; Wei et al. 2003). Understanding and preserving the genetic diversity of wild licorice populations is crucial (Liu et al. 2019). Several investigations have revealed notable variations in the amount of its active ingredients as well as obvious discrepancies in the effectiveness and quality of various basal licorice varieties (Zhang and Ye 2009; Yang et al. 2016a, b). Resources for licorice are becoming more scarce because to the enhanced rate of anthropogenic activities causing habitat loss and unrestrained exploitation. Artificial culture technology has evolved in response to the expanding market needs, and cultivated licorice variants are now the main source of licorice-based medicines. Despite this, the variety' quality falls short of expectations for a number of reasons (Yu et al. 2011; Zheng et al. 2015). First, the three licorice types do not experience gamete isolation during interspecific hybridization, which results in a high hybridization affinity (Commission 2010; Liang-Bi et al. 2014). Second, the patrilineal nature of licorice chloroplast DNA enhances intraspecific genetic variety and adds complexity to interspecific hybridization and genetic variance (Chen et al. 2017). Additionally, licorice germplasm research is still in its early stages. For instance, there are few licorice planting bases with expertise, and the selection of different resources is primarily relied on breeding, which causes the genetic diversity within the population to be complex (Li et al. 2021; Liu et al. 2021; Wang et al. 2020).

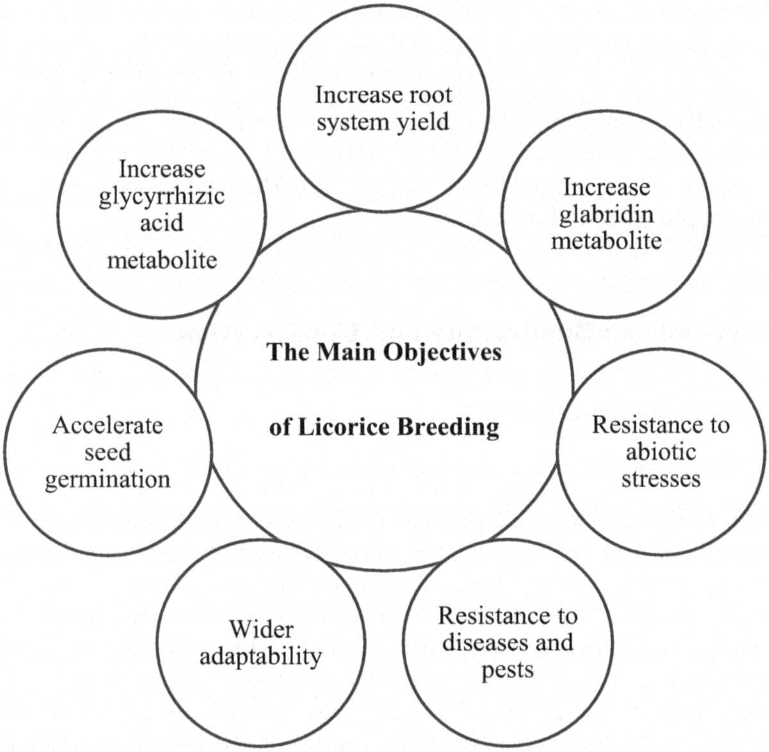

Fig. 2.5 The main objectives of licorice breeding

Table 2.1 Species belonging to the genus *Glycyrrhiza*

No.	Species	No.	Species
1.	*G. acanthocarpa* (Lindl.) J. M. Black	12.	*G. gontscharovii* Maslenn
2.	*G. aspera* Pall	13.	*G. iconica* Hub
3.	*G. astragalina* Hook. & Arn	14.	*G. inflata* Batalin
4.	*G. bucharica* Regel	15.	*G. korshinskyi* Grig
5.	*G. echinata* L	16.	*G. lepidota* Pursh
6.	*G. eglandulosa* X. Y. Li	17.	*G. pallidiflora* Maxim
7.	*G. eurycarpa* P. C. Li	18.	*G. squamulosa* Franch
8.	*G. foetida* Desf	19.	*G. triphylla* Fisch. & C. A. Mey
9.	*G. foetidissima* Tausch	20.	*G. uralensis* Fisch
10.	*G. frearitis* (Boiss.) Orph. ex Beck	21.	*G. yunnanensis* P. C. Li
11.	*G. glabra* L		

2.3.2 Cultivar's Characterization and Phylogeny

There are about 30 taxa in the genus *Glycyrrhiza* (Fabaceae-Leguminosae), which comprise 21 species. These are mentioned in Table 2.1 (Öztürk et al. 2018).

G. glabra L.: Erect perennial, 30–60 cm. Leaflets 5–9-paired, elliptic. Inflorescence 5–9 cm, lax, elongate. Corolla blue to violet. Legume red-brown, glandular or eglandular. 1–6-seeded. Fl. 6–7. Cultivated ground, alluvial river valleys, sand dunes, etc., *G. glabra* also includes three varieties (Altay et al. 2016; Sofia and Walter 2009).

1. *G. glabra var. glabra* (Persian licorice): This plant has violet flowers, a legume, and glandular ovarian tissue.
2. *G. glabra var. glandulifera* (Russian licorice): It doesn't bear stolons but has a large root stock and a few lengthy roots. A leguminous plant with an ovary and at least a few stalked or sessile glands.
3. *G. glabra var. typica* (Spanish and Italian licorice): These plant's papilionaceous blossoms are a shade of purplish blue. It releases a lot of stolons (Öztürk et al. 2018).

2.3.3 Genetic Resources Conservation Approaches

Ex situ and in situ conservation are the two primary methods for protecting plant genetic resources. The following definitions for these categories are provided by the Convention on Biological Diversity: Ex situ conservation refers to the preservation of biological diversity components away from their natural settings. In situ conservation refers to the maintenance and recovery of healthy populations of species in their natural habitats and, in the case of domesticated or cultivated species, in the environments where they have evolved their distinctive characteristics. It also refers to the conservation of ecosystems and natural habitats (Engelmann 2000). There are numerous distinct ways that fall within the two main conservation strategies. Field genebanks, in vitro storage facilities for pollen, DNA, and seeds, as well as botanical gardens are all examples of ex situ conservation, whereas the *in-situ* method focuses on genetic reserves, on-farm conservation, and home gardening (Braverman 2014; Engelmann 2000).

Through assessing the NCBI, EBI, and DDBJ databases, it can be stated that the whole genome of licorice has not been sequenced yet, but the whole EST data of rhizome and SRA data of different licorice organs and many genomics and metabolic studies on different species including *G. glabra*, *G. uralensis*, *G. inflata*, *G. lepidota*, are available in NCBI and DDBJ databases. A metagenomes study has also been uploaded to the EBI database. Various genomics studies have identified about 110 genes, 76 of which are protein-coding licorice. Investigating the largest international germplasm databases in Central Europe, EURISCO, and GENESYS, the licorice genetic resources were found in different genetic resource centers all over the world (the Australian Pastures Genebank, Leibniz Institute of Plant Genetics and Crop Plant Research, Germany, Genetic Resources Institute, Azerbaijan, Agricultural Research Organization, International Center for Agricultural Research in Dry Areas, Lebanon, Western Regional Plant Introduction Station, USDA-ARS, Washington State University, United States, Millennium Seed Bank Project, Seed Conservation Department, Royal Botanic Gardens, Kew, Wakehurst Place, United Kingdom, N.I. Vavilov Research Institute of Plant Industry, Russia, Faculty of Horticulture, etc.) available as seed collection. However, licorice genetic materials are also collected in some botanical gardens whose main purpose is to collect and preserve plant germplasm, such as the Missouri Botanical Garden.

2.3.4 Cytogenetics

A subfield of cytogenetics called "cytotaxonomy" compares karyological characteristics for both systematic and evolutionary reasons (Siljak-Yakovlev and Peruzzi 2012). Chromosome studies can now be used to gather a variety of information, including information about the number of chromosomes, karyotype structure, karyotype asymmetry, chromosome banding, fluorescence in situ hybridization, genomic in situ hybridization, and chromosome painting (Gill et al. 1991; Gokhman 2022; Zhou et al. 2020). Among these, karyotype asymmetry is one of the most common, affordable, and commonly employed method especially by botanists (Peruzzi and Eroğlu 2013). According to karyological investigations on various species of *Glycyrrhiza* (Table 2.2), their chromosome number is $2n = 16$. However, one species, *G. pallidiflora*, has been shown to have $2n = 16$ and $2n = 18$ chromosome numbers (Chuanji et al. 1989; Gurzenkov 1973). Chromosome numbers in *G. echinata*, *G. glabra*, *G. flavescens* subsp. *flavescens* and *G. flavescens* subsp. *antalyensis* are $2n = 16$, consistent with previously stated values for this genus. The chromosome number of *G. asymmetrica* is $2n = 14$.

Table 2.2 Chromosome counts in some taxa of the genus *Glycyrrhiza* (Çetin et al. 2015)

Taxon	2*n*	Literature
G. glabra L	16	Ashurmetov and Karshibaev (1982a, b, 1989), Astanova (1981), Ma et al. (1985), Magulaev (1980), Nafanailova (1989), Verma and Nadkarni (1985), Barghi and Siljak-Yakovlev (1990), Hong (2006), Li et al. (1992), Li and Wei (1991) Nazarova et al. (2004)
G. alalensis X.Y. Li	16	Li and Wei (1992)
G. aspera Pall	16	Li and Wei (1991)
G. echinata L	16	Barghi and Siljak-Yakovlev (1990), Ghaffari (1985), Nafanailova (1989)
G. eglandulosa X.Y. Li	16	Li and Wei (1992)
G. eurycarpa P.C. Li	16	Li and Wei (1991), (1992)
G. foetida Desf	16	Ashurmetov and Karshibaev (1982b)
G. inflata Batalin	16	Hong (2006), Li et al. (1992), Li and Wei (1991), Ma et al. (1985), Yan (1989)
G. korshinskyi Grig	16	Ashurmetov and Karshibaev (1982a), Li et al. (1992), Li and Wei (1991), Nafanailova (1989)
G. laxiflora X.Y. Li and D.C. Feng	16	Li and Wei (1992)
G. lepidota Pursh 16 Löve and Löve	16	Ashurmetov and Karshibaev (1982b), Barghi and Siljak-Yakovlev (1990), Löve (1982), Mulligan (1984), Siljak-Yakovlev and Peruzzi (2012)
G. macedonica Boiss. and Orphanides	16	Ashurmetov and Karshibaev (1982b), Magulaev (1980), Verma and Nadkarni (1985)
G. macrophylla X.Y. Li	16	Li et al. (1992), Li and Wei (1991)
G. nutantiflora X.Y. Li	16	Li and Wei (1992)
G. pallidiflora Maxim	16	Ashurmetov and Karshibaev (1982b), Gurzenkov (1973), Hong (2006), Li et al. (1992), Li and Wei (1991), Nafanailova (1989), Pavlova et al. (1989), Yeh (1986)
G. pallidiflora Maxim	18	Chuanji et al. (1989
G. prostrata X.Y. Li	16	Li et al. (1992), Li and Wei (1991)
G. shiheziensis X.Y. Li	16	Li et al. (1992), Li and Wei (1991)
G. triphylla Fisch. & C. A. Mey	16	Sheidai et al. (2008)
G. uralensis Fisch	16	Ashurmetov and Karshibaev (1982a), Krasnikov (1990), Li et al. (1992), Li and Wei (1991), Ma et al. (1985), Nafanailova (1989), Rostovtseva (1977), Stepanov (1994), Verma and Nadkarni (1985), Xiaochu et al. (1991), Yan (1989)
G. yunnanensis Cheng f. and L.K. Dai ex P.C. Li	16	Li et al. (1992), Li and Wei (1991)

2.4 Traditional Breeding

2.4.1 Traditional Breeding Methodologies

Plant breeding might be characterized as modifications made to herbs as a consequence of their usage by humans (Singer et al. 2021). Three categories can be used to categorize the enormous variety of breeding techniques: (i) Plant selection based on natural variations seen in nature or within conventional types, or plant breeding based on observed variation; (ii) controlled breeding of plants through the selection of seedlings that exhibit the recombination of desired genes from various parents; and (iii) plant breeding that uses molecular techniques to analyze within-genome variation and monitored recombination by choosing certain genes or marker profiles. Continuous use of conventional breeding techniques in a particular species may result in a reduction cultivars that are created from gene pools that contain variations that make crops more vulnerable to biotic and abiotic stresses and hinder development. There are several techniques for adding exotic variety to elite germplasm without having negative impacts (Haroon et al. 2022; Liang et al. 2021). Mass selection, pure line selection, pedigree, bulk population, single seed descent, backcrossing, multiline, and composite are typical techniques for breeding self-pollinated species. The use of mass selection, recurrent selection, family selection, and synthetics are techniques for breeding cross-pollinated species (Priyadarshan 2019).

2.4.2 Traditional Breeding Limitations

Traditional ideas of using captive reproduction to protect biodiversity have limits that are now widely acknowledged. All but a few traditional breeding operations fall short of their ultimate objective of protecting the targeted species due to a lack of space and low knowledge of how to build ideal artificial conditions (Rani et al. 2021). To achieve reasonable success, modernize assisted reproduction with new technology must be prepared. For a small number of critically endangered species, techniques including cryogenics, artificial insemination, and the transfer of embryos created in vivo and in vitro have already been shown effective (Ferré et al. 2020). Despite being a very useful tool, conventional plant breeding has some drawbacks. First, breeding is only possible between two plants that are capable of sexual reproduction. This restricts how many more features can be added to a species' existing traits. Main limitations of breeding for quality traits are given below: (i) In cereals and pulses, there is negative association between grain yield and seed protein. (ii) There is negative association between seed oil and seed protein. (iii) Sometimes selection for high lysine leads to reduction in grain size as in barley (Fig. 2.6) (Amegbor et al. 2022; Loskutov and Khlestkina 2021; Xingzhu et al. 2021).

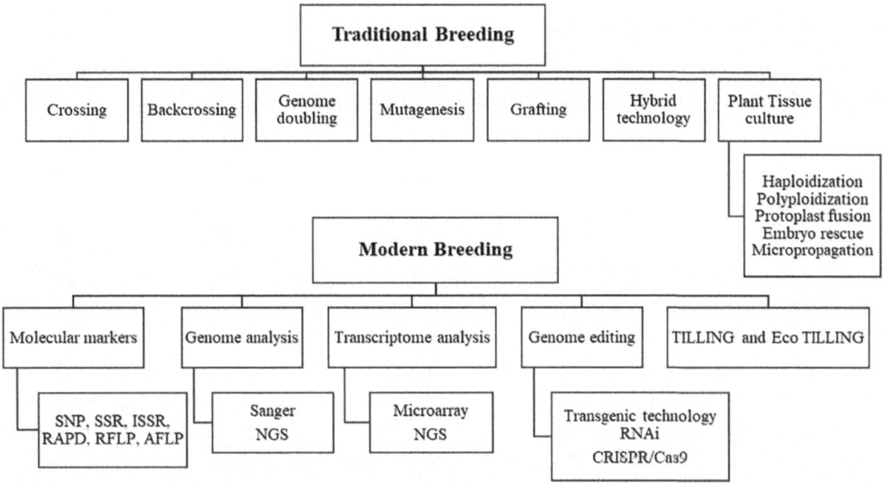

Fig. 2.6 Technics of Traditional and modern breeding

2.4.3 Role of Biotechnology

Biotechnology has been used in a variety of ways that have helped mankind either directly or indirectly. Worldwide, licorice species' roots and stolons have long been utilized as herbal remedies and natural sweeteners. Glycyrrhizin, the primary component responsible for sweetness, is a sweetening factor found in many commercial and herbal products. To now, tissue culture technology has made it possible to mass produce the elite population in less time than with traditional methods. The most crucial component, glycyrrhizin, is improved via in vitro propagation, elicitation, genetic modifications, and bioconversions. High performance liquid chromatography, capillary electrophoresis, UV spectroscopy, counter-current chromatography, and thin layer chromatography have all proved effective in the identification and quantification of glycyrrhizin and its related substances. The genetic variety of plant populations and species has been revealed thanks to advances in molecular markers, which have also helped to identify individual genes, create genetic linkage maps, and study genetic relationships. Numerous contributions have been made recently regarding the assessment of genetic variety, mass production, in vitro production, and genetic improvement of *G. glabra* for increased production of glycyrrhizin (Pandey and Ayangla 2018).

2.5 Molecular Breeding

2.5.1 Molecular Marker-Assisted Breeding

The biodiversity of licorice is currently under threat in Iran as a result of rising demand and a sharp loss in its natural habitats. Therefore, domestication would be required in the near future, and the introduction of cultivable genotypes would be facilitated through molecular breeding. Since phenotypic criteria are frequently scarce and can be greatly impacted by the environment, molecular markers are more appropriate for estimating genetic diversity (Bajpai et al. 2014). Utilizing a variety of techniques, the genetic diversity of *Glycyrrhiza* species has been investigated, including restriction fragment length polymorphism (RFLP) (Shimomura et al. 1994), inter-simple sequence repeat (ISSR) (Yang et al. 2008), random amplified polymorphic DNA (RAPD) (Khan et al. 2009), amplified fragment length polymorphism (AFLP) (Dashzeveg et al. 2018; Zhang et al. 2010) and simple sequence repeat (SSR) (Esmaeili et al. 2020; Liu and Lin 2012; Liu et al. 2015) markers. In a study on Iranian licorice accessions, based on UPGMA cluster analysis was performed, and the 22 populations under study were divided into four major groupings (Fig. 2.7). Only one population belongs to each of the groupings I, III, and IV, which correspond to the Taft, Kashmar, and Bajgah accessions, respectively. With merge value, Taft accession was distinct in GA quantity. However, the aerial part of this population exhibited very high levels of antioxidant potential and phenolics. Kashmar and Bajgah acessions were distinguished in glabridin and GA contents compared to other populations. The second category contained the remaining populations (Esmaeili et al. 2019).

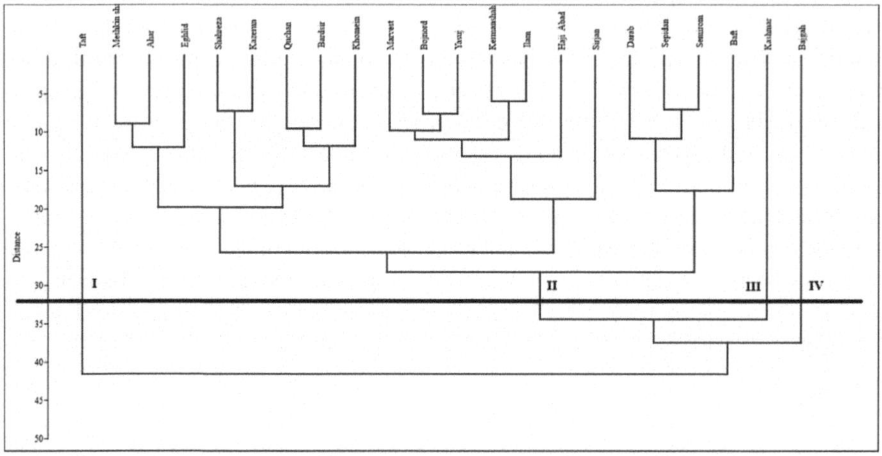

Fig. 2.7 The UPGMA dendrogram for the investigated traits of *G. glabra* populations using Euclidean distance (Esmaeili et al. 2019)

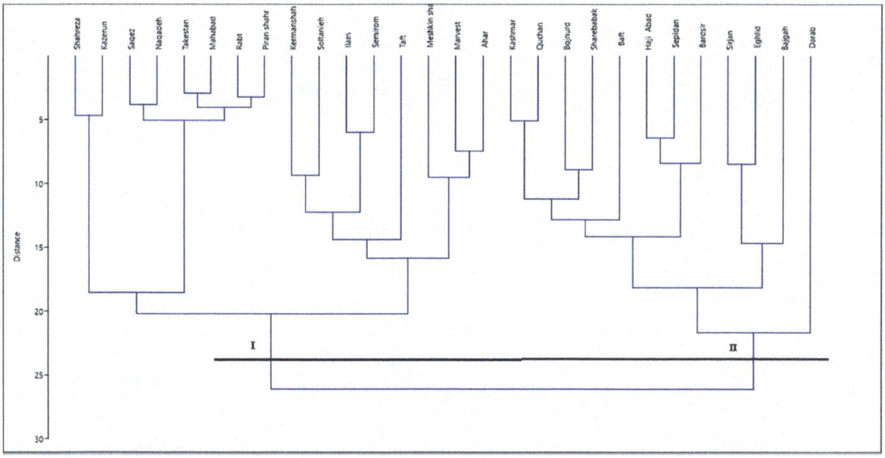

Fig. 2.8 UPGMA (unweighted pair group method with arithmetic mean) cluster analysis for studied morphological traits of cultivated *G. glabra* (Esmaeili et al. 2020)

UPGMA-based morphological cluster analysis revealed two distinct major clusters (Figs. 2.8 and 2.9). The first category included all of the populations taken from Iran's west and northwest. Although some of the populations gathered from Iran's north, east, and south fell into the first category, the majority were grouped into the second one (II). It can be inferred that there was significant phenotypic variety within the population because several populations collected from the same geographic regions did not belong to the same group (Esmaeili et al. 2020).

Recently, research was done on licorice species to examine genetic variation and undertake population analysis using SSR (Erayman et al. 2014; Liu et al. 2019; Esmaeili et al. 2020), RAPD (Jiang et al. 2020), and ISSR (Liu et al. 2015) (Table 2.3). The research of wild populations' genetic, morphological, and phytochemical traits is the first step in the cultivation and domestication process (Hadian et al. 2015). As licorice is endangered, its domestication would be necessary, and molecular breeding would help to introduce genotypes suitable for cultivation. A study was conducted using 170 unique licorice plants that were collected in the wild from 59 locations across 21 provinces of Iran. Six phenotypic phytochemical traits were examined in relation to the 436 polymorphic AFLP markers obtained by 15 primer combinations (EcoRI/MseI). Gene diversity was revealed by the AMOVA analysis both between and within localizations. Two primary sub-populations with high genetic diversity were found through an investigation of population structure. Significant associations were identified between three markers (E3/M40-4, E34/M4-12 and E12/M31-15) and glycyrrhizin concentration, and between four markers (E11/M34-12, E11/M34-15, E9/M7-29, and E9/M7-30) and phenolic compounds contents. Both breeding efforts and licorice domestication can benefit from the markers found. The glycyrrhizin and antioxidant content of licorice sampled from four localizations (KBA1, KBA2,

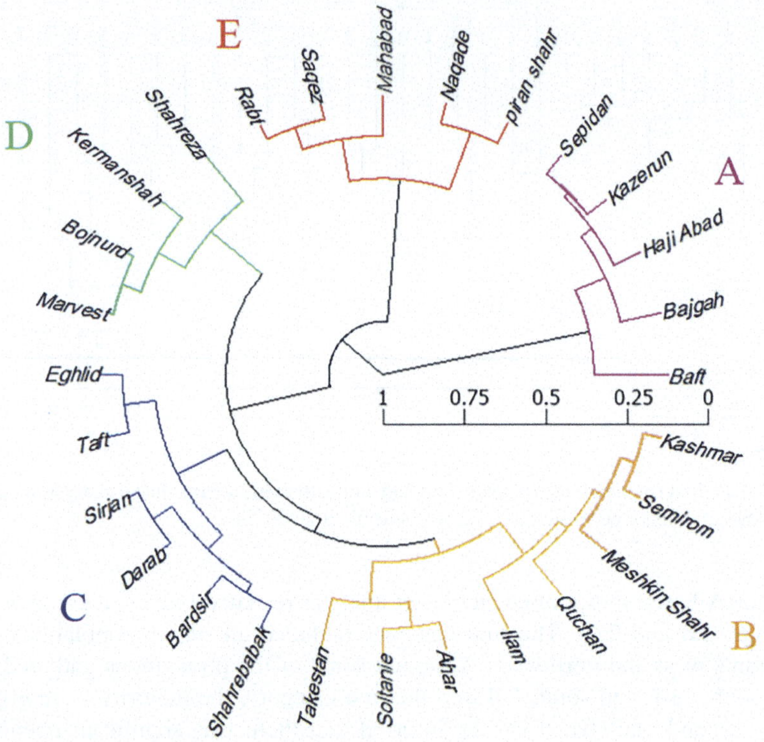

Fig. 2.9 Genetic relationships of the twenty-eight populations of *Glycyrrhiza glabra* from Iran based on microsatellite markers (Esmaeili et al. 2020)

SKh2, and Fa1) was shown to be superior; as a result, they might be regarded as elite genotypes that would be used in the domestication process (Hosseini et al. 2021).

2.5.2 *Functional Genomics*

In order to understand the role and relationships between diverse biological components, including genes, proteins, and metabolites, functional genomics integrates genomics, transcriptomics, proteomics, and metabolomics (Yamazaki et al. 2018). A plant species' chromosome-level de novo whole genome assemblies have been accelerated thanks to the combination of lower sequencing costs, cutting-edge sequencing techniques like long reads sequencing (PacBio, NanoPore), linkedread sequencing (10x-genomics), optical genome mapping (Bionano), and Hi-C genome linked sequencing, as well as smart sequence assembling informatics (Yamazaki et al. 2018). However, given the price and resources needed, full genome sequencing and assembly for all plant species may not be practicable. Therefore, sequencing RNAs

Table 2.3 Genetic diversity of licorice with different molecular markers

Species	Marker	References
Glycyrrhiza glabra L	RAPD	Ahangaran et al. (2021)
G. glabra L	AFLP	Hosseini et al. (2021)
G. glabra L	SSR	Esmaeili et al. (2020)
G. glabra L	SSR	Sorkheh et al. (2019)
G. glabra L	SSR	Liu et al. (2019)
G. glabra L	SSR	Lee et al. (2019)
G. glabra L	SSR	Li et al. (2018)
G. glabra L	AFLP	Hakimi et al. (2017)
G. glabra L	SSR	Erayman et al. (2014)
G. uralensis Fisch	AFLP	Dashzeveg et al. (2018)
G. uralensis Fisch	SSR	Um et al. (2016)
G. uralensis Fisch	SSR	Liu et al. (2015)
G. uralensis Fisch	ISSR	Liu et al. (2015)
G. uralensis Fisch	AFLP	Ge et al. (2009)
G. uralensis Fisch	ISSR	Yao et al. (2008)
G. inflata	SSR	Jiang et al. (2020)
G. lepidota L	SSR	Bang et al. (2019)
G. lepidota L	SSR	Raveendar et al. (2017)
G. inflata	SSR	Yang et al. (2016a, b)
G. echinata L	SSR	Erayman et al. (2014)

has become the method of choice for plant biologists if the goal is to uncover functional and structural aspects of a genome (mRNAs, non-coding RNAs). Numerous studies have successfully used RNA-sequencing and de novo deep transcriptome assemblies and analysis, also known as RNA-seq analysis, to pinpoint the genes encoding potential proteins involved in the manufacture of distinctive specialized compounds in a variety of medicinal herbs (Han et al. 2016; Tripathi et al. 2016). The genomic sequences of a few representative medicinal herbs have also been investigated in recent years, in addition to de novo transcriptome analysis of a variety of therapeutic plants. In two different studies, the genome and transcriptome of Chinese licorice have been studied, respectively (Mochida et al. 2017; Ramilowski et al. 2013).

2.5.3 Bioinformatics

Understanding the genetic and molecular underpinnings of biological activities in herbs that are significant to the specie is the aim of plant genomics. This knowledge is essential for the effective use of herbs as biological resources in the creation of new cultivars with higher standard and lower financial expenses. Additionally, the creation of new plant diagnostic tools depends on this information. Resistance to pathogens and abiotic stress, plant quality features, and reproductive factors that affect yield are traits that are thought to be of the utmost importance. Now, a genome program might be thought of as a crucial tool for enhancing plants. A "quantum leap" in plant improvement will be made possible by this strategy of identifying important genes and comprehending their function. Furthermore, being able to look at gene expression will help us comprehend how herbs respond to and engage with their surroundings and horticultural techniques. Future breeding decision management systems may incorporate this information along with the proper technologies to produce predictive measurements of plant health and quality. The vast amount of data produced by current genome programs will need to be processed, stored, and distributed to the international research community. In addition to sequence information, the data also include details on mutations, markers, maps, newly discovered functions, etc. supplying logical annotation for genes, proteins, and phenotypes, and promoting the public release of all sequence data through repositories, and elaborating relationships both within the data of the plants and between herbs and other organisms are some of the main goals of plant bioinformatics (Imran et al. 2018; Pazhamala et al. 2021; Vassilev et al. 2005).

2.6 Tissue Culture Applications

2.6.1 Micropropagation Approaches

There are primarily two methods for plant in vitro propagation. The plants can either reproduce by developing apical and axillary buds, adventitious shoots, or somatic embryos. It is common practice to encourage pre-existing axillary buds to develop into many shoots, the number of which grows exponentially with subsequent subcultures (Malakar et al. 2022). Organogenesis and somatic embryogenesis are similar micropropagation methods. Adventitious shoot formation is synonymously regarded as direct organogenesis. It lacks the suspension phase and callus, where morphogenesis takes place. Through the process of callus initiation, tissues are formed during indirect organogenesis. There are two types of somatic embryogenesis: direct (when somatic embryos develop directly on intact explants) and indirect (when the same is raised through callus or suspension phase from explants cells). The researchers emphasize the efficacy of micropropagation after shoot tip or nodal cultures most often among the several studies of licorice in vitro multiplication that are now accessible (Table 2.4). Stolon division is the process used for *G. glabra*'s conventional proliferation, but it is extremely slow and produces few offspring. Furthermore, it is necessary to maintain a sizable number of stock plants. *G. glabra* can also be propagated by the use of seeds, however these have poor viability and germination, and the seedlings grow slowly (Shaheen et al. 2020). In vitro propagation can be employed to produce raw plant material to meet the rising demands of the pharmaceutical industry as well as planting material for *G. glabra* production. Prior research on *G. glabra* in vitro propagation mostly concentrated on indirect regeneration through callus development (Chunhua et al. 2010; Mousa et al. 2006; Neelima et al. 2017; Sarkar and Roy 2014; Wongwicha et al. 2008). Direct regeneration of *G. glabra* has been achieved via axillary shoot proliferation using nodal explants (Arya et al. 2009; Mehrotra et al. 2009; Mousa et al. 2006; Shah and Dalal 1980; Thengane et al. 1998; Yadav and Singh 2012), in vitro stolon (Gupta et al. 2013), shoot tip culture, and apical and axillary buds (Kohjyouma et al. 1995; Sawaengsak et al. 2011). Limited research, meanwhile, has focused on the impact of *G. glabra* culture establishment on the type of nodal explants (location on the main stem) (Yadav and Singh 2012). For *G. glabra* to develop and morphologically mature to its full potential, media composition and PGR content are essential (Jaiswal et al. 2017). Additionally, the effectiveness of in vitro regeneration is dependent on a number of steps, each of which has unique requirements. Although, traditionally *G. glabra* is propagated by seed and stolon cuttings, but commercial cultivation is constrained because of poor seed viability and a small seed set. Additionally, the typical propagation process is relatively slow, and only 40–50% of the cuttings establish in the field. To cater the growing demand for pharmaceutical based industries, it is essential to multiply this species through tissue culture (Sharma et al. 2010). Bud break response in its nodal segments was achieved within 21 days of inoculation on Murashige and Skoog (MS) medium supplemented with 2.0–4.0 mg/l BAP. One to two shoots were produced on 2.0 mg/l BAP added to MS medium. The regenerated shoots were isolated and subcultured on MS medium supplemented with 6-Benzyleaminopurine (BAP) (1.0–5.0 mg/l), Kn (1.0–5.0 mg/l) and adenine sulphate (0–100 mg/l) for in vitro shoot multiplication. 8–9 folds shoot multiplication rate was reported on MS medium with 2.0 mg/l BAP, 0.5 mg/l Kn and 50 mg/l adenine sulphate. Rooting of these in vitro raised stems was reported on half strength MS medium with 0.1–4.0 mg/l IAA. Tissue culture raised plantlets were hardened and adapted in mist chamber and shade house conditions for two months (Arya et al. 2009).

2.7 Genetic Engineering and Gene Editing

Producing transformed hairy roots with the T-DNAs of the helper Ri plasmid and of a vector based on a disarmed mini-Ti plasmid is possible with the help of the binary vector system based on the Agrobacterium-Ri plasmid (Saito et al. 1990). This method relies on the ability of vir gene products from the Ri plasmid to mobilize T-DNA produced from the Ti plasmid. Given that double transformants may be anticipated in 60% of the cases, this technique offers a straightforward way to create transgenic tissues carrying any desired foreign genes (Rehman et al. 2021).

Table 2.4 Some research related to tissue culture of different organs of licorice (from 2010 to 2021)

No	Species of glycyrrhiza	Explant	References
1	*G. glabra* L	Stem and leaf	Jaiswal et al. (2021a, b)
2	*G. glabra* L	Intermediate nodal explants	Shaheen et al. (2020)
3	*G. glabra* L	Shoot tip	Manekar and Jadhav (2019)
4	*G. glabra* L	Leaf	El-Hameid et al. (2018)
5	*G. glabra* L	Nodal explant	Nimbalkar et al. (2018)
6	*G. glabra* L	Leaf and nodal	Neelima et al. (2017)
7	*G. glabra* L	Shoot tips, young leaf and nodal	Badkhane et al. (2016)
8	*G. glabra* L	Shoot tip	Manisha et al. (2014)
9	*G. glabra* L	Shoot tip	Srivastava et al. (2013)
10	*G. glabra* L	Nodal explant	Yadav and Singh (2012)
11	*G. glabra* L	Shoot tip	Sawaengsak et al. (2011)
12	*G. glabra* L	Nodal explant	Sharma et al. (2010)

2.8 Mutation Breeding

2.8.1 Conventional Mutagenesis (Seeds)

Mutation breeding is centered on the induction of mutations in organisms through physical, chemical, or spatial processes, as opposed to selective breeding, which depends on natural genetic mutation in plants. One of the most successful methods for acquiring fresh germplasm resources is physical mutagenesis. The various radiation sensitivity of plants can have a substantial impact on breeding (Wang et al. 2020). For instance, it has been discovered that a low dose of gamma irradiation given to the seeds of *G. uralensis* causes varying increases in the width, length, and biomass of rhizomes. Additionally, it was discovered that the amount of glycyrrhizic acid, one of the plant's primary active components that helps clear mucus and alleviate coughing, had grown from 6.57 to 8.54%. Researchers have discovered that *G. uralensis* seeds contain more glycyrrhizic acid and liquiritin when exposed to the physical conditions of spaceflight (Zhang et al. 2011). Mutation breeding is a technique of creating heritable variability through modifying genes by introducing mutations and employing sophisticated selection processes to improve species for desired traits. Physical mutagens (X-rays, gamma rays, thermal neutrons, and fast neutrons) have been used in mutation breeding to create approximately 89% of the mutant kinds, with gamma rays alone responsible for the generation of approximately 60% of the mutant varieties. A major contribution to the countries' national economies has been made by mutation breeding. India has also developed into a significant acknowledged center with the publication of more than 305 mutant cultivars from 56 plant species, it is the second largest distributor of mutant varieties in the world and a center for research on induced mutations (Koli et al. 2021). In medicinal plants examples of successful breeding by mutation induction are mutant variety Sujata (LL34) in *Opium poppy* with opium less, very low alkaloids (Nonnarcotic), oil-seed crop, Aela (Y-17) and Aekla (Ub) varieties of Black henbane with Un-branched, more biomass, and high tropane alkaloid yield and in *Plantago ovata* mutant varieties Niharika (M20-22) and Mayuri (CIMAP/S 5) with high seed and seed husk yield was developed by mutation breeding at CIMAP, Lucknow (India) (Koli et al. 2021).

2.9 Transcriptomes Analysis

A essential regulating enzyme for the biosynthesis of various significant groups of chemicals, such as sterols and triterpenoids, is squalene epoxidase (SQE). Researchers looked at the transcriptome-wide discovery of squalene epoxidase genes from *G. glabra* L. They identified and characterized five SQE genes (*GgSQE1* to *GgSQE5*) from *G. glabra* through transcriptome data mining and homology-based cloning, for the first time. Their functional divergence was revealed by the phylogenetic study. Following transient and stable expressions, one of the five SQEs, *GgSQE1*, was cloned for its function in a heterologous system. Following the deduced protein's anticipated molecular mass, transient expression and GgSQE1-encoding protein purification showed a roughly 58.0-kDa protein. The qRT-PCR gene expression profiling revealed that the roots that were 10 months old had the highest expression (6.4-folds). Furthermore, ABA and GA_3 upregulated the expression of *GgSQE1* in the shoots, which was also reflected in glycyrrhizin accumulation. Functional study of the GgSQE1 promoters using the GUS fusion construct verified the inductive effects of ABA and GA_3 over GgSQE1 expression (Manzoor et al. 2021). In another study, combined analysis of pharmaceutical active ingredients and transcriptomes of *G. uralensis* under PEG6000-induced drought stress were studied. Under drought stress, it was determined how pharmacological active compounds accumulated and were regulated in *G. uralensis*. To determine phytohormones and pharmacological active components, RNA sequencing was performed on *G. uralensis* seedlings that had been exposed to 10% PEG6000 for 0, 2, 6, and 24 h. For content detection and gene expression analysis, *G. uralensis* seedlings were treated with PEG6000, ibuprofen (IBU), and jasmonic acid (JA). The findings demonstrated that under drought stress, *G. uralensis* underground parts considerably accumulated glycyrrhizic acid, glycyrrhetinic acid, and flavonoids (licochalcone A, glabridin, liquiritigenin, isoliquiritigenin, and liquiritin). Kyoto genes and genomes encyclopedia analysis of the *G. uralensis* transcriptome data after drought treatment revealed that up-regulated differentially expressed genes (UDEGs) involved in the production of flavonoids in both the aerial and underground portions and glycyrrhizic acid in the underground regions were highly enriched. It's interesting to note that both aerial and underground portions of the body include UDEGs that are involved in the signal transduction of jasmonic acid (JA). Additionally, JA accumulation under drought stress was particularly pronounced in aerial and underground sections. Additionally, drought stress elevated the expression of genes controlling the synthesis and transduction pathway, which in turn stimulated the levels of flavonoids, glycyrrhizic acid, and JA. Additionally, it was found that *G. uralensis* treated with PEG6000 and JA accumulated considerable amounts of glycyrrhizic acid and flavonoids and increased expression of the associated genes in these pathways. However, IBU, a JA signaling inhibitor, greatly blocked these increases. PEG treatment also dramatically boosted JA content as well as gene expression levels for JA production and signal transduction. Their research establishes the basis for increasing the medicinal component of *G. uralensis* using genetic engineering techniques by suggesting that drought stress may encourage the accumulation of pharmacological active components via JA-mediated signaling pathway (Yao et al. 2022). Research on the aerial portions of *G. uralensis* under salt stress included an integrated investigation of the pharmacological active component and transcriptome also done. The results showed that salinity enhanced the amount of *G. uralensis*'s active components. In a study, the main genes and metabolic pathways regulating the accumulation of pharmacological active components were identified by performing RNA-sequencing (RNA-seq) on the aerial parts of *G. uralensis* that had been exposed to 150 mM NaCl for 0, 2, 6, and 12 h. Liquiritin was identified as the primary constituent and had a considerable rise of more than 10 times during the 6 h NaCl treatment, according to the main active component detection. Two screening profiles, which included the significantly upregulated DEGs (UDEGs) at various treatment points, were identified by temporal expression analysis of the 4245 differentially expressed genes (DEGs) collected by RNA-seq. The most significantly enriched pathways according to a Kyoto Encyclopedia of Genes and Genomes (KEGG) analysis of these UDEGs were phenylpropanoids metabolism and flavonoid biosynthesis. Interestingly, the ABA synthesis pathway associated with carotenoid biosynthesis was also identified, and the ABA content was markedly increased after a 6-h NaCl treatment. After 2 h of treatment, the concentration of liquiritin showed a significant and immediate increase in response to ABA stimulation. Genes involved in the pathways of ABA signal transduction and flavonoid biosynthesis were similarly consistently expressed, but not those involved in the pathway of glycyrrhizic acid biosynthesis. As a result of their research, which suggests that salinity may encourage liquiritin accumulation via the ABA-mediated signaling pathway, *G. uralensis* can now be improved genetically and fully utilized (Bi et al. 2022).

2.10 Enhanced Traits and Improved Cultivars

In various studies, different populations or accessions were studied under various stresses. One of the primary environmental factors preventing plant growth and development is salinity stress. In a study, the effects of six NaCl concentrations (0, 100, 200, 400, 600, and 800 mM) were examined in licorice (*G. glabra*) rhizomes from two populations. The findings demonstrated that the salt stress raised the proline concentrations in the rhizomes while gradually reducing the growth characteristics. When compared to the controls, the K^+/Na^+ ratio under salinity significantly decreased. According to high levels of lipid peroxidation, hydrogen peroxide, and antioxidant enzyme activity, salt exposure led to oxidative stress on the rhizomes. Only the 100 and 200 mM NaCl treatments resulted in an increase in glycyrrhizin content. The expression of the genes bAS, CYP88D6, and CYP72A154 in the Fars population all followed the same pattern. Glycyrrhizin content was discovered to be higher in the Fars population than the Khorasan population. However, compared to the Fars population, the Khorasan population showed higher loss in growth, glycyrrhizin concentration, and glycyrrhizin biosynthesis genes. According to the findings, adding 100 mM NaCl to a solution increased the expression of crucial genes involved in the manufacture of triterpenoid saponins and thus increased the amount of glycyrrhizin produced. Therefore, *G. glabra* can be introduced as a halophyte plant (Behdad et al. 2020). In the other study, sixteen Iranian accessions of *G. glabra* L., inoculated with *Azotobacter* sp., and growing under salinity stress (NaCl 200 mM) were studied. The findings revealed that Bardsir and Bajgah accessions considerably shown higher resistance to salinity, for instance by increasing crown diameter (11.05 and 11 cm, respectively) compared to an average diameter of 9.5 cm in other accessions. Plant height and crown diameter significantly increased after Azotobacter inoculation. Kashmar (46.21%) and Ilam (44.95%) showed the highest rates of membrane stability index among the accessions examined. Evaluation of enzyme activity revealed that the presence of salt in the environment caused bacteria to produce more polyphenol oxidase (0.21 U mg^{-1} protein), and peroxidase (3.09 U mg^{-1} protein), activity. Comparing all research accessions, the Darab accession demonstrated the largest increase in antioxidant potential (6.45%) after Azotobacter inoculation. Principal component analysis (PCA) revealed that Meshkinshahr's accession demonstrated a more notable capacity to activate its enzymatic defense system under salinity (Mousavi et al. 2022).

2.11 Hybridization

2.11.1 Conventional Hybridization

In hybrid breeding, a hybrid is created by mating two homozygous parental lines with different genetic backgrounds. Higher yields occur from the heterosis effect, which makes the ensuing heterozygous offspring ("hybrids") significantly more productive than both parents (Cui et al. 2020). Hybrid varieties are uniform and high performance by hybrid vigor. Genic-cytoplasmic pollen sterility should be used for hybrid breeding or outbreeding MAP to minimize the expense of emasculation. Separate lines are needed for hybrid varieties development with the following characteristics;

(a) Male sterile.
(b) Male fertile with maintainer properties to maintain the male sterile line.

(c) Male pollinator that is fertile and has good combining skills to produce F1 seeds.
(d) A male fertile restorer must be given to the F1 seeds if generative organs are collected and the pollinator parent lacks restorer qualities in order to guarantee an adequate seed set in the production field (Rakshit and Bellundagi 2019).

The process of hybrid breeding involves fusing the beneficial qualities of two or more kinds. Given that the progeny of crossing will fully depend on parent plant genetics in this method of breeding, parent selection is one of the most crucial aspects in the breeding program. Additionally, hybridization takes use of heterosis, also known as hybrid vigor, which occurs when a generation outperforms either of its parents in some way (Fridman 2015). Crossbreeding has thus far produced a few promising variants. For instance, strain C-2 of *G. uralensis*, which was produced by mating strain A-19, which had a high glycyrrhizin content, and strain G-6, which had robust growth, showed vigorous growth as well as high glycyrrhizin and total flavonoid concentrations (Ozaki et al. 2010). Glycyrrhizin, the primary active component of licorice, a traditional Korean medicinal herb. Chinese licorice (*G. uralensis*) has less glycyrrhizin than the reference amount (2.5%) established by the Korean Pharmacopoeia, which lowers its value as a therapeutic herb. In a study, researchers tried to address this issue by creating 32 F1 lines of an interspecific licorice hybrid by mating European licorice (*G. glabra*) with Chinese licorice. Genetic features were compared and showed that glycyrrhizin content varies among lines, ranging from 1.5 to 5.6%, with a mean value of 3.2%; these values are higher than those of the parental plants. Additionally, 25 lines (78.1%) exhibited glycyrrhizin contents greater than 2.5%, above the Korean Pharmacopoeia's reference threshold. Glycyrrhizin concentrations in four of these lines were greater than the WHO-recommended level of 4%. The leaves of the hybrids exhibited all the traits of both European and Chinese licorice, according to phenotypic comparisons; nevertheless, the stems of most hybrids had traits of European licorice. Finally, the research determined the genetic distances of 34 samples of *Glycyrrhiza* plants (parents, 32 F1 lines) by random amplified polymorphic DNA (RAPD); the F1 lines showed a close genetic distance. They plan to develop a cultivar using five of these lines (glycyrrhizin content < 4.0%) (Table 2.5) (Lee et al. 2017; Wang et al. 2020).

Glycyrrhizin quantity of F1 Lines for 25 (78.1%) of the 32 F1 lines, the glycyrrhizin content was > 2.5%; that of four lines (Ggu 8, Ggu 15, Ggu 30, and Ggu 31) even exceeded the 4.0% glycyrrhizin level established by the WHO. The average amount of glycyrrhizin in the F1 lines was 3.2%, which was more than the parental herbs' levels. Four hybrid lines (Ggu3, Ggu13, Ggu14, and Ggu28) were unable to meet the Korean Pharmacopoeia's glycyrrhizin threshold level of 2.5% (Table 2.5). They also examined the glycyrrhizin content of domestically grown and imported licorice to compare the glycyrrhizin concentration amongst the F1 lines. The glycyrrhizin concentration of licorice plants grown in China and Inner Mongolia (Neimenggu) was higher than that of domestically grown herbs, and these plants typically met the standard threshold of 2.5% established by the Korean Pharmacopoeia. One Chinese sample revealed glycyrrhizin levels that was higher than the World Health Organization's suggested threshold level of 4.0% (Lee et al.

Table 2.5 Root characteristics and glycyrrhizin contents of F1 lines and their parents

Parent/F1 line	Root diameter (mm)	Root weight		Glycyrrhizin content (% dry weight)
		Fresh weight (g)	Dry weight (g)	
G. glabra (♀)	11.6*	36.4	20.1	2.29 ± 0.02
G. uralensis (♂)	9.5	29.8	16.4	0.86 ± 0.01
Ggu 1	16.9	51.7	28.5	3.24 ± 0.01
Ggu 2	10.6	34.8	19.1	3.67 ± 0.02
Ggu 3	12.9	48.1	26.8	2.46 ± 0.02
Ggu 4	13.8	40.8	22.3	4.29 ± 0.03
Ggu 5	14.3	69.8	36.1	2.74 ± 0.02
Ggu 6	12.9	51.1	29.5	3.63 ± 0.01
Ggu 7	13.2	62.3	35.5	2.61 ± 0.04
Ggu 8	14.3	42.5	24.5	3.51 ± 0.04
Ggu 9	12.7	57.5	33.5	3.61 ± 0.02
Ggu 10	13.9	38.5	23.8	4.43 ± 0.02
Ggu 11	13.2	38.3	23.3	5.29 ± 0.04
Ggu 12	10.3	30.0	17.5	3.63 ± 0.01
Ggu 13	9.5	34.8	21.5	1.56 ± 0.01
Ggu 14	10.2	47.8	27.8	1.92 ± 0.03
Ggu 15	14.1	80.5	49.0	3.22 ± 0.02
Ggu 16	10.9	53.3	31.8	4.93 ± 0.02
Ggu 17	12.3	34.0	20.1	2.61 ± 0.02
Ggu 18	13.9	51.5	31.3	3.27 ± 0.01
Ggu 19	14.2	57.5	35.3	2.97 ± 0.01
Ggu 20	13.8	62.5	40.3	3.52 ± 0.01
Ggu 21	11.1	32.3	20.8	3.33 ± 0.02
Ggu 22	15.3	59.8	35.8	3.39 ± 0.03
Ggu 23	13.7	126.5	76.0	3.73 ± 0.02
Ggu 24	13.1	46.0	27.5	4.17 ± 0.02
Ggu 25	13.3	47.5	28.5	2.80 ± 0.01
Ggu 26	13.0	44.1	26.8	2.94 ± 0.01
Ggu 27	13.5	42.8	27.5	2.08 ± 0.02
Ggu 28	17.0	99.8	60.8	1.97 ± 0.01
Ggu 29	15.7	78.8	47.0	3.24 ± 0.02
Ggu 30	12.6	35.5	21.5	3.91 ± 0.05
Ggu 31	13.6	36.3	22.0	2.96 ± 0.02
Ggu 32	14.5	80.0	50.3	2.95 ± 0.01

* Data are presented are the mean ± SD of more than three replicates (Lee et al. 2017)

2017). The glycyrrhizin levels in the F1 lines varied, ranging from 1.56% (Ggu13) to 5.29% (Ggu11), indicating that heterosis altered both the amounts of the secondary metabolic product, glycyrrhizin, as well as the phenotypic traits of the F1 plants. Additionally, the hybridization trials changed other aspects of the plant, such as the production and shape of the roots, which may help in choosing the best types for future propagation. For instance, among the F1 lines, the hybrid line Ggu11 had the lowest dry weight and the greatest glycyrrhizin content (5.29%), indicating low productivity (Lee et al. 2017). By using conventional breeding techniques to cross European and Chinese licorice, they created an interspecific hybrid and then grew the F1 lines that resulted. The hybrid lines had leaves with a variety of forms that resembled both of their parental plants, however they were typically larger than their parents' leaves (Lee et al. 2017).

2.12 Somatic Cell Hybridization

2.12.1 Hybrid Cultivars

The herbal crop team at the National Institute of Horticultural and Herbal Science, Rural Development Administration, Korea created a new interspecific hybrid cultivar, Wongam, as a high-yield licorice from a *G. glabra* × *G. uralensis* (*G. korshinski* Grig.) cross. The leaf shape is oval, and the plant type is erect. Compared to the control species *G. uralensis*, it possesses well-developed vascular bundles (Table 2.6). In comparison to the control species, *G. uralensis*, "Wongam" had dried root weights of 54.6 g per plant as opposed to 22.7 g per plant in control. Additionally, "Wongam's" average yield was 359 kg/10 a, which was 127% higher than *G. uralensis*' average yield (Table 2.7). It demonstrates resistance to lodging and the brown spot disease. 'Wongam' contains 3.96% of glycyrrhizin and 0.8% of liquiritigenin, which complies with Korean Pharmacopoeia Standards (Table 2.8) (Lee et al. 2020). The Rural Development Administration created the new interspecific hybrid licorice cultivar Wongam as a cross between *G. glabra* and *G. uralensis* (*G. korshinski* Grig). Compared to *G. uralensis*, wongam produces greater yield and is more resistant to infection. Various examinations have done on the biological activities of *G. uralensis*, but Wongam, a novel cultivar, has only been the subject of limited research on its anti-ulcer, anti-allergic, and immunomodulatory properties (Kim et al. 2021; Lee et al. 2020).

Table 2.6 Inherent characteristics of *G. glabra* (♀) x *G. uralensis* (♂) hybrid cultivar 'Wongam'

Cultivar	Plant type	Leaf		Root color		Root shape	
Wongam	Erect type	Oval	Green	Brown	Light yellow	Thick	Developed
G. uralensis	Semi-runner type	Lanceolate	Green	Reddish brown	White	Thin	Undeveloped

Table 2.7 Agronomical characteristics of *G. glabra* (♀) x *G. uralensis* (♂) hybrid cultivar 'Wongam'

Cultivar	Plant length (cm)	Stem diameter (mm)	Branch number (ea/plant)	Root length (cm)	Main root diameter (mm)	Root dry weight (g/plant)	Yield (kg/10a)	Index
Wongam	102.9	8.40	6.0	51.3	15.2	54.6	359	227
G.uralensis	54.5	4.3	4.7	35.1	11.3	22.7	158	100

Table 2.8 Ingredient content (%) of *G. glabra* (♀) x *G. uralensis* (♂) hybrid cultivar 'Wongam' (Lee et al. 2020)

Cultivar	Glycyrrhizin	Liquiritigenin
Wongam	3.96 ± 0.02	0.80 ± 0.03
G. uralensis	1.90 ± 0.01	0.57 ± 0.10

2.13 Conclusions and Prospects

Legumes are essential for maintaining the world's food supply and for enhancing soil quality and fixing nitrogen. To meet the demands of the world's expanding population, higher genetic gains must be accelerated. Due to improvements in next-generation sequencing and high-throughput genotyping technology, the field of legume genomics has recently seen rapid breakthroughs. In the last five years, reference genome sequences for numerous legumes have been revealed. The ability to find structural changes at a wide scale has been made possible by the availability of the draft genome sequences and the re-sequencing of elite genotypes for a number of significant legume crops. Large-scale genomic resources and low-cost, high-throughput genotyping technologies are making genetic mapping and marker-trait association research more effective and precise. Most crucially, the use of molecular breeding techniques has led to the creation of better lines. Breeder-friendly genomics and decision support technologies must be implemented in breeding programs in underdeveloped countries to facilitate genomics-driven agricultural development at a rapid pace. This chapter gave a summary of cutting-edge genomics and informatics techniques and methods that will be the main force behind increasing genomics-assisted breeding in the *Glycyrrhiza* genus.

References

Ahangaran M, Gharaviri M, Masoumian M (2021) A study on the genetic variation in licorice (*Glycyrrhiza glabra*) Iran by molecular DNA markers (RAPD). Health Food Biotechnol 2(4):89–97

Altay V, Karahan F, Öztürk M, Hakeem KR, Ilhan E, Erayman M (2016) Molecular and ecological investigations on the wild populations of Glycyrrhiza L. taxa distributed in the East Mediterranean Area of Turkey. J Plant Res 129(6):1021–1032

Amanifar S, Khodabandeloo M, Mohseni Fard E, Askari MS, Ashrafi M (2019) Alleviation of salt stress and changes in glycyrrhizin accumulation by arbuscular mycorrhiza in liquorice (*Glycyrrhiza glabra*) grown under salinity stress. Environ Exp Bot 160:25–34. https://doi.org/10.1016/j.envexpbot.2019.01.001

Amegbor IK, van Biljon A, Shargie N, Tarekegne A, Labuschagne MT (2022) Heritability and associations among grain yield and quality traits in quality protein maize (QPM) and non-QPM hybrids. Plants 11(6):713

Arya S, Rathi N, Arya ID (2009) Micropropagation protocol for *Glycyrrhiza glabra* L. Phytomorphology 59(1):71

Ashurmetov AA, Karshibaev KK (1989) Morfometricheskij i polykaryogramnyj analiz chromosom Nekotorykh Predstavitelej Semejstva Fabaceae Lindl. Tesizy II Symp. Plant Karyology 68–70

Ashurmetov AA, Karshibaev KK (1982a) Kariologicheskoe issledovanie vidov roda Glycyrrhiza L. (Sekt. EuGlycyrrhiza Boiss.). Dokl Akad Nauk Uzbeksk SSR Tashkent 3:44–45

Ashurmetov AA, Karshibaev KK (1982b) Kariologicheskoe issledovanie vidov roda Glycyrrhiza L. (Sekt. PseudoGlycyrrhiza Krug.). Uzbeksk Biol Zurn 1:36–39

Astanova SB (1981) Chromosome numbers of Leguminosae of flora in Tajikistan. Doklady Akadamii Nauk Tadzikskoj SSR 24:61–63

Badkhane Y, Yadav AS, Bajaj A (2016) Effect of explant source and different concentrations of plant growth regulators on in vitro micropropagation of *Glycyrrhiza glabra* L. Indo Amer J Pharma Res 6:5830–5840

Bajpai PK, Warghat AR, Sharma RK, Yadav A, Thakur AK, Srivastava RB, Stobdan T (2014) Structure and genetic diversity of natural populations of Morus alba in the Trans-Himalayan Ladakh Region. Biochem Genet 52(3):137–152

Bang JH, Jo IH, Chung JW (2019) Molecular chracterization of new gDNA SSR markers for glycyrrhiza lepidota and cross-amplification of other Glycyrrhiza Species. 한국육종학회 학술발표회지, 242

Barghi N, Siljak-Yakovlev S (1990) Karyological study in three species of Glycyrrhiza genus (G. glabra, G. lepidota and G. echinata). Caryologia 43(3–4):223–234

Behdad A, Mohsenzadeh S, Azizi M, Moshtaghi N (2020) Salinity effects on physiological and phytochemical characteristics and gene expression of two *Glycyrrhiza glabra* L. populations. Phytochemistry 171:112236

Bi Q, Yao H, Wang F, He D, Xu W, Xie S, Chen X, Li Y, Liu H, Shen H (2022) Integrative analysis of the pharmaceutical active ingredient and transcriptome of the aerial parts of *Glycyrrhiza uralensis* under salt stress reveals liquiritin accumulation via ABA-mediated signaling. Mol Genet Genomics 297(2):333–343

Braverman I (2014) Conservation without nature: the trouble with in situ versus ex situ conservation. Geoforum 51:47–57

Breseghello F, Coelho ASG (2013) Traditional and modern plant breeding methods with examples in rice (Oryza sativa L.). J Agric Food Chem 61(35):8277–8286

Çetin Ö, Duran A, Martin E, Küçüködük M (2015) Karyological studies in some Glycyrrhiza (Fabaceae) taxa from Turkey. Caryologia 68(3):254–264

Chapman SC, Chakraborty S, Dreccer MF, Howden SM (2012) Plant adaptation to climate change—opportunities and priorities in breeding. Crop Pasture Sci 63(3):251–268

Chen C, Lu J, Li X, Zhou L, Xie L, Li X, Song F (2017) Inheritance analysis and discovery of chloroplast paternal inheritance in interspecific crossing of Glycyrrhiza. Guangxi Zhiwu/guihaia 37(2):162–168

Chen KZ, Song H, Ruyu C (2014) Licorice industry in China: implications for licorice producers in Uzbekistan. Intl Food Policy Res Inst

Chuanji G, Yankun L, Yue Z, Pingsheng H (1989). Observations on the chromosome numbers of medicinal plants of Shandong. Chin Tradit Herb Drugs (中草药) 20(6):34–35

Chunhua F, Lei C, Gan L, Li M, Yang Y, Yu L (2010) Optimization of embryogenic-callus induction and embryogenesis of *Glycyrrhiza glabra*. Afr J Biotechnol 9(36)

Commission CP (2010) Pharmacopoeia of the People's Republic of China, vol 1. Chinese Medical Science and Technology Press, Beijing, p 292

Cui Y, Li R, Li G, Zhang F, Zhu T, Zhang Q, Ali J, Li Z, Xu S (2020) Hybrid breeding of rice via genomic selection. Plant Biotechnol J 18(1):57–67

Dashzeveg N, Vornam B, Nergui S, Buerkert A (2018) Differentiation of *Glycyrrhiza uralensis* Fisch. Ex DC. Populations in the bulgan river basin, western Mongolia. Genet Resour Crop Evol 65(7):1857–1865

Dowarah B, Gill SS, Agarwala N (2021) Arbuscular mycorrhizal fungi in conferring tolerance to biotic stresses in plants. J Plant Growth Regul 1–16

El-Hameid AA, El-Kheir ZA, Abdel-Hady M, Helmy W (2018) In vitro and molecular characterization using ISSR markers of *Glycyrrhiza glabra* L. BioTechnologia J Biotechnol Comput Biol Bionanotechnol 99(4)

Engelmann F (2000) Importance of cryopreservation for the conservation of plant genetic resources

Erayman M, Ilhan E, Güzel Y, Eren AH (2014) Transferability of SSR markers from distantly related legumes to Glycyrrhiza species. Turk J Agric for 38(1):32–38

Esmaeili H, Karami A, Hadian J, Saharkhiz MJ, Nejad Ebrahimi S (2019) Variation in the phytochemical contents and antioxidant activity of *Glycyrrhiza glabra* populations collected in Iran. Ind Crops Prod 137:248–259. https://doi.org/10.1016/j.indcrop.2019.05.034

Esmaeili H, Karami A, Hadian J, Ebrahimi SN, Otto LG (2020) Genetic structure and variation in Iranian licorice (*Glycyrrhiza glabra* L.) populations based on morphological, phytochemical and simple sequence repeats markers. Ind Crop Prod 145:112140. https://doi.org/10.1016/j.indcrop.2020.112140

Ferré LB, Kjelland ME, Strøbech LB, Hyttel P, Mermillod P, Ross PJ (2020) Recent advances in bovine in vitro embryo production: reproductive biotechnology history and methods. Animal 14(5):991–1004

Fiore C, Eisenhut M, Ragazzi E, Zanchin G, Armanini D (2005) A history of the therapeutic use of liquorice in Europe. J Ethnopharmacol 99(3):317–324

Fridman E (2015) Consequences of hybridization and heterozygosity on plant vigor and phenotypic stability. Plant Sci 232:35–40

Ge S, Li G, Ma Z, Wu X, Meng Y, Huo Y (2009) Analysis on genetic diversity of wild populations of licorice (*Glycyrrhiza uralensis* Fisch.) with AFLP markers. Sci Agric Sin 42(1):47–54

Ghaffari SM (1985) Chromosome counts of some angiosperms from Iran. Iran J Bot 3:67–73

Gill BS, Friebe B, Endo TR (1991) Standard karyotype and nomenclature system for description of chromosome bands and structural aberrations in wheat (*Triticum aestivum*). Genome 34(5):830–839

Gokhman VE (2022) Comparative karyotype analysis of parasitoid Hymenoptera (Insecta): major approaches, techniques, and results. Genes 13(5):751

Gupta S, Pandotra P, Gupta AP, Verma MK, Ahuja A, Vishwakarma RA (2013) Direct rhizogenesis, in vitro stolon proliferation and high-throughput regeneration of plantlets in *Glycyrrhiza glabra*. Acta Physiol Plant 35(9):2699–2705

Gurzenkov NN (1973) Studies of chromosome numbers of plants from the south of the Soviet Far East. Komarov Lectur 20:47–62

Hadian J, Karami A, Azizi A, Khadivi-Khub A (2015) Ubiquitous genetic diversity among and within wild populations of Satureja rechingeri assessed with ISSR markers. Plant Syst Evol 301(3):923–930
Haghighi TM, Saharkhiz MJ (2022) Mycorrhizal colonization and silicon nutrition mitigates drought stress in Licorice (*Glycyrrhiza glabra* L.) with morphophysiological and biochemical perspectives. Ind Crop Prod 178:114650
Hakimi A, Zolfaghari M, Sorkheh K (2017) RETRACTED ARTICLE: genetic structure and diversity analysis revealed by AFLP markers on different *Glycyrrhiza glabra* L. an endangered medicinal species from South of Iran and implications for conservation. Biochem Genet 55(4):345
Han R, Rai A, Nakamura M, Suzuki H, Takahashi H, Yamazaki M, Saito K (2016) De novo deep transcriptome analysis of medicinal plants for gene discovery in biosynthesis of plant natural products. Methods Enzymol 576:19–45. https://doi.org/10.1016/bs.mie.2016.03.001
Haroon M, Wang X, Afzal R, Zafar MM, Idrees F, Batool M, Khan AS, Imran M (2022) Novel plant breeding techniques shake hands with cereals to increase production. Plants 11(8):1052
Hong K (2006) Karyotypes of three species in Glycyrrhiza. Zhi Wu Yan Jiu 26(5):560–562
Hosseini MS, Ebrahimi M, Samsampour D, Abadía J, Khanahmadi M, Amirian R, Ghafoori IN, Ghaderi-Zefrehei M, Gogorcena Y (2021) Association analysis and molecular tagging of phytochemicals in the endangered medicinal plant licorice (*Glycyrrhiza glabra* L.). Phytochemistry 183:112629
Huang M, Wang W, Wei S (2010) Investigation on medicinal plant resources of *Glycyrrhiza uralensis* in China and chemical assessment of its underground part. Zhongguo Zhong Yao Za Zhi= Zhongguo Zhongyao Zazhi= China J Chin Mater Med 35(8):947–952
Husain I, Bala K, Khan IA, Khan SI (2021) A review on phytochemicals, pharmacological activities, drug interactions, and associated toxicities of licorice (Glycyrrhiza sp.). Food Front 2(4):449–485
Imran M, Rizwan M, Kerkhi SA, Ali N (2018) Bioinformatics for crop breeding and sustainable agriculture. Journal 6(4):1–5
Jaiswal N, Verma Y, Misra P (2021b) High frequency in vitro callogenesis and plant regeneration of *Glycyrrhiza glabra* L. Vegetos 34(3):495–504
Jaiswal N, Verma Y, Misra P (2017) Micropropagation and in vitro elicitation of licorice (Glycyrrhiza spp.). Vitr Cell & Dev Biol-Plant 53(3):145–166
Jaiswal N, Verma Y, Misra P (2021a) Enhanced production of bioactive compound in the callus culture of *Glycyrrhiza glabra* L. with the improved biomass accumulation
Jiang W, Tan W, Gao H, Yu X, Zhang H, Bian Y, Wang Y, Tian X (2020) Transcriptome and complete chloroplast genome of Glycyrrhiza inflata and comparative analyses with the other two licorice species. Genomics 112(6):4179–4188
Khan S, Mirza KJ, Tayaab M, Abdin MZ (2009) RAPD profile for authentication of medicinal plant *Glycyrrhiza glabra* Linn. Internet J Food Saf 11:24–28
Kim D-G, Lee J, Kim W, An H-J, Lee J-H, Chang J, Kang S-H, Song Y-J, Jeon Y-D, Jin J-S (2021) Assessment of general toxicity of the Glycyrrhiza new variety extract in rats. Plants 10(6):1126
Kohjyouma M, Kohda H, Tani N, Ashida K, Sugino M, Yamamoto A, Horikoshi T (1995) In vitro propagation from axillary buds of *Glycyrrhiza glabra* L. Plant Tissue Cult Lett 12(2):145–149
Koli GK, Arya R, Nimbal S, Kumar D (2021) Chapter-4 genetic improvement for immunity boosting traits in medicinal plants. MEDICINAL 59
Krasnikov AA (1990) Chromosome numbers in representative of some families of vascular plants in the flora of the Novosibirsk region. I. Bot Z 75:116–118
Lee KJ, Raveendar S, Choi JS, Gil J, Lee JH, So Y-S, Chung J-W (2019) Development of chloroplast microsatellite markers for identification of Glycyrrhiza species. Plant Genet Resour 17(1):95–98
Lee J-H, Oh M-W, Lee S-H, Park C-G, Jeong J-T, Han J-W, Ma K-H, Chang J-K (2020) 'Wongam', a licorice interspecific hybrid cultivar with high yield. Korean Soc Breed Sci 52(4):454–459

Lee SH, Kim YB, Lee JM, Lee JH, Park CB, Bang JW, Choi HW, Hur Y, Park CG (2017) Cultivation and characteristics of licorice F 1 (*Glycyrrhiza glabra* × G. uralensis) Lines. Hortic Sci Technol 35(2):276–286

Li XY, Lu JH (2015) Taxonomy and experimental biology of the genus Glycyrrhiza L. Fudan University Press, Shanghai

Li XY, Wei LJ (1991) Karyotype analysis of Glycyrrhiza L. China. Bull Bot Res Harbin 11(3):45–53

Li XY, Wei LJ (1992) The karyotype analyses of some new taxa of Glycyrrhiza. Advances Pl Taxon North W China 1:101–107

Li XY, Feng DC, Wei LJ, Lu YF (1992) Studies on germplasm of Glycyrrhiza by using different taxonomic methods. Adv Plant Taxon Northwest China 1:7–24

Li J, Xiong C, He X, Lu Z, Zhang X, Chen X, Sun W (2018) Using SSR-HRM to identify closely related species in herbal medicine products: a case study on licorice. Front Pharmacol 9:407

Li T, Xuan F, Shen X, Jiang D, Ren G, Liu C (2021) Special genotypes of ABA key functional genes NCEDs members in the glycyrrhizic acid synthesis regulatory network promote glycyrrhizic acid synthesis in licorice

Liang Y, Liu H-J, Yan J, Tian F (2021) Natural variation in crops: realized understanding, continuing promise. Annu Rev Plant Biol 72(090632):10–1146

Liang-Bi XIE, Jia-Hui LU, Xiao-Lan LI, Yu Z, Tao WEI, Xue-Yu LI (2014) The cross compatibility and hybrid seed vigor among three Glycyrrhiza species. Plant Diversity 36(03):342

Lim TK (2016) *Glycyrrhiza glabra*. In: Edible medicinal and non-medicinal plants. Springer, pp 354–457

Liu T, Lin HM (2012) Preliminary assessment of genetic diversity in cultivated *Glycyrrhiza uralensis*, G. inflate and G. glabra by chemical fingerprint and inter-simple sequence repeat markers. Adv Mater Res 347:1318–1325

Liu J, Wu L, Wei S, Xiao X, Su C, Jiang P, Song Z, Wang T, Yu Z (2007) Effects of arbuscular mycorrhizal fungi on the growth, nutrient uptake and glycyrrhizin production of licorice (*Glycyrrhiza uralensis* Fisch). Plant Growth Regul 52(1):29–39

Liu H, Tan Y, Nell M, Zitter-Eglseer K, Wawscrah C, Kopp B, Wang S, Novak J (2014) Arbuscular mycorrhizal fungal colonization of Glycyrrhiza glabra roots enhances plant biomass, phosphorus uptake and concentration of root secondary metabolites. J Arid Land 6(2):186–194

Liu Y, Zhang P, Song M, Hou J, Qing M, Wang W, Liu C (2015) Transcriptome analysis and development of SSR molecular markers in *Glycyrrhiza uralensis* fisch. PLoS ONE 10(11):e0143017

Liu Y, Geng Y, Song M, Zhang P, Hou J, Wang W (2019) Genetic structure and diversity of Glycyrrhiza populations based on transcriptome SSR markers. Plant Mol Biol Report 37(5):401–412

Liu Q, Guo S, Zheng X, Shen X, Zhang T, Liao B, He W, Hu H, Cheng R, Xu J (2021) Licorice germplasm resources identification using dna barcodes inner-variants. In: Plants, vol 10, issue 10. https://doi.org/10.3390/plants10102036

Loskutov IG, Khlestkina EK (2021) Wheat, barley, and oat breeding for health benefit components in grain. Plants 10(1):86

Löve Á (1982) IOPB chromosome number reports LXXV. Taxon 342–368

Ma XH, Qin RL, Xing WB (1985) Chromosome observation of twenty species of drug plants in Xingjiang. Acta Bot Bor-Occid Sinica 5:149–154

Magulaev A (1980) Chromosome numbers of some legumes in the Northern Caucasus. Botanicheskii Zhurnal

Malakar M, Beruto M, Barba-Gonzalez R (2022) Biotechnological approaches to overcome hybridization barriers and use of micropropagation tool for further improvement in Heliconia: a review. Plant Cell, Tissue Organ Cult (PCTOC) 1–20

Manekar R, Jadhav D (2019) In vitro shoot induction in endangered medicinal plant *Glycyrrhiza glabra* L. using shoot tip explants. Plant Cell Biotechnol Mol Biol 890–895

Manisha G, Poonam D, Subhash K, Ashok C (2014) Optimization of media for establishment of *Glycyrrhiza glabra* micropropagation. Ann Agri Bio Res 19(2):197–201

Manzoor MM, Goyal P, Pandotra P, Dar MS, Dar MJ, Misra P, Gupta AP, Vishwakarma RA, Ahuja A, Dhar MK (2021) Transcriptome-wide identification of squalene epoxidase genes from *Glycyrrhiza glabra* L.: expression analysis and heterologous expression of GgSQE1 suggest important role in terpenoid biosynthesis. Protoplasma 258(5):991–1007
Mehrotra S, Mishra BN, Kukreja AK (2009) Efficiency of liquid system for mass multiplication of *Glycyrrhiza glabra* and evaluation of genetic fidelity of micropropagated plants. Int J Biotechnol Biochem 5(2):157–170
Mochida K, Sakurai T, Seki H, Yoshida T, Takahagi K, Sawai S, Uchiyama H, Muranaka T, Saito K (2017) Draft genome assembly and annotation of *Glycyrrhiza uralensis*, a medicinal legume. Plant J 89(2):181–194
Mousa N, Siaguru P, Wiryowidagdo S, Wagih ME (2006) Rapid clonal propagation of licorice (*Glycyrrhiza glabra*) byin vitro shoot culture. Sugar Tech 8(4):292–298
Mousavi SS, Karami A, Saharkhiz MJ, Etemadi M, Ravanbakhsh M (2022) Microbial amelioration of salinity stress in endangered accessions of Iranian licorice (*Glycyrrhiza glabra* L.). BMC Plant Biol 22(1):1–17
Mulligan GA (1984) Chromosome numbers of some plants native and naturalized in Canada. Le Naturaliste Canadien 111:447–449
Nafanailova II (1989) Examination, caryologica comparativa Glycyrrhiza L. et Meristotropis. Fisch. et Mey. Botanicheskie Materialy Gerbariia Instituta Botaniki Akademii Nauk Uzbekskoi SSR 2:51–68
Nazarova E, Ghukasyan A, Gabrielian E (2004) Chromosome numbers of flowering plants of Armenian flora. Institute of Botany, National Academy of Sciences of Ra
Neelima R, Rao NN, Seema D, Sarita A, Arya ID (2017) Optimization of media constituents for regeneration via callus cultures of *Glycyrrhiza glabra* L. an endangered plant. Biochem Cell Arch 17(1):389–398
Nimbalkar RD, Nanoti SA, Bhoite AG (2018) In vitro studies on jeshtamadh (Glycyrrhiza glabral). BIOINFOLET-A Q J Life Sci 15(2):217–222
Ozaki K, Shibano M, Kusano G, Watanabe H (2010) Aim for production of Glycyrrhizae Radix in Japan (2). Selection of pharmaceutically fine strains from *Glycyrrhiza uralensis* Fisher. Shoyakugaku Zasshi 64(2):76–82
Öztürk M, Altay V, Hakeem KR, Akçiçek E (2018) Liquorice: from botany to phytochemistry. Springer
Pandey DK, Ayangla NW (2018) Biotechnological aspects of the production of natural sweetener glycyrrhizin from Glycyrrhiza sp. Phytochem Rev 17(2):397–430
Pavlova NS, Probatova NS, Sokolovskaja AP (1989) Taksonomicheskij obzor semejstva Fabaceae, chisla khromosom i rasprostranenie na Sovetskom Dal'nem Vostoke. Komarovskie Čtenija (Vladivostok) 36:20–47
Pazhamala LT, Kudapa H, Weckwerth W, Millar AH, Varshney RK (2021) Systems biology for crop improvement. The Plant Genome 14(2):e20098
Peruzzi L, Eroğlu HE (2013) Karyotype asymmetry: again, how to measure and what to measure? Comp Cytogenet 7(1):1
Pohtam I, Anuradha KN, Devraj KC (2019) Need for sustainable utilization and conservation of *Glycyrrhiza glabra* Linn: a review
Priyadarshan PM (2019) Breeding self-pollinated crops. In: Plant breeding: classical to modern. Springer, pp 223–241
Purugganan MD (2019) Evolutionary insights into the nature of plant domestication. Curr Biol 29(14):R705–R714
Purugganan MD, Fuller DQ (2009) The nature of selection during plant domestication. Nature 457(7231):843–848
Rakshit S, Bellundagi A (2019) Conventional breeding techniques in sorghum. In: Breeding sorghum for diverse end uses. Elsevier, pp 77–91

Ramilowski JA, Sawai S, Seki H, Mochida K, Yoshida T, Sakurai T, Muranaka T, Saito K, Daub CO (2013) *Glycyrrhiza uralensis* transcriptome landscape and study of phytochemicals. Plant Cell Physiol 54(5):697–710

Rani A, Kashyap R, Azmi W (2021) Conservation of biodiversity by biotechnology. In: Basic concepts in environmental biotechnology. CRC Press, pp 149–175

Raveendar S, So Y-S, Lee KJ, Lee D-J, Sung J, Chung J-W (2017) The complete chloroplast genome sequence of Glycyrrhiza lepidota (Nutt.) pursh-an American wild licorice. J Crop Sci Biotechnol 20(4):295–303

Rehman S, Rehman IU, Jan B, Rashid I, Reshi ZA, Ganie AH (2021) Genome editing: applications for medicinal and aromatic plants. In: Medicinal and aromatic plants. Elsevier, pp 119–144

Rizzato G, Scalabrin E, Radaelli M, Capodaglio G, Piccolo O (2017) A new exploration of licorice metabolome. Food Chem 221:959–968. https://doi.org/10.1016/j.foodchem.2016.11.068

Rostovtseva TS (1977) Chromosome numbers of some plant species from the south of Siberia. II. Botanicheskii Zhurnal

Saito K, Kaneko H, Yamazaki M, Yoshida M, Murakoshi I (1990) Stable transfer and expression of chimeric genes in licorice (*Glycyrrhiza uralensis*) using an Ri plasmid binary vector. Plant Cell Rep 8(12):718–721

Sangwan S, Prasanna R (2021) Mycorrhizae helper bacteria: unlocking their potential as bioenhancers of plant–arbuscular mycorrhizal fungal associations. Microb Ecol 1–10

Sarkar M, Roy SC (2014) Rapid scale micropropagation of *Glycyrrhiza glabra* L. (Leguminoseae) a valuable medicinal herb. Int J Sci Res 3:20–22

Sawaengsak W, Saisavoey T, Chuntaratin P, Karnchanatat A (2011) Micropropagation of the medicinal herb *Glycyrrhiza glabra* L., through shoot tip explant culture and glycyrrhizin detection. Int Res J Plant Sci 2(5):129–136

Shah RR, Dalal KC (1980) In vitro multiplication of Glycyrrhiza [drug plant, India]. Current Sci (India)

Shaheen A, Ali M, Ahmad N, Hassan Dewir Y, El-Hendawy S, Gawad A-E, Ahmed M (2020) Micropropagation of licorice (*Glycyrrhiza glabra* L.) by using intermediate nodal explants. Chil J Agric Res 80(3):326–333

Sharifi-Rad J, Quispe C, Herrera-Bravo J, Belén LH, Kaur R, Kregiel D, Uprety Y, Beyatli A, Yeskaliyeva B, Kirkin C, Özçelik B, Sen S, Acharya K, Sharopov F, Cruz-Martins N, Kumar M, Razis AFA, Sunusi U, Kamal RM, Suleria HAR (2021) Glycyrrhiza genus: enlightening phytochemical components for pharmacological and health-promoting abilities. Oxidative Med Cell Longev. https://doi.org/10.1155/2021/7571132

Sharma P, Tripathi MK, Tiwari G, Tiwari S, Baghel BS (2010) Regeneration of liquorice (*Glycyrrhiza glabra* L.) from cultured nodal segments. Indian J Plant Physiol 15(1):1

Sharma V, Katiyar A, Agrawal R (2017) *Glycyrrhiza glabra*: chemistry and pharmacological activity. Sweeten, Biotechnol, Appl 87–100

Sheidai M, Iraj S, Karamian R, Ranjbar M (2008) Cyto-morphological studies of the genus Glycyrrhiza in Iran. Cytologia 73(3):333–339

Shibata S (2000) A drug over the millennia: pharmacognosy, chemistry, and pharmacology of licorice. Yakugaku Zasshi 120(10):849–862

Shimomura K, Saito K, Murakoshi I (1994) Genetic relationships among glycyrrh1za plants determined by rapd and rflp analyses. Biol Pharm Bull 17(11):1529–1531. https://doi.org/10.1248/bpb.17.1529

Siljak-Yakovlev S, Peruzzi L (2012) Cytogenetic characterization of endemics: past and future. Plant Biosyst-An Int J Deal All Asp Plant Biol 146(3):694–702

Singer SD, Laurie JD, Bilichak A, Kumar S, Singh J (2021) Genetic variation and unintended risk in the context of old and new breeding techniques. Crit Rev Plant Sci 40(1):68–108

Sofia HN, Walter TM (2009) Review of *Glycyrrhiza glabra*, Linn. Siddha Papers Med J 2:1–7

Sorkheh K, Zolfaghari M, Ercisli S (2019) Reliability authentication of *Glycyrrhiza glabra* L. Populations from South Iran Using SSR and SNP-based markers. Proc Natl Acad Sci, India Sect B: Biol Sci 89(4):1283–1294

Srivastava M, Purshottam DK, Srivastava AK, Misra P (2013) In vitro conservation of *Glycyrrhiza glabra* by slow growth culture. Int J Biotechnol Res 3(1):49–58

Stepanov NV (1994) Chromosome numbers of some higher plant taxa of the flora of the Krasnoyarsk region. Botanicheskiĭ Zhurnal 79(2):135–139

Thengane SR, Kulkarni DK, Krishnamurthy KV (1998) Micropropagation of licorice (*Glycyrrhiza glabra* L.) through shoot tip and nodal cultures. In Vitr Cell Dev Biol-Plant 34(4):331–334

Tohma HS, Gulçin I (2010) Antioxidant and radical scavenging activity of aerial parts and roots of Turkish liquorice (*Glycyrrhiza glabra* L.). Int J Food Prop 13(4):657–671

Tripathi S, Jadaun JS, Chandra M, Sangwan NS (2016) Medicinal plant transcriptomes: the new gateways for accelerated understanding of plant secondary metabolism. Plant Genetic Resources 14(4):256–269

Um Y, Jin M-L, Lee Y, Hur M, Cha SW, Jung CS, Kim SM, Lee J-H (2016) Genetic diversity analysis of *Glycyrrhiza uralensis* using 8 novel polymorphic microsatellite markers. J Plant Biotechnol 43(2):174–180

Vassilev D, Leunissen J, Atanassov A, Nenov A, Dimov G (2005) Application of bioinformatics in plant breeding. Biotechnol Biotechnol Equip 19(sup3):139–152

Verma S, Nadkarni RS (1985) Chromosome number and karyotypic studies in Glycyrrhiza. Curr Sci (Bangalore) 54(1):45–47

Wahab S, Annadurai S, Abullais SS, Das G, Ahmad W, Ahmad MF, Kandasamy G, Vasudevan R, Ali MS, Amir M (2021) *Glycyrrhiza glabra* (Licorice): a comprehensive review on its phytochemistry, biological activities, clinical evidence and toxicology. In: Plants (vol 10, issue 12). https://doi.org/10.3390/plants10122751

Wang WQ (2000) Studies on ecological characters of *Glycyrrhiza uralensis* Fisch and effects of ecological environment on its medicinal material quality. Beijing Forestry University, Beijing

Wang W, Xu J, Fang H, Li Z, Li M (2020) Advances and challenges in medicinal plant breeding. Plant Sci 298:110573

Wei S, Wang W, Wang H (2003) Study on licorice resources and their sustainable utilization in center and western area of China. Zhongguo Zhong Yao Za Zhi= Zhongguo Zhongyao Zazhi= China J Chin Mater Med 28(3):202–206

Wei S, Wang W, Liu C, Wang J, Zhang R, Sun M (2012) Analysis of broad-sense heritability and genetic correlation of production and content of glycyrrhizin of annual Glycyrrhiza uralensis. Zhongguo Zhong Yao Za Zhi= Zhongguo Zhongyao Zazhi= China J Chin Mater Med 37(5):553–557

Wongwicha W, Tanaka H, Shoyama Y, Tuvshintogtokh I, Putalun W (2008) Production of glycyrrhizin in callus cultures of licorice. Zeitschrift Für Naturforschung C 63(5–6):413–417

Xiaochu C, Benkun L, Zuosheng J (1991) Observations of chromosome numbers of several medical plants. J Hunan Agric Coll (China)

Xie W, Hao Z, Yu M, Wu Z, Zhao A, Li J, Zhang X, Chen B (2019) Improved phosphorus nutrition by arbuscular mycorrhizal symbiosis as a key factor facilitating glycyrrhizin and liquiritin accumulation in Glycyrrhiza uralensis. Plant Soil 439(1):243–257

Xingzhu D, Qingrui Z, Keren C, Yuxi L, Yunpeng H, Shien Z, Xiangwei F (2021) Cryopreservation of porcine embryos: recent updates and progress. Biopreservation Biobanking 19(3):210–218

Yadav K, Singh N (2012) Factors influencing in vitro plant regeneration of Liquorice (*Glycyrrhiza glabra* L.)

Yamazaki M, Rai A, Yoshimoto N, Saito K (2018) Perspective: functional genomics towards new biotechnology in medicinal plants. Plant Biotechnol Rep 12(2):69–75

Yan GX (1989) Chromosome numbers and geographical distribution of 68 species of forage plants. Grassl China 4:53–60

Yang RW, Zhou YH, Ding CB, Zheng YL, Zhang L (2008) Relationships among Leymus species assessed by RAPD markers. Biol Plant 52(2):237–241

Yang R, Li W, Yuan B, Ma Y, Zhou S, Liu C, Liu Y (2016b) Simultaneous determination of 18 α-glycyrrhizic acid and 18 β-glycyrrhizic acid in three licorice samples from different origin by HPLC∗. Chin J Pharm Anal 36(6):1065–1071

Yang L, Chen J, Hu W, Yang T, Zhang Y, Yukiyoshi T, Zhou Y, Wang Y (2016) Population genetic structure of Glycyrrhiza inflata B.(Fabaceae) is shaped by habitat fragmentation, water resources and biological characteristics. PloS One 11(10):e0164129

Yao H, Zhao Y, Chen DF, Chen JK, Zhou TS (2008) ISSR primer screening and preliminary evaluation of genetic diversity in wild populations of Gycyrrhiza uralensis. Biol Plant 52(1):117–120

Yao H, Wang F, Bi Q, Liu H, Liu L, Xiao G, Zhu J, Shen H, Li H (2022) Combined analysis of pharmaceutical active ingredients and transcriptomes of glycyrrhiza uralensis under PEG6000-induced drought stress revealed glycyrrhizic acid and flavonoids accumulation via JA-mediated signaling. Front Plant Sci 13

Yeh MS (1986) Chromosome nnumbers in the Leguminosae. Sci Rep Res Inst Evol Biol 3:57–71

Yu F, Fang Y, Wang W, Wang Q, Liu F (2011) Genetic variability and interrelationships of mainly quantitative traits in Glycyrrhiza uralensis cultivated population. Zhongguo Zhong Yao Za Zhi= Zhongguo Zhongyao Zazhi= China Journal of Chinese Materia Medica, 36(18):2457–2461

Zhang Q, Ye M (2009) Chemical analysis of the Chinese herbal medicine Gan-Cao (licorice). J Chromatogr A 1216(11):1954–1969

Zhang J T, Xu B, Li M (2010) Genetic diversity of populations of an endangered medicinal plant species (Glycyrrhiza uralensis) in different environments of North China. J Med Plants Res 4(9):830–836

Zhang J-Z, Gao W-Y, Gao Y, Liu D-L, Huang L-Q (2011) Analysis of influences of spaceflight on chemical constituents in licorice by HPLC–ESI-MS/MS. Acta Physiol Plant 33(6):2511–2520

Zheng Y, Wei J, Leng K, Tao W, Fang S, Peng G, Duan J (2015) Research advances in resource chemistry and utilization of genus Glycyrrhiza. Mod. Chin. Med 17:1096–1108

Zhou Y-P, Wang Z-X, Du Y-P, Li J-W, He H, Jia G-X (2020) Fluorescence in situ hybridization of 35S rDNA sites and karyotype of wild Lilium (Liliaceae) species from China: taxonomic and phylogenetic implications. Genet Resour Crop Evol 67(6):1601–1617

Chapter 3
Advances in Marjoram (*Origanum majorana* L.) Breeding Strategies

Amira A. Ibrahim, Sawsan Abd-Ellatif, El-Sayed S. Abdel Razik, and Khaled F. M. Salem

Abstract Marjoram (*Origanum majorana* L.) is a highly valued herb known for its aromatic and medicinal properties. There has been growing interest in developing improved marjoram varieties through innovative breeding strategies in recent years. This chapter provides an overview of the recent advancements in marjoram breeding techniques and explores the potential for enhancing various desirable traits in this herb. The chapter discusses the importance of Marjoram as a versatile herb with numerous applications in the culinary, pharmaceutical, and cosmetic industries. It highlights the need for innovative breeding approaches to overcome challenges such as low yields, disease susceptibility, and limited genetic diversity. Next, the chapter explores conventional breeding methods, including selection, hybridization, backcrossing, and their applicability in marjoram improvement. It delves into identifying and evaluating desirable traits such as yield, essential oil content and

A. A. Ibrahim (✉)
Department of Botany and Microbiology, Faculty of Science, Arish University, Al-Arish, Egypt
e-mail: amiranasreldeen@sci.aru.edu.eg

S. Abd-Ellatif
Bioprocess Development Department, Genetic Engineering and Biotechnology Research Institute (GEBRI), City of Scientific Research and Technology Applications (SRTA-City), Borg EL-Arab, Alexandria, Egypt
e-mail: sabdellatif@srtacity.sci.eg

E.-S. S. A. Razik
Livestock Research Department, Arid Lands Cultivation Research Institute (ALCRI), City of Scientific Research and Technology Applications (SRTA-City), Borg EL-Arab, Alexandria, Egypt
e-mail: eshabaan@srtacity.sci.eg

K. F. M. Salem
Department of Biology, College of Science and Humanitarian Studies, Shaqra University, Qwaieah, Saudi Arabia

Department of Plant Biotechnology, Genetic Engineering and Biotechnology Research Institute (GEBRI), University of Sadat City, Sadat City, Egypt

K. F. M. Salem
e-mail: khaled.salem@gebri.usc.edu.eg; khaledfathi@su.edu.sa

J. M. Al-Khayri et al. (eds.), *Biodiversity and Genetic Improvement of Medicinal and Aromatic Plants II*, Advances in Plant Breeding Strategies 10,
https://doi.org/10.1007/978-3-031-81857-8_3

composition, disease resistance, and adaptability to different environmental conditions. Genetic mapping and molecular markers in marjoram breeding initiatives are also examined. Genomic and transcriptomic data are used to uncover trait-related genes and expedite breeding with marker-assisted and genomic selection. Genome editing, gene stacking, and genomic prediction could manipulate Marjoram's phenotype and improve genetics quickly. This chapter concludes with marjoram breeding's future and the need for eco-friendly methods. It promotes safeguarding marjoram germplasm, studying wild relatives for distinctive traits, and breeding with genomic and molecular approaches. This chapter summarises marjoram breeding strategy developments, helping researchers, breeders, and stakeholders increase cultivation, quality, and production.

Keywords Biodiversity · Biotechnology · Breeding · Marjoram (*Origanum majorana* L.)

3.1 Introduction

The Lamiaceae family has been the subject of extensive research owing to its copious secondary metabolites. The metabolites encompass diverse volatile and nonvolatile compounds found in complex blends. Intricate blends in plants have been discovered to exhibit noteworthy biological properties, rendering them advantageous for application in diverse sectors, including food, cosmetics, and pharmaceuticals (Ramos da Silva et al. 2021; Oalde Pavlovic et al. 2021; Elsherbini et al. 2022).

Marjoram (*Origanum majorana* L.), commonly known as sweet Marjoram, belongs to the Lamiaceae family, a vast group of plants comprising 230 genera and almost 7000 species (Tamokou et al. 2017). This annual herbaceous plant is renowned for its fragrant characteristics and exhibiting a recurring growth pattern. The taxonomically accepted nomenclature for the plant in question is Majorana hortensis, and its synonymous name also recognizes it. The species *O. majorana* has been extensively researched within the Lamiaceae plant family. *O majorana* has been taxonomically categorized as a botanical specimen possessing significant pharmacological properties owing to its diverse chemical composition, enc2014ompassing the essential oil fraction or the plant extracts. Multiple studies, such as those carried out by Bina and Rahimi (2017), Hossain et al. (2014), Sahranavard et al. (2014), Amaghnouje et al. (2020), Gutiérrez et al. (2022) and Soliman et al. (2016), have documented this phenomenon. Numerous investigations have been conducted to evaluate the physiological effects of the volatile oil derived from the above-ground portion of the botanical specimen. *O. majorana* commonly known as sweet Marjoram, is a member of the Lamiaceae family, a group of plants known for their aromatic properties. Bina and Rahimi (2016) observed that the plant is grown in various Mediterranean areas, including Asian, European, and North African nations.

The *O. vulgare* plant is a botanical source for extracting essential oil, frequently used in fragrances for personal care and wellness applications such as bathing and sauna. Oregano, scientifically known as *O. vulgare*, is a perennial herb belonging to the Lamiaceae family. It exhibits a range of bioactive compounds such as phenols, flavonoids, and terpenes that contribute to its antibacterial, antioxidant, antiviral, anti-inflammatory, analgesic, and immune regulatory properties. The properties mentioned above have led to its utilization in traditional medicine for the treatment of various conditions such as sunstroke, fever, acute gastroenteritis, and dysentery, among others, as documented in studies carried out by Sharifi-Rad et al. (2021) and Kosakowska et al. (2021). The essential oil derived from oregano is a botanical extract known to be devoid of adverse effects. This botanical substance has the potential to serve as a sustainable and ecologically sound substitute for synthetic food additives, chemical preservatives, and animal feed supplements. This has been evidenced by several investigations, such as those carried out by Carvalho et al. (2020), Lages et al. (2021), Amer et al. (2021), Guo et al. (2022), and Li et al. (2021). The essential oil derived from oregano has been acknowledged worldwide as a natural feed supplement that holds promise for the preservation of chicken, fish, and rabbit, as per recent studies (Amer et al. 2021; Guo et al. 2022; Li et al. 2021).

The genetic diversity analyses of plant species have predominantly employed using morphological (ALshamrani et al. 2022; Alqahtani et al. 2024; El-Absawy et al. 2015; EL-Mansy et al. 2021; El-Zanaty et al. 2011; Mattar et al. 2009; Salem et al. 2008, 2021d) and PCR-based molecular markers, including AFLP, RAPD, SRAP, SSRs and ISSR (Abdelsalam et al. 2021; El Rabey et al. 2013; El-Sheikh et al. 2020; Eltaher et al. 2018; Essa et al. 2024; Ghazy et al. 2020; Ibrahim et al. 2024a, b; Mattar et al. 2016; Salem 2009; 2015; Salem and Ibrahim 2024a, b; Salem and Mattar 2014a; Salem and Sallam 2016; Salem and Mattar 2014a, b; Salem et al. 2007; 2010; 2015, 2021c, 2022, 2024). A recent investigation was carried out to establish a collection of greatly variable genomic SSR markers through NGS sequencing of genomic DNA extracted from *O. vulgare* ssp. hirtum. A set of genetically diverse *O. vulgare* ssp. hirtum lines, previously characterized with SRAP markers, were subjected to limited testing. Our study's results suggest that the developed SSR set has the potential to be applied to *O. vulgare* ssp. Furthermore, we will evaluate and analyze the efficacy of the recently developed genomic SSRs compared to the previously utilized SRAP markers. The markers were employed to assess the genetic diversity and genetic structure of indigenous populations of *O. vulgare* ssp. hirtum in Bulgaria. This chapter provides recent developments in marjoram breeding techniques, providing researchers, breeders and other stakeholders with useful information for boosting marjoram production, quality, and cultivation.

3.2 Taxonomy and Origin

The genus *Origanum* has several popular members, including Greek oregano (*Origanum vulgare* ssp. hirtum), which is known for its high essential oil content, common oregano (*O. vulgare* L. ssp. vulgare), and sweet marjoram (*O. majorana* L.). This information is according to Alekseeva et al. (2023).

Kingdom: Plantae—plantes, Planta, Vegetal, plants
Subkingdom: Viridiplantae—green plants
Infrakingdom: Streptophyta—land plants
Superdivision: Embryophyta
Division: Tracheophyta—vascular plants, tracheophytes
Subdivision: Spermatophytina—spermatophytes, seed plants, phanérogames
Class: Magnoliopsida
Superorder: Asteranae
Order: Lamiales
Family: Lamiaceae—mints, menthes
Genus: *Origanum* L.
Species: *Origanum majorana* L.
Common names: Sweet Marjoram

Origanum majorana is a botanical species that commonly occurs in the regions of Cyprus and Antalya, Turkey. The plant species are widely distributed across diverse regions in the Mediterranean countries, including Serbia, Italy, Corsica, southern Spain, Portugal, Morocco, and Algeria, as documented by Ietswaart (1980). This particular plant species is widely grown in diverse geographical locations, spanning Europe (France and Hungary), the United States, Asia, and various regions of India.

3.3 Botanical Description

Marjoram (*O. majorana* L.) is a herbaceous perennial plant with a bushy growth habit and attains a vertical stature ranging from 30 to 60 cm. The plant exhibits a sprawling growth habit with square-shaped, reddish-hued stems that branch out extensively, resulting in a mounding form. A straight, weak, hairy structure with a round cross-section characterizes the plant's stem morphology. The stem coloration is predominantly green with red speckles. The plant exhibits diminutive foliage, with leaf dimensions measuring 0.5–1.5 cm in length and 0.2–0.8 cm in width. The specimens in question exhibit herbaceous characteristics and possess a globose shape. Their structure is simple, and petioles support them. The leaves are ovate to oblong-ovate in shape. The foliage exhibits quadrigonus-cylindrical morphology and bears a distinctly glabrous and pubescent surface. The plant's bracts exhibit a clear differentiation from its foliage. The structures exhibit high density, a grey-green hue, and a length equivalent to the calyx.

Furthermore, they exhibit an opposite phyllotaxy on a quadrangular stem. The inflorescence of sweet Marjoram consists of burr-like heads that are 1.3 cm long and bear small flowers that measure 0.3 cm in length. The floral morphology of this plant species exhibits the potential for hermaphroditic or female reproductive structures. The floral morphology exhibits a tubular form and comprises two lips, characterized by one or more upper lips that are either undivided or have tooth-like projections. The plant exhibits flattened calyces and bears white or pale pink flowers accompanied by grey-green bracts arranged in a spike from June through September. Seed maturation occurs in late summer, typically between August and September. The specimens exhibit an ovate morphology and display a deep hue of brown. The subterranean organs exhibit a sub-cylindrical morphology and display longitudinal wrinkles and transverse fissures. The diameter of the measured specimens ranges from 0.2 to 0.6 mm. According to Ietswaart (1980) findings, the exocarp of the specimen appears dark brown, whereas the endocarp exhibits a lighter brown hue. Additionally, the presence of numerous elongated rootlets can be observed.

3.4 Economic Importance of Marjoram

In modern plant biology and commerce, the scope of economically important plants has expanded to include not just fodder, staple foods, and textile crops but also botanical specimens that possess valuable fragrant or medicinal attributes or that can serve as raw materials for the fragrance, culinary, and chemical sectors (Safaei et al. 2023). The expansion phenomenon has been observed and documented by notable plant biologists (Abd El-Wahab 2013; Balandrin and Klocke 1988; Banchio et al. 2008). *Origanum majorana* L. is a member of the Lamiaceae family, commonly known as Marjoram. This particular botanical specimen is frequently cultivated within the regions of the Mediterranean.

3.4.1 Health Benefits

Marjoram leaves smell wonderful. Marjoram, *Origanum majorana*, is used in cooking to improve taste, aroma, and preservation (Yousefi et al. 2021). Marjoram's leaves and blossoming tips, fresh or dried, are used for this because of their strong scent, according to Shahidi et al. (1992). Medical and aromatic characteristics make this botanical specimen valuable to plant biologists. Strong antibacterial and antioxidant effects against many bacteria and mycotoxigenic fungi (Baydar et al. 2004; Bina and Rahimi 2017; Charles 2013; Fanourakis et al. 2022). Tahraoui et al. (2007) say its antioxidants can prevent and treat glaucoma. Marjoram extracts prevent lipid oxidation, color deterioration, and microbial proliferation in various types of meat (Fasseas et al. 2007; El-Raghi et al. 2023; Camo et al. 2008; Djenane et al. 2003). This plant or plant derivative is appreciated for its industrial and medicinal potential.

Studies have connected *O. majorana* secondary metabolites to antioxidant, antibacterial, anti-inflammatory, antiacetylcholinesterase, anticancer, antidepressant, and analgesic effects. Besides biological properties, the plant's essential oil repels and kills insects.

Essential oil and plant extracts of marjoram (*O. marjorana* L.) vary. Several studies (Amaghnouje et al. 2020; Bina and Rahimi 2017; Gutiérrez et al. 2022; Hossain et al. 2014; Sahranavard et al. 2014; Soliman et al. 2016) *O. majorana* has therapeutic value. Many investigations have explored the physiological effects of the botanical specimen's aboveground volatile oil.

Antibacterial, antifungal, and antioxidant properties are present in marjoram. Aureli et al. (1992), El-Ashmawy et al. (2007), Muller et al. (1995), and Vera and Chane-Ming (1999) studied these features.

O. majorana treats asthma, paralysis, and diuresis, per Yadava and Khare (1995). The leaves may treat diabetes and stomach ulcers, say Pimple et al. (2012). Marjoram oils are antimicrobial, say Abdel-Massih et al. (2010). These oils are anti-fungal, according to Souza et al. (2010) and Kunicka-Styczy'nska (2011). In 2001, Chung et al. (2001) reported *O. majorana* ethanol extracts inhibited Acetylcholinesterase. Al Dhaheri et al. (2013) showed these extracts suppressed metastasis, cell cycle, and neoplasm proliferation.

Pharmacological experiments on *O. majorana* methanol extracts and essential oils showed biological action (Fig. 3.1). This chemical is antibacterial against *Bacillus subtilis, Enterococcus faecalis, Escherichia coli, Klebsiella pneumonia, Salmonella choleraensius*, and *Serratia* spp. Xu et al. (2008) say OMEO's carvacrol and thymol increase membrane permeability, exude critical cellular components, and block quorum sensing, making it antimicrobial. *O. majorana* inhibits *Aspergillus niger, Trichoderma viride, Penicillium cyclopium, Phytophthora infestants, Candida* sp., and other pathogens. This study contains Vagi et al. (2005), Hajlaoui et al. (2016), Della Pepa et al. (2019), and Thanh et al. (2019) *O. majorana* is insecticidal, larvicidal, and antiparasitic, research shows. El-Akhal et al. (2016) and (2019) found larvicidal properties, while Barbouche et al. (2001) found insecticidal properties. Besides its anti-inflammatory, analgesic, and antipyretic properties, *O. majorana* may prevent nephrotoxicity (Pimple et al. 2012). Antidiabetic phenols in *O. majorana.* It stimulates hepatic glycogen production, plasma insulin, glucokinase, and G6Pase/PEPCK downregulation. Research supports hepatoprotective, antimutagenic, and gastrointestinal advantages (Mossa et al. 2013; Qari 2008; Khatab and Elhaddad 2015). According to Seoudi et al. (2009), toxicological studies have revealed *O. majorana* ethanolic extract is harmless. *O. majorana* ethanolic extract did not kill rodents, according to Selim et al. (2013).

3.4.2 Phytochemical Profile of **Origanum Majorana**

The phytochemical components of plants, such as essential oils, which are synthesized during secondary metabolism, are responsible for the medicinal properties exhibited by plants (Kabera et al. 2014; Ramadan and Ibrahim 2021) Kakouri et al. 2022). Plant-derived essential oils are complex mixtures of volatile organic compounds that exhibit strong fragrances and are insoluble in water. As observed by Dhifi et al. (2013) and Hyldgaard et al. (2012), these are volatile chemical compounds originating from plants. As per the findings of Xylia et al. (2021), the marjoram herb synthesizes a yellowish essential oil, which usually generates a yield ranging from 0.5 to 3% (Sarfaraz et al. 2023). The major constituents of essential oil include carvacrol,

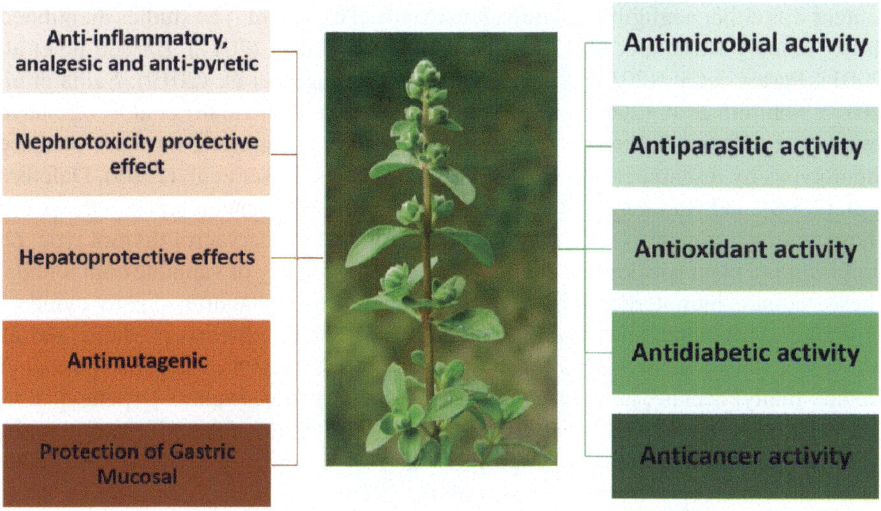

Fig. 3.1 Pharmacological properties of Marjoram (*Origanum majorana L.*) Source: Bouyahya et al. 2021)

thymol, terpinene-4-ol, trans-caryophyllene, gamma-terpinene, and p-cymene, as reported in various studies (Baatour et al. 2010; Novak et al. 2010; Sellami et al. 2009; Zgheib et al. 2016). The growing inclination towards natural substances in developed nations has resulted in the development of a distinct market for medicinal and aromatic plants.

Typically, the essential oils comprise terpenoids, particularly monoterpenes that contribute to the aroma, and both oxygenated and non-oxygenated sesquiterpenes. The occurrence of additional components may be ascribed to the monoterpenoid derivatives. Salakhutdinov et al. (2017) reported that these compounds exhibit diverse functional groups, such as esters, acetates, and alcohols. Monoterpenes and sesquiterpenes are classified as terpenoids and possess a molecular formula of $(C_5H_8)n$. Monoterpenes are composed of two isoprene units; hence n equals 2. It can be alternatively stated that sesquiterpenes are tricyclic compounds composed of three isoprene units, which gives them an n value of 3. *O. majorana* is recognized for its notable presence of distinct volatile organic compounds, particularly hydrocarbons and oxygenated monoterpenes belonging to the class of monoterpenoids. Minor constituents such as sesquiterpenes have been detected, exhibiting varying degrees of oxygenation. As noted earlier, the essential oil of *O. majorana*, which is sourced from different geographic locations, is primarily composed of oxygenated monoterpenes. Among these compounds, terpinene-4-ol is the most abundant, as reported by Jelali et al. (2011), Sellami et al. (2009), and Vagi et al. (2005).

The terpinene-4-ol, cis-sabinene hydrate, and g-terpinene compounds were highly prevalent in most investigations. In certain instances, the composition of the essential oil exhibited a high abundance of carvacrol and thymol, while the terpinene-4-ol content was either negligible or equivalent to half of carvacrol. The studies mentioned above, namely Busatta et al. (2008), Erdogan and Ozkan (2017), Ezzeddine et al. (2001), Partovi et al. (2018), Prabu et al. (2020); Ragab et al. (2019); Salha et al. (2017); Schmidt et al. (2008) and Verma et al. (2016), Abd-Ellatif et al. (2022)have reported significant findings. *O. majorana* has been categorized into two primary chemotypes by researchers using qualitative measures. Baser et al. (1993), Daferera et al. (2008), and Hajlaoui et al. (2016) have identified two chemotypes in the plant. These are the terpinene-4-ol/cis-sabinene hydrate and the carvacrol/thymol type. *O. majorana* is a plant species that is known to contain a variety of valuable compounds. These include monoterpenoids such as β-pinene, α-phellandrene, para cymene, terpinolene, sabinene, and myrcene, sesquiterpenoids like β-caryophyllene and α-humulene, phenol-methyl-ethers such as trans-anethole, terpenic esters like terphenyl acetate, linalyl acetate, and geranyl acetate, monoterpenols such as terpene-1-ol-4/3, trans/cis-thuyanol-4, and linalool, hydrocarbons, tannins, phenolic compounds, acids, aroma compounds like a-terpineol and 1, 8-cineol, and flavonoids. This information was reported by Saxena et al. in (2016).

The primary constituents of *O. marjorana* essential oil have been identified by Zargary (2012) as thymol, myrcene, carvacrol, gamma-terpinene, P-cymene, and α-pinene. The essential oil of MO is primarily composed of phenolic and alcoholic compounds. Phytochemicals like alkaloids, terpenoids, and phenolics are present in medicinal plants and are known to have antimicrobial properties. Phytochemicals are vital in enhancing animal productivity without cytotoxicity (Soliman et al. 2017).

3.5 Current Cultivation Practices

Beyond consumer needs, essential oil success depends on other factors. It also demands the ability to create high-quality plant specimens in large quantities and at a fair price. These plants thrive best in perfect conditions. Crop growth, essential oil output, and chemical makeup depend on many things. Genetics, environment, and cultivation affect plant development. Plant growth is affected by cultivation, according to Sotiropoulou and Karamanos (2010), Tibaldi et al. (2011), Tuttolomondo et al. (2016), and Virga et al. (2020). Environment affects plant development, according to La Bella et al. (2015), Murillo-Amador et al. (2015), and Vokou et al. (1993). Bakkali et al. (2008), Murillo-Amador et al. (2013), and Ninou et al. (2017a, b) found that genetic variables affect plant growth and development. Numerous research has explored how water deficiency and nitrogen supplementation impact *Origanum* taxonomy development, biomass, essential oil output, and characteristics. The studies were undertaken by Azizi et al. (2009), Ninou et al. (2017b), Khater (2019), Said-Al Ahl et al. (2009), and Thaner Dos Santos et al. (2020). Water stress affects photosynthetic and transpirational systems, reducing plant growth. Stress affected essential oil amount and quality. Azizi et al. (2009), Khater (2019), Ninou et al. (2017a), Said-Al Ahl et al. (2009), and Thaner Dos Santos et al. (2020) agree. Increased irrigation intervals harmed *O. vulgare* morphology, Gerami et al. (2016) discovered. However, essential oil output and content remained stable. Németh-Zámbori et al. (2016) observed that soil moisture impacts essential oil yield and composition in four Lamiaceae species, including Marjoram. Alsafar and Al-Hassan (2009), Dasha et al. (2006), and Usha and Patra (2003) discovered nitrogen fertilizer boosts oil crop yields. Fresh and dry leaf output, volatile oil production, and vegetative growth have improved. *Origanum vulgare* herbage yield was tested after nitrogen application (0, 40, 80, and 120 kg/ha). The maximum output was 80 kg/ha N, demonstrating nitrogen use affected plant development. This investigation was done by Sotiropoulo et al. in 2010. According to Omer (1999), nitrogen fertilizer boosts *O. syriacum* essential oil production.

Manukyan and Schnitzler found in 2006 that air temperature influences plant growth, herbage yield, essential oil content, and oil crop productivity. Bettray and Vömel (1992), Mathe et al. (1992), and Dorais et al. (2001) examined this. Mortensen (2014) found that raising air temperature from 18 to 27 °C boosted plant height and herbage yield in *Ocimum basilicum*, *Salvia officinalis*, Rosmarinus, and Eruca vesicaria. Soil characteristics have not been fully examined on medicinal and fragrant plant growth, yield, and chemical composition. Biomass output, essential oil yield, and chemical composition of *O. syriacum* were modified by soil type, according to El-Alam et al. (2019). Aziz and colleagues (2008) showed soil type affected Thymus vulgaris growth, essential oil output, and primary components. Calcareous soil promoted plant growth and yield most, the study found. Clay and sandy loam fared moderately. Soil type impacts *Rosmarinus officinalis* essential oil, according to Hendawy et al. (2017). Because essential oil depends on flowering, two harvest times were chosen. The harvest region was carefully selected to maintain *Origanum majorana* cultivation exposure intensity (Fig. 3.2).

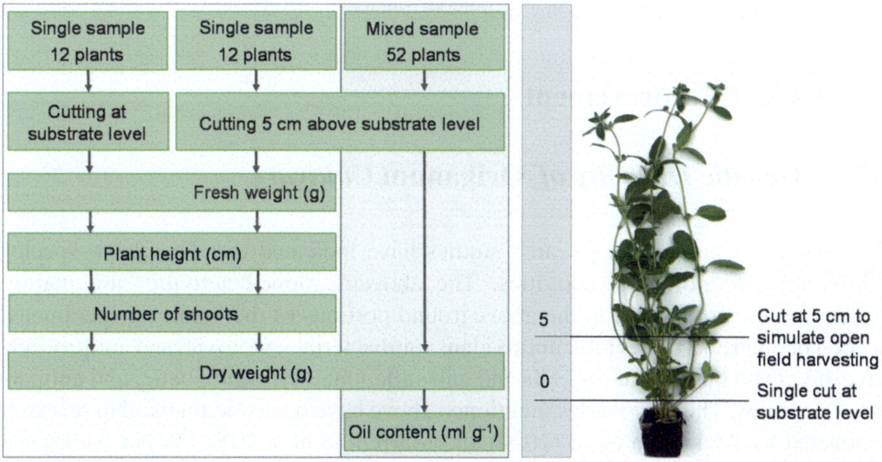

Fig. 3.2 A flow chart that provides an exemplary representation of the data collection process during the harvest of *Origanum majorana*. *Source* Wittmann et al. (2020)

3.6 Current Agricultural Problems and Challenges

Despite the potential benefits of Marjoram, there has been an insufficient assessment of its effects on plant growth, physiology, essential oil production, and chemical makeup across various agricultural methods. This observation implies that the present comprehension of its agricultural practices is insufficient and uncertain. Insufficient data impairs the optimal exploitation of this economically significant plant species, which exhibits fragrant and therapeutic characteristics. This study was undertaken to fill the previously identified research and development gaps. The study aimed to assess the physiological and biochemical reactions of Marjoram, as Malaka et al. (2023) reported. The impact of air temperature on plant growth, herbage yield, essential oil composition, and yield of essential oil crops has been studied by various researchers (Bettray and Vömel 1992; Dorais et al. 2001; Manukyan and Schnitzler 2006; Mathe et al. 1992). Mortensen (2014) reported that an increase in air temperature from 18 to 27 °C resulted in a significant increase in plant height and herbage yield for *Ocimum basilicum, Salvia officinalis, Rosmarinus officinali*s, and *Eruca vesicaria.* The impact of soil composition on various medicinal and fragrant plant species' development, yield, and chemical makeup has yet to be thoroughly investigated. The study conducted by El-Alam et al. (2019) revealed that the growth rate, production of essential oil, and chemical composition of essential oil in *O. syriacum* were significantly impacted by the soil type. Aziz and colleagues (2008) observed that the growth, essential oil yield, and primary components of Thymus vulgaris were impacted by soil type. The study revealed that calcareous soil exhibited the highest efficacy in enhancing plant growth and yield, while clay loam and sandy loam followed suit in descending order. Hendawy et al. (2017) reported that the soil type affected the essential oil composition of cultivated *Rosmarinus officinalis.*

3.7 Genetic Improvement

3.7.1 Genetic Diversity of Origanum *Cultivars*

Over the years, numerous research studies have indicated that *Origanum* species exhibit diverse biological activities. The abovementioned activities are mainly ascribed to the volatile oil in the aboveground portions of the botanical specimens. Several plant properties exhibit antioxidant, antibacterial, cytotoxic, and antiproliferative effects on human cancer cells and anti-inflammatory, antidiabetic, and antiparasitic activities. The discoveries mentioned above have been documented in research conducted by Alekseeva et al. (2020) and Marrelli et al. (2018). As per Alekseeva et al. (2020), the genetic diversity analyses of *Origanum* species have predominantly employed dominant PCR-based molecular markers, including AFLP, RAPD, SRAP, and ISSR. Using PCR-based dominant markers is convenient in plant biology as it eliminates the need for prior knowledge of the loci's sequence. However, the information provided is limited concerning the allelic state of the loci (EL-Mansy et al. 2021; ALshamrani et al. 2022).

Furthermore, replicating data produced by dominant markers may present difficulties in diverse laboratories due to several factors that can influence the PCR-derived banding profile. This is due to the non-specific nature of the primers utilized (Abd El-Moneim et al. 2021). The genus *Origanum* poses considerable challenges in identifying breeding lines and cultivars using DNA markers.

Additionally, the implementation of marker-assisted selection (MAS) in breeding programs is not without its difficulties. Simple sequence repeats (SSRs), commonly known as microsatellite markers, are a co-dominant molecular marker type that relies on PCR-based techniques (Salem 2004; Salem and Matter 2014b; Ibrahim et al. 2024a, b). Vieira et al. (2016) observed that microsatellites had been extensively employed as the preferred marker system for genotype identification in diverse plant species. The SSR analysis protocols exhibit ease of transferability between diverse laboratories and are acknowledged for their remarkable data reproducibility. As per Maul and Töpfer (2015) findings, the SSR profiles acquired from the analysis can be conveniently organized in digital databases and employed for scrutinizing genetic fidelity and discerning cultivars. Thus far, establishing SSR markers within the *Origanum* genus has solely relied on expressed sequence tags (ESTs). It has been observed that the polymorphic information content or PIC was relatively low, which resulted in restricted applicability in genetic diversity studies (Alekseeva et al. 2021; Novak et al. 2008).The establishment of genetic resource collections that mirror the diversity in natural populations, in terms of genetics and metabolism, has become necessary due to the considerable economic value and high interest in these species (Börner et al. 2010). Furthermore, breeding superior plant varieties that exhibit exceptional performance is crucial (Aboshama et al. 2015; Emara et al. 2011, 2013; Nawara et al. 2016, 2017; Saleh et al. 2020a; 2020b; Salem 1997; 2004; Salem et al. 2021a, b, d, e, 2023). The cited works were authored by Alekseeva et al. (2021) and Sarrou et al. (2017). The Fitch-Margoliash method is a widely used computational algorithm in molecular phylogenetics. Its primary function is to estimate the evolutionary distances between different species based on their genetic makeup. The clustering analysis was performed to model the evaluation of genetic variation within a pool of genetic resources. Typically, botanical assemblages exhibit considerable genetic heterogeneity, and conventional phylogenetic methodologies, such as constructing a phylogenetic dendrogram, are frequently employed. The phylogenetic tree depicted in Fig. 3.3 demonstrates the remarkable discriminatory power of the genomic SSR markers in distinguishing all the studied plants within the population. As a plant biologist, this finding is noteworthy. The dendrogram exhibited a distinct grouping of the specimens that precisely correlated with the coloration pattern of the specimens belonging to population 3 in the genetic composition analysis. The SSR marker's probability of identity (PI) was determined for different cultivars, namely R-40C, R-6M, R-103C, R-38C, R-105C, and R-115C. The PI values were 0.21, 0.17, 0.1, 0.13, 0.06, and 0.12, respectively. The total PI for all the cultivars was calculated to be 3.126384 × 10–6. The cumulative polymorphic information value indicates that the selected set of simple sequence repeat markers is suitable for discerning distinct plants within the given population. The findings of Marina Alekseeva et al. (2023) suggest that the selected genomic SSR markers are effective in characterizing the genetic resource collections of oregano species, as indicated by the results of the phylogenetic analysis. Furthermore, these genetic markers have the potential to be utilized in traditional phylogenetic methodologies to investigate genetic variation.

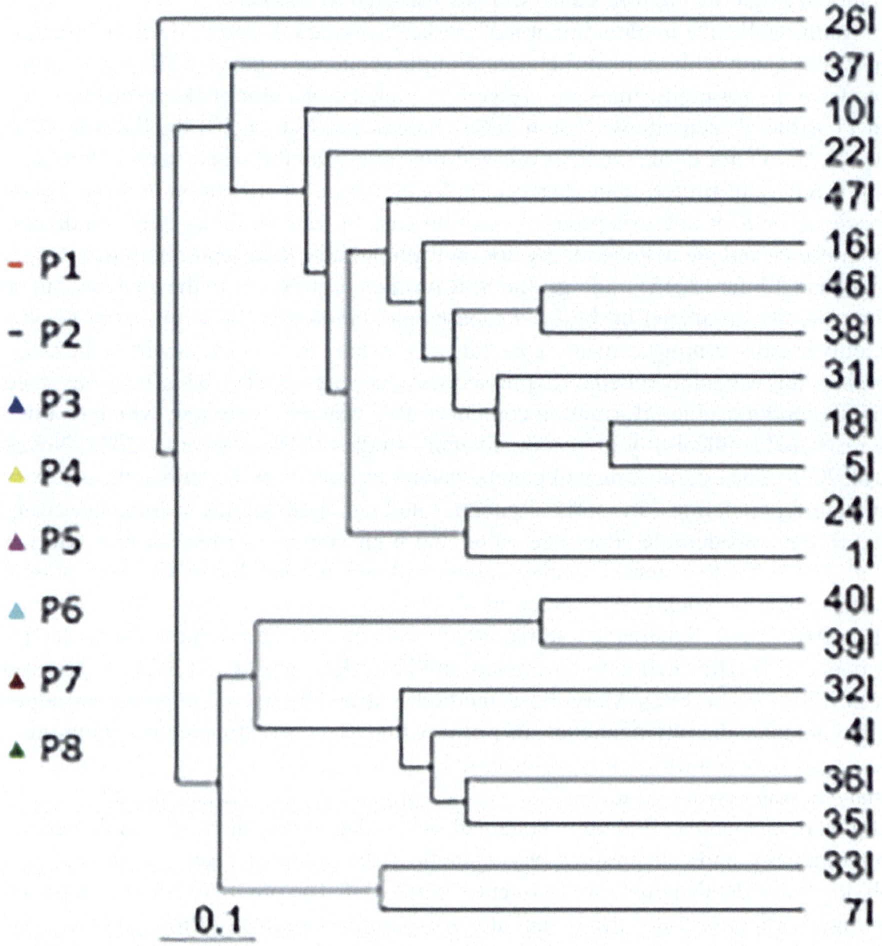

Fig. 3.3 Phylogenetic tree constructed by the Fitch-Margoliash method representing the genetic similarities among the plants in population 3. *Source* Alekseeva et al. (2023)

3.8 Traditional Breeding

3.8.1 Traditional Breeding and Limitations

The practice of cultivating this particular plant species is observed in diverse regions spanning Europe, North America, and Asia. The cultivation of Marjoram involves two methods: directly sowing seeds into the soil or planting greenhouse-grown seedlings. However, the latter technique is not widely used compared to the former (Zawi'slak and Dzida 2010). Plant growth, quality, yield, and oil content significantly vary due to unfavorable and frequently unpredictable environmental and soil factors. As per the findings of Nurzy et al. (2009), using seedlings can potentially augment the crop yield and guarantee consistent plant dimensions. This is because climatic factors relatively less influence the early growth phase. The optimal time for Marjoram harvest is when the initial signs of flower bud formation become visible. Cutting the aboveground biomass at a height of 5 cm is recommended to minimize contamination. The data above is corroborated by studies conducted by Anbauverfahren Majoran (2020), Nurzy et al. (2009), and Zawi'slak (2008).

The essential oils derived from Marjoram plants are composed primarily of thymol, carvacrol, terpinene-4-ol, and sabinene. The distinct aroma and taste of Marjoram can be attributed to the presence of terpinene-4-ol and cis-sabinene hydrate in their chemical composition. Several studies have documented these results, such as Vági et al. (2005) and Sellami et al. (2009). Furthermore, Vera and Chane-Ming (1999) identified noteworthy levels of p-cymene and -terpinene. The accumulation and final composition of essential oils are significantly influenced by cultivation conditions, climate, and growth stage during harvest, as noted by Murillo-Amador et al. (2013), Sellami et al. (2009), and Trivino and Johnson (2000).

Indoor vertical farming is a viable strategy to maintain the uniform quality of crops throughout all seasons (Saleh et al 2021a, b). Multi-layer hydroponic systems employ artificial lighting to grow plants within a controlled environment, enabling autonomous functioning irrespective of external factors. As per Kozai et al. (2020), this facilitates absolute regulation of abiotic parameters. As per the findings of Graamans et al. (2018), Ohyama et al. (2020), and Pennisi et al. (2019), this particular cultivation method has been observed to result in a noteworthy enhancement in crop yield while also ensuring the retention of optimal resource utilization efficiencies. Additionally, it is feasible to augment the secondary metabolite accumulation in botanical specimens such as *Coriandrum sativum* L. (Nguyen et al. 2020) and *Ocimum basilicum* (Rihan et al. 2020) by accurately manipulating the environmental factors throughout the cultivation procedure.

Furthermore, the potential for cross-contamination can be mitigated, as Kalantari et al. (2017) reported. Various studies have explored the impact of distinct light spectral qualities provided by LEDs on the volatile compounds of plants, including *Ocimum basilicum*, as reported by Carvalho et al. (2016), Pennisi et al. (2019) and Schwend et al. (2016). The practice of indoor vertical farming offers promising prospects for the sustainable and reliable cultivation of premium herbs for the fresh market, along with a constant supply of raw materials throughout the year for the food, chemical, or perfume industry. It is important to acknowledge that the present capital and running costs associated with an indoor vertical farming system are substantial, as stated in the studies conducted by Graamans et al. (2018) and Barbosa et al. (2015).

3.8.2 Molecular Breeding

In recent years, the application of DNA markers in examining and recognizing therapeutic plants has gained significant traction (Gostimsky et al. 2005; Safhi et al. 2022; Ibrahim et al. 2024a, b). Identifying the target gene in individual plants can be achieved by detecting closely linked molecular markers. Molecular markers offer a significant advantage in precision and speed over traditional breeding techniques. In the realm of plant biology, the commonly employed molecular markers include simple sequence repeat (SSR), random amplified polymorphic DNA (RAPD), restriction fragment length polymorphism (RFLP), sequence-related amplified polymorphism (SRAP), amplified fragment length polymorphism (AFLP), DNA barcoding, and single nucleotide polymorphism (SNP) (Hasan et al. 2021; Nadeem et al. 2018; Varshney et al. 2008). Pandey et al. (2018) highlighted that the increasing use of molecular marker technology had enabled plant biologists to perform various tasks, including genetic diversity analysis, genetic map construction, parental analysis, and molecular breeding (Eltaher et al. 2021; 2022; Ismail et al. 2013; Mourad et al. 2019; Salem et al. 2023, b; Abdel-Hamid and Salem 2021; Abu-Ellail et al. 2021; Aldahak et al. 2021; Hassan et al. 2021; Ibrahim et al. 2023a, b, 2023c). The genetic resources of plant species' wild relatives are highly valued in plant biology. Research on genetic diversity is carried out to advance the conservation of resources and the promotion of sustainable utilization. The abovementioned phenomenon has been elucidated in studies by Hashemifar and Rahimmalek (2018) and Poland and Rife (2012).

3.8.3 Tissue Culture Studies

There are various drawbacks associated with conventional propagation techniques. For instance, large-scale seed propagation is hindered by low germination, low viability, seed sterility, repeated vegetative propagation that leads to progressive yield loss, and low rooting capacity of cuttings (Goleniowski et al. 2003). Exploring alternative pathways through in vitro cultivation of these species is a promising biotechnological strategy for conserving the species and achieving large-scale production. It enables quick and efficient mass propagation and maintains the product's durability.

3.8.3.1 In Vitro Micropropagation of *O. majorana*

Savitikadi et al. (2020) state that the in vitro propagation technique is important in preserving scarce and precious medicinal plant species that exhibit restricted reproductive efficacy. In their recent publication, Manokari et al. (2020) present a novel approach to conserving a diverse array of plant species for future utilization. In vitro regeneration methodology is employed for large-scale propagation of uncommon and endangered plant species to augment their populace and reinforce in situ conservation initiatives (Chokheli et al. 2020). Somaclonal variation is a well-known phenomenon in plants propagated in vitro and has been observed since the early days of plant tissue culture (Anandan et al. 2018). The phenomenon of somaclonal variation is subject to the influence of various factors, including but not limited to the nature of explants employed, the constitution of the medium, the frequency of explant sub-culturing, and the concentration of phytohormones. The abovementioned topic has been examined in research conducted by Govindaraju and Arulselvi (2018) and Jogam et al. (2020).

The herbaceous plant known as Marjoram has been selectively extracted from open forest ecosystems due to its notable botanical characteristics and commercial significance attributed to its essential oil content (Khadhri et al. 2019). Marjoram is propagated through both sexual and asexual means, specifically through seeds and root suckers, respectively. The seed propagation method exhibits low efficiency, and the poor rooting of stem cuttings results in a low survival rate, as observed by Yashoda Bai et al. (2011). In recent years, several PCR-based molecular marker techniques have been established to evaluate the genetic homogeneity of micro-propagated plant specimens. Inter simple sequence repeats (ISSR) PCR primers are widely used in plant biology to evaluate the genetic consistency between in vitro propagated plantlets and their maternal plants (Bhattacharyya et al. 2016; Sadhu et al. 2020; Alsharari et al. 2023; Soliman et al. 2023).

The efficacy of the micropropagation protocol is contingent upon the capacity of tissue-cultured plantlets to endure ex-vitro conditions. The morpho-anatomical and physiological characteristics of the plantlets are influenced by the specific physicochemical conditions present in their in vitro microenvironments. In vitro-derived plantlets frequently encounter challenges adapting to ex-vitro and in vivo conditions, leading to their demise due to varying environmental factors (Chandra et al. 2010; Mesfer ALshamrani et al. 2022). Scientists have endeavored to enhance the survival rates and reduce mortality by manipulating the micro-environments during the hardening and acclimatization protocols. The phenomenon in question has been recorded and analyzed in scientific research by Lavanya et al. (2009) and Deb and Imchen (2010). Despite the existing literature on in vitro regeneration of *O. majorana*, including studies by Iyer and Pai (1998, 2000), Korkar et al. (2017), and Tejavathi and Padma (2012), there has been no assessment of the genetic and micro-morphological fidelity of the micro propagated plants. Micropropagation is a technique that can lead to somaclonal variation, as well as micro-morpho-anatomical and physiological abnormalities in plantlets that are propagated through tissue culture. It has been documented by researchers such as Chandra et al. (2010) and Haisel et al. (2001).

Assessment of clonal homogeneity is imperative for micro-propagated *O. majorana* specimens. In addition, performing a foliar micro-morphological examination on the regenerated plants and the parent plant may offer valuable information on the anatomical ability of the young plants to survive in their native habitat. Sandhya and colleagues (2021) performed ISSR molecular marker analysis on Origanum majorana plants propagated in vitro and compared them to the original mother plants. As depicted in (Fig. 3.4), the visual representation illustrates the data in a graphical format.

3.9 Conclusions and Prospects

Recent studies have described an effective in vitro clonal propagation procedure for *O. majorana*, a highly traded aromatic plant. The studies on genetic homogeneity using ISSR markers have shown no genetic variations between the mother plant and the plants propagated clonally. Additionally, the evaluation of foliar micromorphology showed that the developed plantlets were more stable than the donor plant. The results obtained from this study could be utilized for the large-scale in vitro synthesis of bioactive compounds from *O. majorana*. Advances in Marjoram (*Origanum majorana* L.) breeding strategies have paved the way for a promising future in the cultivation and improvement of this valuable herb. Marjoram is widely used in culinary, medicinal, and aromatic applications, making it a significant crop that demands continuous development. It has been established that Marjoram possesses considerable genetic diversity through various breeding techniques, such as conventional breeding, molecular markers, and genomic studies. This diversity is a valuable resource for breeders to develop improved cultivars with desirable traits, including yield, quality, disease resistance, and adaptability. Trait improvement through breeding programs has focused on enhancing specific characteristics of Marjoram. Some traits targeted for enhancement are yield improvement, oil content, aroma, flavor profile, and resistance to pests and diseases. Integrating conventional breeding methods with modern biotechnological tools, such as marker-assisted selection and genetic engineering, holds immense potential to accelerate the breeding process and achieve desired trait improvements. Stress tolerance, like other crops, Marjoram faces challenges related to abiotic and biotic stresses. Breeding efforts have been directed towards developing marjoram cultivars with improved tolerance to various stress factors, including drought, salinity, temperature fluctuations, and fungal infections. Identification and incorporation of stress-responsive genes and quantitative trait loci (QTLs) into breeding programs can aid in developing resilient cultivars. The nutritional and medicinal properties of Marjoram possess several medicinal properties attributed to its bioactive compounds. Future breeding strategies could focus on enhancing the concentration of these compounds, such as phenolics, flavonoids, and essential oils, to elevate the medicinal value of Marjoram. Improving the herb's nutritional composition, including essential minerals and secondary metabolites, can also be a target of breeding programs. With increasing

Fig. 3.4 Marjoram (*Origanum majorana*), in its many forms after micropropagation. A) *O. majorana* nodal explants can directly induce shoot growth 10 days following vaccination in experiment **a**. **b** After a month in culture, several shoots have developed from the explants. **c** Proliferation and elongation of shoots following the first subculture on MS medium with 0.5 mg/L of BAP and 0.25 mg/L of NAA. **d** After the third cultural movement, new shoots proliferate rapidly. **e** Half-strength MS medium + 1.0 mg/L IBA yielded rooted shoots in vitro. **f** Plant seedlings in paper cups with a mixture of soilrite®, manure, and sand (2: 1: 1). **g** Field-transferred plant of micro-propagated *O. majorana*. *Source* Dulam Sandhya et al. (2021)

consumer demand for organically grown products, breeding marjoram cultivars suitable for organic farming becomes crucial for organic production and sustainability. Breeding efforts can aim to develop varieties that exhibit natural resistance to pests and diseases, reducing the reliance on chemical inputs.

Additionally, optimizing cultivation practices and improving resource-use efficiency contribute to the sustainability of marjoram production.

In conclusion, advances in marjoram breeding strategies offer exciting prospects for developing improved cultivars with enhanced traits, stress tolerance, medicinal properties, and sustainability. Integrating conventional breeding methods with cutting-edge technologies will expedite the breeding process, providing solutions to the marjoram industry's challenges. Continued research and collaboration among breeders, geneticists, agronomists, and other stakeholders are essential for realizing the full potential of Marjoram as a valuable crop.

References

Abd El-Moneim D, ELsarag EIS, Aloufi S et al (2021). Quinoa (*Chenopodium quinoa* Willd.): genetic diversity according to ISSR and SCoT markers, relative gene expression, and morpho-physiological variation under salinity stress. Plants 10:2802

Abd-Ellatif S, Ibrahim AA, Safhi FA, Abdel Razik ES, Kabeil SSA, Aloufi S, Alyamani AA, Basuoni MM, ALshamrani SM, Elshafie HS (2022) Green synthesized of thymus vulgaris chitosan nanoparticles induce relative WRKY-genes expression in solanum lycopersicum against fusarium solani, the causal agent of root rot disease. Plants 11:3129. https://doi.org/10.3390/plants11223129

Abd El-Wahab MA (2013) Productivity of marjoram (*Origanum majoranum* L.) in variable ecosystem locations in Egypt. J Appl Sci 3:83–89

Abdel-Hamid AME, Salem KFM (2021) Breeding strategies of garden pea (*Pisum sativum* L.). In: Al-Khayri JM, Jain SM, Johnson DV (eds) Advances in plant breeding strategies: vegetable crops. Springer, Cham. https://doi.org/10.1007/978-3-030-66969-0_9

Abdel-Massih R, Abdou E, Baydoun E, Daoud Z (2010) Antibacterial activity of the extracts obtained from *Rosmarinus officinalis, Origanum majorana* and *Trigonella foenum-graecum* on highly drug-resistant gram-negative bacilli. J Botany 2010:464087

Abdelsalam NR, Hasan ME, Rabie SMA et al (2021) Endorsement and phylogenetic analysis of some fabaceae plants based on DNA barcoding gene MatK. https://doi.org/10.1101/2021.07.27.454001

Aboshama HM, Salem KFM, Hassan HMF, Tahoun MH (2015) Estimating heterosis and combining ability for *in vivo* and *in vitro* traits using diallel cross on *tomato* (*Solanum lycopersicum* L.). Minufiya J Agric Res 40(3):757–770

Abu-Ellail FFB, Salem KFM, Saleh MM et al (2021) Molecular breeding strategies of beetroot (*Beta vulgaris* ssp. vulgaris var. conditiva Alefeld). In: Al-Khayri JM, Jain SM, Johnson DV (eds) Advances in plant breeding strategies: vegetable crops. Springer, Cham. https://doi.org/10.1007/978-3-030-66965-2_4

Al Dhaheri Y, Eid A, AbuQamar S et al (2013) Mitotic arrest and apoptosis in breast cancer cells induced by *Origanum majorana* extract: upregulation of TNF-α and downregulation of survivin and mutant p53. PLoS ONE 8(2):e56649

Aldahak L, Salem KFM, Al-Salim SHF, Al-Khayri JM (2021) Advances in chicory (*Cichorium intybus* L.) breeding strategies. In: Al-Khayri JM, Jain SM, Johnson DV (eds) Advances plant breeding strategies: vegetable crops. Springer, Cham. https://doi.org/10.1007/978-3-030-669 69-0_1

Alsharari SF, Ibrahim AA, Okasha SA (2023). Combining ability for yield, oil content, and physio biochemical characters of canola (Brassica napus L.) Under salt stress conditions. SABRAO J Breed Genet 55(4): 1003–1024. https://doi.org/10.54910/sabrao2023.55.4.1

Alekseeva M, Rusanova M, Rusanov K, Atanassov I (2023) A set of highly polymorphic microsatellite markers for genetic diversity studies in the genus *Origanum*. Plants 12:824. https://doi.org/10.3390/plants

Alekseeva M, Zagorcheva T, Atanassov I, Rusanov K (2020) *Origanum vulgare* L.—a review on genetic diversity, cultivation, biological activities and perspectives for molecular breeding. Bulg J Agric Sci 26:1183–1197

Alekseeva M, Zagorcheva T, Rusanova M et al (2021) Genetic and flower volatile diversity in natural populations of *Origanum vulgare* subsp. hirtum (Link) Ietsw. In Bulgaria: toward the development of a core collection. Front Plant Sci 12:679063

Alqahtani MM, Saleh MM, Alwutayd KM et al (2024) Performance and genotypic variability in diverse date palm (*Phoenix dactylifera* L.) cultivars for fruit characteristics. Genet Resour Crop Evol. https://doi.org/10.1007/s10722-023-01719-0

Alsafar MS, Al-Hassan YM (2009) Effect of nitrogen and phosphorus fertilizers on growth and oil yield of indigenous mint (*Mentha longifolia* L.). Biotechnol J 8:380–384

ALshamrani M, Safhi S, Alshaya FA et al (2022) Genetic diversity using biochemical, physiological, karyological and molecular markers of *Sesamum indicum* L. Front Genet 13:1035977

Amaghnouje A, Mechchate H, Es-safi I et al (2020) Anxiolytic, antidepressant-like proprieties and impact on the memory of the hydro-ethanolic extract of *Origanum majorana* L. on mice. Appl Sci 10:8420

Amer SA, Tolba SA, AlSadek DMM et al (2021) Effect of supplemental glycerol monolaurate and oregano essential oil blend on the growth performance, intestinal morphology, and amino acid digestibility of broiler chickens. BMC Vet Res 17:312

Anandan R, Prakash M, Deenadhayalan T et al (2018) Efficient *in vitro* plant regeneration from cotyledon-derived callus cultures of sesame (*Sesamum indicum* L.) and genetic analysis of true-to-type regenerants using RAPD and SSR markers. South Afr J Bot 119:244–251

Anbauverfahren Majoran (2020) Available online: https://publikationen.sachsen.de/bdb/artikel/13856

Aureli P, Costantini A, Zole S (1992) Antimicrobial activity of some essential oils against *Listeria monocytogenes*. J Food Prot 55:344–348

Aziz EE, Hendawy ST, Azza ED, Omer EA (2008) Effect of soil type and irrigation intervals on plant growth, essential oil yield and constituents of *Thymus vulgaris* plant. Am Eur J Agric Environ Sci 4:443–450

Azizi A, Yan F, Honermeier B (2009) Herbage yield, essential oil content and composition of three oregano (*Origanum vulgare* L.) populations as affected by soil moisture regimes and nitrogen supply. Ind Crop Prod 29:554–561

Baatour O, Kaddour R, Wannes WA et al (2010) Salt effects on growth, mineral nutrition, essential oil yield and composition of Marjoram (*Origanum majorana*). Acta Physiol Plant 32:45–51

Bakkali F, Averbeck S, Averbeck D, Idaomar M (2008) Biological effects of essential oils—a review. Food Chem Toxicol 46:446–475

Balandrin MF, Klocke JA (1988) Medicinal, aromatic, and industrial materials from plants. In: Bajaj YPS (ed) Medicinal and aromatic plants I, Biotechnology in agriculture and forestry. Springer, Berlin

Banchio E, Bogino P, Zygadlo J, Giordano W (2008) Plant growth promoting rhizobacteria improve growth and essential oil yield in *Origanum majorana* L. Biochem Syst Ecol 36:766–771

Barbosa GL, Gadelha FDA, Kublik N et al (2015) Comparison of land, water and energy requirements of lettuce grown using hydroponic vs. conventional agricultural methods. Int J Environ Res Public Health 12:6879–6891

Barbouche N, Hajjem B, Lognay G, Ammar M (2001) Contribution a letude de l'activite biologique d'extraits de feuilles de Cestrum parqui L'Herit. (Solanaceae) sur le criquet p'elerin Schistocerca gregaria (Forsk.). BASE

Baser KHC, Kirimer N, Tümen G (1993) Composition of the essential oil of *Origanum majorana* L. from Turkey. J Essent Oil Res 5:577–579

Baydar H, Sagdic O, Ozkan G, Karadogan T (2004) Antibacterial activity and composition of essential oils from Origanum, Thymbra and Satureja species with commercial importance in Turkey. Food Control 15:169–172

Bettray G, Vömel A (1992) Influence of temperature on yield and active principles of Chamomilla recutita (L.) Rausch. Under controlled conditions. Acta Hortic 306:83–87

Bhattacharyya P, Kumaria S, Tandon P (2016) High-frequency regeneration protocol for *Dendrobium nobile*: a model tissue culture approach for propagation of medicinally important orchid species. South Afr J Bot 104:232–243

Bina F, Rahimi R (2016) Sweet Marjoram: a review of ethnopharmacology, phytochemistry, and biological activities. J Evid Based Complement Altern Med 22:175–185

Bina F, Rahimi R (2017) Sweet Marjoram: a review of ethnopharmacology, phytochemistry, and biological activities. J Evid Based Complement Altern Med 22:175–185

Börner A, Landjeva S, Salem KFM et al (2010) Plant genetic resources—a prerequisite for drought tolerance breeding in wheat. Tagungsband 60. Tagung der Vereinigung Pflanzenzüchter und Saatgutkaufleute Österreichs, Gumpenstein, Österreich, 24–26 November 2009

Bouyahya A, Chamkhi I, Benali T et al (2021) Traditional use, phytochemistry, toxicology, and pharmacology of Origanum majorana L. J Ethnopharm 265:113318. https://doi.org/10.1016/j.jep.2020.113318

Busatta C, Vidal RS, Popiolski AS et al (2008) Application of *Origanum majorana* L. essential oil as an antimicrobial agent in sausage. Food Microbiol 25:207–211

Camo J, Beltrán JA, Roncalés P (2008) Extension of the display life of lamb with an antioxidant active packaging. Meat Sci 80:1086–1091

Carvalho CB, Madrona GS, Mitcha JG et al (2020) Effect of active packaging with oregano oil on beef burgers with low sodium content. Acta Sci Technol 42:e42892

Carvalho SD, Schwieterman ML, Abrahan CE et al (2016) Light quality dependent changes in morphology, antioxidant capacity, and volatile production in sweet basil (*Ocimum basilicum*). Front Plant Sci 7:1328

Chandra S, Bandopadhyay R, Kumar V, Chandra R (2010) Acclimatization of tissue cultured plantlets: from laboratory to land. Biotechnol Lett 32(9):1199–1205

Charles DJ (2013) Marjoram sweet. In: Charles DJ (ed) Antioxidant properties of spices, herbs and other sources. Springer, New York, pp 393–399

Chokheli VA, Dmitriev PA, Rajput VD et al (2020) Recent development in micropropagation techniques for rare plant species. Plants 9(12):1733

Chung YK, Heo HJ, Kim EK et al (2001) Inhibitory effect of ursolic acid purified from *Origanum majorana* L. on the Acetylcholinesterase, 2. In: Molecules & cells, vol 11. Springer

Daferera DJ, Ziogas BN, Polissiou MG (2008) GC-MS analysis of essential oils from some Greek aromatic plants and their fungitoxicity on *Penicillium digitatum*. J Agric Food Chem 48:2576–2581

Dasha R, Muni R, Ranjeet S (2006) Optimization of water and nitrogen application to menthol mint (*Mentha arvensis* L.) through sugarcane trash mulch in a sandy loam soil of semi-arid subtropical climate. Bioresour Technol 97:886–893

Deb CR, Imchen T (2010) An efficient *In vitro* hardening technique of tissue culture-raised plants. Biotechnol 9(1):79–83

Della Pepa T, Elshafie HS, Capasso R et al (2019) Antimicrobial and phytotoxic activity of *Origanum heracleoticum* and *O. majorana* essential oils growing in cilento (Southern Italy). Molecules 24:2576

Dhifi W, Bellili S, Jazi S et al (2013) Essential oils' chemical characterization and investigation of some biological activities: a critical review. Medicines 3:25

Djenane D, Sánchez-Escalante A, Beltrán JA, Roncalés P (2003) Extension of the shelf life of beef steaks packaged in a modified atmosphere by treatment with rosemary and displayed under UV-free lighting. Meat Sci 64:417–426

Dorais M, Papadopoulos AP, Luo X et al (2001) Soilless greenhouse production of medicinal plants in northeastern Canada. Acta Hortic 554:297–303

El Rabey H, Salem KFM, Mattar MZ (2013) The genetic diversity and relatedness of eight rice (*Oryza sativa* L.) cultivars as revealed by AFLP and SSRs markers. Life Sci J 10(1):1471–1479

El-Absawy EA, Nowar AA, Salem KFM et al (2015) Estimation of genetic diversity among some bread wheat (*Triticum aestivum* l.) genotypes using molecular markers and agronomic traits. Minufiya J Agric Res 40(3):223–233

El-Akhal F, Guemmouh R, Maniar S et al (2016) Larvicidal activity of essential oils of *Thymus vulgaris* and *Origanum majorana* (Lamiaceae) against of the malaria vector *Anopheles labranchiae* (Diptera: Culicidae). Int J Pharm Pharmaceut Sci 8:372–376

El-Alam I, Zgheib R, Iriti M et al (2019) *Origanum syriacum* essential oil chemical polymorphism according to soil type. Foods 8:90

El-Ashmawy IM, Saleh A, Osama M (2007) Effects of marjoram volatile oil and grape seed extract on ethanol toxicity in male rats. Basic Clin Pharmacol Toxicol 101:320–327

EL-Mansy AB, Abd El-Moneim D, ALshamrani SM et al (2021) Genetic diversity analysis of tomato (Solanum lycopersicum L.) with morphological, cytological, and molecular markers under heat stress. Horticulturae 7:65. https://doi.org/10.3390/horticulturae7040065

El-Raghi AA, Hassan MAE, Hashem NM Abdelnour SA (2023) Struggling thermal stress impacts on growth performance and health status of newly weaned rabbits using nanoemulsion of origanum majorana considering the economic efficiency of supplementation. Animals 13:1772. https://doi.org/10.3390/ani13111772

El-Sheikh MH, Ghareeb RY, Ibrahim AA, Hafez EE (2020) Genetic diversity among Juglans regia and Prunus dulcis genotypes using RAPD and ISSR molecular markers. Plant Archives 20(2)

Elsherbini DMA, Almohaimeed HM, El-Sherbiny M et al (2022) *Origanum majorana* L. extract attenuated benign prostatic hyperplasia in rat model: effect on oxidative stress, apoptosis, and proliferation. Antioxidants 11:1149. https://doi.org/10.3390/antiox11061149

Eltaher S, Baenziger S, Belamkar V et al (2021) GWAS revealed effect of genotype × environment interactions for grain yield of Nebraska winter wheat. BMC Genomics 22:1–14

Eltaher S, Sallam A, Belamkar V et al (2018) Genetic diversity and population structure of F3:6 Nebraska winter wheat genotypes using genotyping-by-sequencing. Front Genet 9:76. https://doi.org/10.3389/fgene.2018.00076

Eltaher S, Sallam A, Emara HA et al (2022) Genome-wide association mapping revealed SNP alleles associated with spike traits in wheat. Agronomy 12:1469. https://doi.org/10.3390/agronomy12061469

El-Zanaty AM, Salem KFM, Esmail RM (2011) Detection of genetic diversity in Egyptian cotton (*Gossypium barbadense* L.) varieties using RAPD markers and morphological traits. J American Sci 7(12):1107–1115

Emara HA, Nower AA, Salem KFM, Eltaher S (2011) Estimating heterosis and combining ability for *in vivo* and *in vitro* traits using diallel cross in Maize (*Zea mays* L*)*. J Prod Dev 16:101–120

Emara HA, Salem KFM, Abdel-Halim AA, Abd El-Kowy RNF (2013) An in vivo and in vitro traits analysis of a diallel cross in rice (*Oryza sativa L.*). Egyptian J Plant Breed 17:591–610

Erdogan A, Ozkan A (2017) Investigatıon of antioxıdative, cytotoxic, membrane-damaging and membrane-protective effects of the essentıal oil of origanum majorana and its oxygenated monoterpene component linalool in human-derived hep G2 cell line. Iran J Pharm Sci 16:24–34

Essa NM, Ibrahim AA, Soliman MI (2024) Association studies between some cultivated species and their wild relatives from Apiaceae, Asteraceae and Brassicaceae families based on molecular and DNA barcoding in Egypt. Genet Resour Crop Evol https://doi.org/10.1007/s10722-023-01681-x

Ezzeddine NBHB, Abdelkefi MM, Aissa RB, Chaabouni MM (2001) Antibacterial screening of *Origanum majorana* L. oil from Tunisia. J Essent Oil Res 13:295–297

Fanourakis D, Paschalidis K, Tsaniklidis G et al (2022) Pilot cultivation of the local endemic Cretan marjoram *Origanum microphyllum* (Benth.) vogel (Lamiaceae): effect of fertilizers on growth and herbal quality features. Agronomy 12(1):94. https://doi.org/10.3390/agronomy12010094

Fasseas MK, Mountzouris KC, Tarantilis PA et al (2007) Antioxidant activity in meat treated with oregano and sage essential oils. Food Chem 106:1188–1194

Gerami F, Moghaddam PR, Ghorbani R, Hassani A (2016) Effects of irrigation intervals and organic manure on morphological traits, essential oil content and yield of oregano (*Origanum vulgare* L.). An Acad Bras Cienc 88:2375–2385

Ghazy MI, Salem KFM, Sallam A (2020) Utilization of genetic diversity and marker-trait to improve drought tolerance in rice (*Oryza sativa* L.). Mol Biol Rep. https://doi.org/10.1007/s11033-020-06029-7

Goleniowski ME, Flamarique C, Bima P (2003) Micropropagation of Oregano (*Origanum vulgare* × applii) from meristem tips. Plant 39:125–128. https://doi.org/10.1079/IVP2002361

Gostimsky SA, Kokaeva ZG, Konovalov FA (2005) Studying plant genome variation using molecular markers. Russ J Genet 41:378–388

Govindaraju S, Arulselvi PI (2018) Effect of cytokinin combined elicitors (l-phenylalanine, salicylic acid and chitosan) on *in vitro* propagation, secondary metabolites and molecular characterization of medicinal herb–*Coleus aromaticus* Benth (L). J Saudi Soc Agric Sci 17(4):435–444

Graamans L, Baeza E, van den Dobbelsteen A et al (2018) Plant factories versus greenhouses: comparison of resource use efficiency. Agric Syst 160:31–43

Guo GX, Li KX, Zhu QH et al (2022) Improvements of immune genes and intestinal microbiota composition of turbot (Scophthalmus maximus) with dietary oregano oil and probiotics. Aquaculture 547:737442

Gutiérrez RMP, Jerónimo FFM, Soto JGC et al (2022) Optimization of ultrasonic-assisted extraction of polyphenols from the polyherbal formulation of *Cinnamomum verum*, *Origanum majorana*, and *Origanum vulgare* and their anti-diabetic capacity in zebrafish (*Danio rerio*). Heliyon 8:e08682

Haisel D, Hofman P, Vagner M et al (2001) *Ex vitro* phenotype stability is affected by *in vitro* cultivation. Biol Plantarum 44(3):321–324

Hajlaoui H, Mighri H, Aouni M et al (2016) Chemical composition and in vitro evaluation of antioxidant, antimicrobial, cytotoxicity and anti-acetylcholinesterase properties of Tunisian *Origanum majorana* L. essential oil. Microb Pathog 95:86–94

Hasan N, Choudhary S, Naaz N et al (2021) Recent advancements in molecular marker-assisted selection and applications in plant breeding programmes. J Genet Eng Biotechnol 19:128

Hashemifar Z, Rahimmalek M (2018) Genetic structure and variation in *Perovskia abrotanoides* Karel and *P. atriplicifolia* as revealed by molecular and morphological markers. Sci Hortic 230:169–177

Hassan MN, Mekkawy SA, Mahdy M et al (2021) Recent molecular and breeding strategies in lettuce (*Lactuca* spp.). Genet Resour Crop Evol 68:3055–3079. https://doi.org/10.1007/s10722-021-01246-w

Hendawy SF, Hussein MS, Amer HM, El-Gohary AE (2017) Effect of soil type on growth, productivity, and essential oil constituents of rosemary, Rosmarinus officinalis. Asian J Agric Biol 4:303–311

Hossain MB, Camphuis G, Aguiló-Aguayo I et al (2014) Antioxidant activity guided separation of major polyphenols of Marjoram (*Origanum majorana* L.) using flash chromatography and their identification by liquid chromatography coupled with electrospray ionization tandem mass spectrometry. J Sep Sci 37:3205–3213

Hyldgaard M, Mygind T, Mayer RL (2012) Essential oils in food preservation: mode of action, synergies, and interactions with food matrix components. Front Microbiol 3:12

Ibrahim AA, Abdel Razik ES, Abd-Ellatif S, Ramadan S (2023a) *Nigella sativa* and stressful conditions. In: Latef AAHA (ed) Medicinal plant responses to stressful conditions. CRC Press. https://doi.org/10.1201/9781003242963

Ibrahim AA, Abd-Ellatif S, Abdel Razik ESS, Salem KFM (2024) Biodiversity of cereal crops and utilization in food and nutritional security. In: Al-Khayri JM, Jain, SM, Penna S (eds) Sustainable utilization and conservation of plant genetic diversity. Sustainable development and biodiversity, vol 35. Springer, Singapore. https://doi.org/10.1007/978-981-99-5245-8_2

Ibrahim AA et al (2024b) Characterization and comparative genomic analyses of complete chloroplast genome on Trema orientalis L. Genet Resour Crop Evol. https://doi.org/10.1007/s10722-023-01678-6

Ibrahim AA, Salem KFM, Abdein MA, Ramadan SM (2023b) Advances in summer squash (*Cucurbita pepo* L.) molecular breeding strategies. In: Singh S, Sharma D, Sharma SK, Singh R (eds) Smart plant breeding for vegetable crops in post-genomics era. Springer, Singapore. https://doi.org/10.1007/978-981-19-5367-5_8

Ibrahim AA, Salem KFM, Ramadan S (2023c) *Allium sativum* and stressful conditions. In: Latef AAHA (ed) Medicinal plant responses to stressful conditions. CRC Press. https://doi.org/10.1201/9781003242963

Ietswaart JH (1980) A taxonomic revision of the genus *Origanum* (Labiatae). Leiden Botanical Ser 4(1):1–153

Ismail M, Salem KFM, El-Zanaty AM, Esmail RM (2013) Association mapping of fiber and agronomic characters in Egyptian cotton (*Gossypium barbadense* L.) genotypes. J Agric Chem Biotech Mansoura Univ 4:205–216

Iyer PV, Pai JS (1998) Micropropagation of sweet Marjoram (*Majorana hortensis* Moech.). J Spices Aromatic Crops 7(1):47–79.

Iyer PV, Pai JS (2000) *In vitro* regeneration of *Majorana hortensis* Moench from callus and nodal stem segments. J Spices Aromatic Crops 9(1):47–50

Jelali N, Dhifi W, Chahed T, Marzouk B (2011) Salinity effects on growth, essential oil yield and composition and phenolic compounds content of Marjoram (*Origanum majorana* L.) leaves. J Food Biochem 35:1443–1450

Jogam P, Dulam S, Shekhawat SS et al (2020) Genetic stability analysis using DNA barcoding and molecular markers and foliar micro-morphological analysis of *in vitro* regenerated and *in vivo* grown plantlets of *Artemisia vulgaris* L. Ind Crop Prod 150(1):112476

Kabera JN, Semana E, Mussa AR, He X (2014) Plant secondary metabolites: biosynthesis, classification, function and pharmacological properties. J Pharm Pharmacol 2:377–392

Kakouri E, Daferera D, Kanakis C, Revelou PK, Kaparakou EH, Dervisoglou S, Perdikis D, Tarantilis PA (2022) *Origanum majorana* essential oil—a review of its chemical profile and pesticide activity. Life 12:1982. https://doi.org/10.3390/life12121982

Kalantari F, Mohd Tahir O, Mahmoudi Lahijani A, Kalantari S (2017) A Review of vertical farming technology: a guide for implementation of building integrated agriculture in cities. AEF Adv Eng Forum 24:76–91

Khadhri A, Bouali I, Aouadhi C et al (2019) Determination of phenolic compounds by MALDI-TOF and essential oil composition by GC-MS during three development stages of *Origanum majorana* L. Biomed Chromatogr 33:e4665

Khatab HA, Elhaddad NS (2015) Evaluation of mutagenic effects of monosodium glutamate using *Allium cepa* and antimutagenic action of *Origanum majorana* L. And *Ruta chalepensis* medical plants. Biotechnol J Int 8:1–11. https://doi.org/10.9734/BBJ/2015/17695

Khater RM (2019) Effect of irrigation with magnetic water and nitrogen fertilizers source on the vegetative growth, essential oil ingredients and productivity of *Origanum majorana*. J Archit Constr 2:27–40

Korkar AM, Mohamed SA, El-Kafie OMA, Gohar AA (2017) Adaptation of the *in vitro* culture of *Origanum majorana* L. for production of phenolic acids. IOSR J Pharm Biol Sci 12:30–38

Kosakowska O, Weglarz Z, Pióro-Jabrucka E, Przybyl et al (2021) Antioxidant and antibacterial activity of essential oils and hydroethanolic extracts of greek oregano (*O. vulgare* L. subsp. hirtum (Link) Ietswaart) and common oregano (*O. vulgare* L. subsp. vulgare). Molecules 26:988

Kozai T, Niu G, Takagaki M (2020) Plant Factory, An indoor vertical farming system for efficient quality food production, 2nd ed, Academic Press, an Imprint of Elsevier: London, UK

Kunicka- Styczy'nska A (2011) Activity of essential oils against food-spoiling yeast: a review. Flavour Fragrance J 26(5):326–328

La Bella S, Tuttolomondo T, Dugo G et al (2015) Composition and variability of the essential oil of the flowers of Lavandula stoechas from various geographical sources. Nat Prod Commun 10:2001–2004

Lages L, Radunz M, Goncalves BT et al (2021) Microbiological and sensory evaluation of meat sausage using thyme (*Thymus vulgaris* L.) essential oil and powdered beet juice (*Beta vulgaris* L., EarlyWonder cultivar). LWT 148:111794

Lavanya M, Venkateshwarlu B, Devi BP (2009) Acclimatization of neem micro shoots adaptable to semi-sterile conditions. Indian J Biotechnol 8:218–222

Li CY, Niu JL, Liu YX et al (2021) The effects of oregano essential oil on production performance and intestinal barrier function in growing Hyla rabbits. Ital J Anim Sci 20:2165–2173

Malaka MJ, Araya NA, Soundy P et al (2023) Biomass, essential oil yield, and composition of Marjoram as influenced by interactions of different agronomic practices under controlled conditions. Plants 12:173. https://doi.org/10.3390/plants12010173

Manokari M, Priyadharshini S, Jogam P, Shekhawat MS (2020) Application of seismo-mechanical stress in attenuation of *in vitro* induced abnormalities in *Scaevola taccada* (Gaertn.) Roxb. Biocatalysis Agric Biotech 29:101821

Manukyan A, Schnitzler W (2006) Influence of air temperature on productivity and quality of some medicinal plants under controlled environment conditions. Eur J Hortic Sci 71:36–44

Marrelli M, Statti GA, Conforti F (2018) *Origanum* spp.: an update of their chemical and biological profiles. Phytochem Rev 17:873–888

Mathe I Jr, Olah L, Mathe A et al (1992) Changes in the essential oil production of Salvia officinalis under climatic conditions of the temperature belt. Planta Med 58:A680

Mattar MZ, Eissa OA, Salem KFM (2009) Genotypic variability in some wheat genotypes (*Triticum aestivum* L.) exposed to salinity at germination stage. Egyptian J Exp Biol (Bot) 5:143–148

Mattar MZ, Salem KFM, El-Abed AB (2016) Identification of candidate microsatellite markers associated with agronomic traits in rice (*Oryza sativa* L.). J Bot 15(5):213–232

Maul E, Töpfer R (2015) Vitis International Variety Catalogue (VIVC): a cultivar database referenced by genetic profiles and morphology. BIO Web Conf 5:01009

Mesfer ALshamrani S, Safhi FA, Alshaya DS, Ibrahim AA, Mansour H, Abd El Moneim D (2022) Genetic diversity using biochemical, physiological, karyological and molecular markers of Sesamum indicum L. Front Genet 13:1035977. https://doi.org/10.3389/fgene.2022.1035977

Mortensen LM (2014) The effect of air temperature on growth of eight herb species. Am Plant Sci 5:1542–1546

Mossa ATH, Refaie AA, Ramadan A, Bouajila J (2013) Amelioration of prallethrin-induced oxidative stress and hepatotoxicity in rat by the administration of *Origanum majorana* essential oil. BioMed Res Int 2013: ID 859085

Mourad AMI, Alomari DZ, Alqudah AM et al (2019) Recent advances in wheat (*Triticum* spp.) breeding. In: Al-Khayri J, Jain S, Johnson D (eds) Advances in plant breeding strategies: cereals. Springer, Cham. https://doi.org/10.1007/978-3-030-23108-8_15

Muller RF, Berger B, Yegen O (1995) Chemical composition and fungi toxic properties to phytopathogenic fungi of essential oils of selected aromatic plants growing wild in Turkey. J Agric Food Chem 43:2262–2266

Murillo-Amador B, Morales-Prado LE, Troyo-Diéguez E et al (2015) Changing environmental conditions and applying organic fertilizers in *Origanum vulgare* L. Front Plant Sci 6:549

Murillo-Amador B, Nieto-Garibay A, López-Aguilar R et al (2013) Physiological, morphometric characteristics and yield of *Origanum vulgare* L. and *Thymus vulgaris* L. exposed to open-field and shade-enclosure. Ind Crop Prod 4:659–667

Nadeem MA, Nawaz MA, Shahid MQ et al (2018) DNA molecular markers in plant breeding: current status and recent advancements in genomic selection and genome editing. Biotechnol Biotec Eq 32:261–285

Nawara HM, Mattar MZ, Salem KFM, Eissa OA (2016) Estimating heterosis and combining ability for agronomic and tissue culture traits using diallel cross on bread wheat (*Triticum aestivum* L.). Egypt J Bot (special issue):6; In: International conference on biotechnology and its applications in botany and microbiology 11–12 May 2016, pp 177–200

Nawara HM, Mattar MZ, Salem KFM, Eissa OA (2017) Diallel study on some in vitro traits of bread wheat (*Triticum aestivum* L.) under salt stress. Inter J Agric Environ Res 3(1):1988–2006

Németh-Zámbori É, Szabó K, Pluhár Z et al (2016) Changes in biomass and essential oil profile of four Lamiaceae species due to different soil water levels. J Essen Oil Res 28:391–399

Nguyen DTP, Kitayama M, Lu N, Takagaki M (2020) Improving secondary metabolite accumulation, mineral content, and growth of coriander (*Coriandrum sativum* L.) by regulating light quality in a plant factory. J Hortic Sci Biotechnol 95:356–363

Ninou E, Paschalidis K, Mylonas I (2017a) Essential oil responses to water stress in Greek oregano populations. J Essent Oil Bear Plants 20:12–23

Ninou EG, Paschalidis KA, Mylonas IG et al (2017b) The effect of genetic variation and nitrogen fertilization on productive characters of Greek oregano. Acta Agric Scand Sect B Soil Plant Sci 67:372–379

Novak J, Lukas B, Bolzer K et al (2008) Identification and characterization of simple sequence repeat markers from a glandular Origanum vulgare expressed sequence tag. Mol Ecol Resour 8:599–601

Novak J, Lukas B, Franz C (2010) Temperature influences thymol and carvacrol differentially in *Origanum* spp. (Lamiaceae). J Essent Oil Res 22:412–415

Oalde Pavlovic OM, Kolarevic S, Dordevic J et al (2021) A study of phytochemistry, genoprotective activity, and antitumor effects of extracts of the selected Lamiaceae species. Plants 10:2306

Ohyama K, Yamaguchi J, Enjoji A (2020) Resource utilization efficiencies in a closed system with artificial lighting during continuous lettuce production. Agronomy 10:723

Omer AE (1999) Response of wild Egyptian oregano to nitrogen fertilization in a sandy soil. J Plant Nutr 22:103–114

Pandey M, Kumar R, Srivastava P et al (2018) WGSSAT: a high throughput computational pipeline for mining and annotation of SSR markers from whole genomes. J Hered 109:339–343

Partovi R, Talebi F, Sharifzadeh A (2018) Antimicrobial efficacy and chemical properties of *Caryophyllus aromaticus* and *Origanum majorana* essential oils against foodborne bacteria alone and in combination. Int J Enteric Pathog 6:95–103

Pennisi G, Blasioli S, Cellini A et al (2019) Unraveling the role of red: Blue LED lights on resource use efficiency and nutritional properties of indoor grown sweet basil. Front Plant Sci 10:305

Pimple BP, Kadam PV, Patil MJ (2012) Ulcer healing properties of different extracts of Origanum majorana in streptozotocin-nicotinamide induced diabetic rats. Asian Pacific J Tropical Disease 2:312–318

Poland JA, Rife TW (2012) Genotyping-by-sequencing for plant breeding and genetics. Plant. Genome 5:92–102

Prabu S, Jing D, Chandran V, Mathew P (2020) Insecticidal activity of *Origanum majorana* L. essential oil as anticholinergic agent. Entomol Res 50:402–413

Qari SH (2008) *In vitro* evaluation of the anti-mutagenic effect of *Origanum majorana* extract on the meristematic root cells of Vicia faba. J Taibah Univ Sci 1:6–10. https://doi.org/10.1016/S1658-3655(12)60028-4

Ragab TI, El Gendy ANG, Saleh IA, Esawy MA (2019) Chemical composition and evaluation of antimicrobial activity of the *Origanum majorana* essential oil extracted by microwave-assisted extraction, conventional hydro-distillation and steam distillation. J Essent Oil-Bear Plants 22:563–573

Ramadan SA, Ibrahim AA (2021) Fruits and vegetables as sources of functional phytochemicals for the management of obesity, diabetes, and cancer. In: Egbuna C, Hassan S (eds) Dietary phytochemicals. Springer. Switzerland. https://doi.org/10.1007/978-3-030-72999-8_8

Ramos da Silva LR, Ferreira OO et al (2021) Lamiaceae essential oils, phytochemical profile, antioxidant, and biological activities. Evid Based Complement Altern Med 2021:6748052

Rihan HZ, Aldarkazali M, Mohamed SJ et al (2020) A novel new light recipe significantly increases the growth and yield of sweet basil (*Ocimum basilicum*) grown in plant factory system. Agronomy 10:934

Sadhu S, Jogam P, Thampu RK et al (2020) High-efficiency plant regeneration and genetic fidelity of regenerants by SCoT and ISSR markers in chickpea (*Cicer arietinum* L.). Plant Cell Tissue Organ Cult 141:465–477

Safaei SMH, Dadpasand M, Mohammadabadi M et al (2023) An *Origanum majorana* leaf diet influences myogenin gene expression, performance, and carcass characteristics in lambs. Animals 13:14. https://doi.org/10.3390/ani13010014

Safhi FA, ALshamrani SM, Jalal AS et al (2022) Genetic characterization of some Saudi Arabia's accessions from *Commiphora gileadensis* using physio-biochemical parameters, molecular markers, DNA barcoding analysis and relative gene expression. Genes 13:2099. https://doi.org/10.3390/genes13112099

Sahranavard S, Ghafari S, Mosaddegh M (2014) Medicinal plants used in Iranian traditional medicine to treat epilepsy. Seizure 23:328–332

Said-Al Ahl HAH, Omer EA, Naguib NY (2009) Effect of water stress and nitrogen fertilizer on herb and essential oil of oregano. Int Agrophysics 23:269–275

Salakhutdinov N, Volcho K, Yarovaya O (2017) Monoterpenes as a renewable source of biologically active compounds. Pure Appl Chem 89:1105–1117

Saleh MM, Ali WA, Alwan MH et al (2020a) Multi-environmental evaluation of wheat tetraploid genotypes for agronomic traits under rainfed conditions in Syria. J Aridland Agric 6:1–5

Saleh MM, Alnaddaf LM, Almuhammady AK et al (2021b) Applications of plant-derived nanomaterials in mitigation of crop abiotic stress. In: Al-Khayri JM, Ansari MI, Singh AK (eds) Nanobiotechnology. Springer, Cham. https://doi.org/10.1007/978-3-030-73606-4_9

Saleh MM, Mahmoud AS, Abbas HS et al (2021b) Nanotechnological approaches for efficient delivery of plant ingredients. In: Faizan M, Hayat S, Yu F (eds) Sustainable agriculture reviews, vol 53, Springer, Cham. https://doi.org/10.1007/978-3-030-86876-5_11

Saleh MM, Salem KFM, Elabd AB (2020b) Definition of selection criterion using correlation and path coefficient analysis in rice (*Oryza sativa* L.) genotypes. Bull Natl Res Cent 44(1):1–6

Salem KFM, Ibrahim AA (2024) Plant biodiversity in the context of food security under climate change. In: Al-Khayri JM, Jain SM, Penna S (eds) Sustainable utilization and conservation of plant genetic diversity. Sustainable development and biodiversity, vol 35. Springer, Singapore. https://doi.org/10.1007/978-981-99-5245-8_1

Salem KFM (1997) Breeding studies on rice (*Oryza sativa* L.). M.Sc thesis, Faculty of Agriculture, Minufiya University, Shibin El-Kom, Egypt

Salem KFM (2004) The inheritance and molecular mapping of genes for post-anthesis drought tolerance (PADT) in wheat. Martin-Luther-University, Halle-Wittenberg, p 124

Salem KFM (2009) Relationship between genetic diversity based on SSRs markers with heterosis and combining ability in diallel cross of bread wheat (*Triticum aestivum* L.). Minufiya J Agric Res 34(6):2159–2178

Salem KFM (2015) Allelic state at the microsatellite locus *Xgwm261* marking the dwarfing gene *Rht8* in Egyptian bread wheat (*Triticum aestivum* L.) genotypes released from 1947 to 2004. Genetika 47:741–750

Salem KFM, Abd-Ellatif S, Abdel Razik ES et al (2023) Effect of nanomaterials on water and solutes translocation in plants. In: Al-Khayri JM, Alnaddaf LM, Jain SM (eds) Nanomaterial interactions with plant cellular mechanisms and macromolecules. Springer, Cham

Salem KFM, Alghuthaymi MA, Elabd AB et al (2022) Prediction of heterosis for agronomic traits in half-diallel cross of rice (*Oryza sativa* L.) under drought stress using microsatellite markers. Plants 11:1532. https://doi.org/10.3390/plants11121532

Salem KFM, Alloosh MT, Saleh MM et al (2021a) Utilization of nanofertilizers in crop tolerance to abiotic stress. In: Al-Khayri JM, Ansari MI, Singh AK (eds) Nanobiotechnology. Springer, Cham. https://doi.org/10.1007/978-3-030-73606-4_11

Salem KFM, Elabsawy A, Aldahak L et al (2021b) Evaluation of stem reserve mobilization in Egyptian bread wheat (*Triticum aestivum* L.) genotypes and F_1 hybrids under post-anthesis chemical desiccation stress. J Genet Environ Resour Conserv 9(1):176–182

Salem KFM, El-Absawy EA, Khidr YA, El-Shamy H (2021c) Assessment genetic diversity among some Egyptian Bread wheat (*Triticum aestivum* L.) cultivars using RAPD markers. J Genet Environ Resour Conserv 9(1):168–175

Salem KFM, El-Zanaty AM, Esmail RM (2008) Assessing wheat (*Triticum aestivum* L.) genetic diversity using morphological characters and microsatellite markers. World J Agric Sci 4(5):538–544

Salem KFM, Ibrahim AA (2024) Plant biodiversity in the context of food security under climate change. In: Al-Khayri JM, Jain, SM, Penna S (eds) Sustainable utilization and conservation of plant genetic diversity. Sustainable development and biodiversity, vol 35. Springer, Singapore. https://doi.org/10.1007/978-981-99-5245-8_1

Salem KFM, Mattar MZ (2014) Genetic diversity in old and modern Egyptian bread wheat (Triticum aestivum L.) varieties revealed by simple sequence repeats. Egypt J Genet Cytol 43:143–156

Salem KFM, Matter MZ (2014b) Identification of microsatellite alleles for salt tolerance at seedling stage in wheat (Triticum aestivum L.). Life Sci J 11(12s):1064–1073

Salem KFM, Röder MS, Börner A (2015) Assessing genetic diversity of Egyptian hexaploid wheat (Triticum aestivum L.) using microsatellite markers. Genet Resour Crop Evol 62:377–385

Salem KFM, Safhi FA, Alwutayd KM et al (2024) Analysis of genetic diversity, population structure and phylogenetic relationships of rice (*Oryza sativa* L.) cultivars using simple sequence repeat (SSR) markers. Genet Resour Crop Evol. https://doi.org/10.1007/s10722-023-01789-0

Salem KFM, Saleh MM, Abu-Ellail FFB et al (2021e) Role of quantum dots, polymeric NPS and dendrimers in emphasizing crops tolerate biotic and abiotic stresses. In: Faizan M, Hayat S, Yu F (eds) Sustainable agriculture reviews, vol 53, Springer, Cham. doi.org/https://doi.org/10.1007/978-3-030-86876-5_1

Salem KFM, Saleh MM, Aldahak L, Elabd AB (2021e) Assessment phenotypic diversity of rice (*Oryza sativa* L.) genotypes by multivariate analysis. J Arid Agric 7:52–59. https://doi.org/10.25081/jaa.2021.v7.6516

Salem KFM, Sallam A (2016) Analysis of population structure and genetic diversity of Egyptian and exotic rice (*Oryza sativa* L.) genotypes using SSR markers. Comptes Rendus Biol 339:1–9. https://doi.org/10.1016/j.crvi.2015.11.003

Salem KFM, Tawfik E, Mahdy M et al (2023) Advances in lettuce (*Lactuca* spp.) molecular breeding strategies. In: Singh S, Sharma D, Sharma SK, Singh R (eds) Smart plant breeding for vegetable crops in post-genomics era. Springer, Cham

Salem KFM, Röder MS, Börner A (2007) Identification and mapping quantitative trait loci for stem reserve mobilisation in wheat (*Triticum aestivum* L.). Cereal Res Common 35:1367–1374

Salem KFM, Varshney RK, Röder MS, Börner A (2010) EST-SSR based estimates on functional genetic variation in a barley (*Hordeum vulgare* L.) collection from Egypt. Genet Resour Crop Evol 57(4):515–521

Salha GB, Díaz RH, Labidi J, Abderrabba M (2017) Deterpenation of *Origanum majorana* L. essential oil by reduced pressure steam distillation. Ind Crops Prod 109:116–122

Sandhya D, Jogam P, Manokari M et al (2021) High-frequency *in vitro* propagation and assessment of genetic uniformity and micro-morphological characterization of *Origanum majorana* L. A highly traded aromatic herb. Biocat Agric Biotec 34:02024. https://doi.org/10.1016/j.bcab.2021.102024

Sarfaraz D, Rahimmalek M, Sabzalian MR et al (2023) Essential oil composition and antioxidant activity of oregano and marjoram as affected by different light-emitting diodes. Molecules 28:3714. https://doi.org/10.3390/molecules28093714

Sarrou E, Tsivelika N, Chatzopoulou P et al (2017) Conventional breeding of Greek oregano (*Origanum vulgare* ssp. hirtum) and development of improved cultivars for yield potential and essential oil quality. Euphytica 213:104

Savitikadi P, Jogam P, Rohela GK et al (2020) Direct regeneration and genetic fidelity analysis of regenerated plants of *Andrographis echioides* (L.)- An important medicinal plant. Ind Crop Prod 155:112766

Saxena D, Jayant SK, Soni K, Neekhra K (2016) *Origanum majorana*, a potential herb for functional food. Europ J Pharm Med Res 3(2):321–325

Schmidt E, Bail S, Buchbauer G et al (2008) Chemical composition, olfactory evaluation and antioxidant effects of the essential oil of *Origanum majorana* L. from Albania. Nat Prod Commun 3(7):1051–1056

Schwend T, Prucker D, Peisl S et al (2016) The rosmarinic acid content of basil and borage correlates with the ratio of red and far-red light. Eur J Hortic Sci 81:243–247

Selim SA, Aziz MA, Mashait MS, Warrad MF (2013) Antibacterial activities, chemical constitutes and acute toxicity of Egyptian *Origanum majorana* L., *Peganum harmala* L. and *Salvia officinalis* L. essential oils. Afr J Pharm Pharmacol 7:725–735

Sellami IH, Maamouri E, Chahed T et al (2009) Effect of growth stage on the content and composition of the essential oil and phenolic fraction of sweet Marjoram (*Origanum majorana* L.). Ind Crops Prod 30:395–402

Seoudi DM, Medhat AM, Hewedi IH et al (2009) Evaluation of the anti-inflammatory, analgesic, and anti-pyretic effects of *Origanum majorana* ethanolic extract in experimental animals. J Radiat Res Appl Sci 2:513–534

Shahidi F, Janitha PK, Wanasundara P (1992) Phenolic antioxidants. Crit Rev Food Sci Nutr 32:67–102

Sharifi-Rad M, Yılmaz YB, Antika G et al (2021) Phytochemical constituents, biological activities, and health-promoting effects of the genus *Origanum*. Phytother Res 35:95–121

Soliman AM, Desouky S, Marzouk M, Sayed AA (2016) *Origanum majorana* attenuates nephrotoxicity of cisplatin anticancer drug through ameliorating oxidative stress. Nutrients 8:264

Soliman MI, Ibrahim AA, El-Metwally MA, Eshak DA (2017) Cytotoxicity effects of biological control and antioxidants using *Vicia faba* chromosomal aberration assay. J Environ Sci Technol 10:230–237. https://doi.org/10.3923/jest.2017.230.237

Soliman MI, Mohammed NS, EL-Sherbeny G, Safhi FA, ALshamrani SM, Alyamani AA, Alharthi B, Qahl SH, Al Kashgry NAT, Abd-Ellatif S, Amira AA (2023) Antibacterial, antioxidant activities, GC-mass characterization, and cyto/genotoxicity effect of green synthesis of silver nanoparticles using latex of Cynanchum acutum L. Plants 12:172. https://doi.org/10.3390/plants12010172

Sotiropoulou DEK, Karamanos AJ (2010) Field studies of nitrogen application on Greek oregano (*Origanum vulgare* ssp. Hirtum (Link) Ietswaart. Ind Crop Prod 32:450–457

Souza NAB, Lima EDO, Guedes DN et al (2010) Efficacy of Origanum essential oils for inhibition of potentially pathogenic fungi. Brazil J Pharm Sci 46(3):499–508

Tahraoui A, El Hilaly J, Israili ZH (2007) Ethnopharmacological survey of plants used in the traditional treatment of hypertension and diabetes in south-eastern Morocco (*Errachidia province*). J Ethnopharmacol 110:105–117

Tamokou JDD, Mbaveng AT, Kuete V (2017) Antimicrobial activities of african medicinal spices and vegetables in medicinal spices and vegetables from Africa. In: Kuete V (ed) Medicinal spices and vegetables from Africa. Elsevier Academic Press, Cambridge, pp 207–237

Tejavathi DH, Padma AV (2012) *In vitro* multiplication of *Majorana hortensis* moench—an aromatic medicinal herb. Indian J Plant Sci 1(1):48–56

Thaner Dos Santos H, Alcarde Sermarini R, Moreno-Pizani MA, Alves Marques PA (2020) Effects of irrigation management and seasonal stages on essential oil content and biomass of Origanum vulgare L. Not Sci Biol 12:42–56

Thanh VM, Bui LM, Bach LG et al (2019) *Origanum majorana* L. Essential oil-associated polymeric nano dendrimer for antifungal activity against *Phytophthora infestans*. Materials 12:1446

Tibaldi G, Fontana E, Nicola S (2011) Growing conditions and postharvest management can affect the essential oil of *Origanum vulgare* ssp. hirtum (Link) Ietswaart. Ind Crop Prod 34:1516–1522

Trivino MG, Johnson CB (2000) Season has a major effect on the essential oil yield response to nutrient supply in *Origanum Majorana*. J Hortic Sci Biotechnol 75:520–527

Tuttolomondo T, La Bella S, Leto C et al (2016) Effect of plant density on the number of glandular trichomes and on yield and quality of essential oils from oregano. Nat Prod Res 11:849–852

Usha K, Patra DD (2003) Medicinal and aromatic plant materials as nitrification inhibitors for augmenting yield and nitrogen uptake of Japanese mint (*Mentha arvensis* L. var. piperascens). Bioresour Technol 6:267–276

Vagi E, Simandi B, Suhajda A, Hethelyi E (2005) Essential oil composition and antimicrobial activity of *Origanum majorana* L. extracts obtained with ethyl alcohol and supercritical carbon dioxide. Food Res Int 38:51–57

Varshney RK, Salem KFM, Röder MS et al (2008) Assessment and comparison of genetic diversity in barley germplasm collection using gSSR, eSSR and SNP markers. Plant Genet Res 6:167–174

Vera RR, Chane-Ming J (1999) Chemical composition of the essential oil of Marjoram (*Origanum majorana* L.) from Reunion Island. Food Chem 66:143–145

Verma RS, Padalia RC, Chauhan A et al (2016) Changes in the essential oil composition of *Origanum majorana* L. during post-harvest drying. J Essent Oil-Bear Plants 19:1547–1552

Vieira MLC, Santini L, Diniz AL, Munhoz CDF (2016) Microsatellite markers: what they mean and why they are so useful. Genet Mol Biol 39:312–328

Virga G, Sabatino L, Licata M et al (2020) Effects of irrigation with different sources of water on growth, yield and essential oil compounds in oregano. Plants 9:1618

Vokou D, Kokkini S, Bessiere JM (1993) Geographic variation of Greek oregano (*Origanum vulgare* spp. Hirtum) essential oils. Biochem Syst Ecol 21:287–295

Wittmann S, Jüttner I, Mempel H (2020) Indoor farming marjoram production—quality, resource efficiency, and potential of application. Agronomy 10:1769. https://doi.org/10.3390/agronomy10111769

Xu J, Zhou F, Ji BP et al (2008) The antibacterial mechanism of carvacrol and thymol against *Escherichia coli*. Lett Appl Microbiol 47:174–179. https://doi.org/10.1111/j.1472-765X.2008.02407.x

Xylia P, Chrysargyris A, Tzortzakis N (2021) The combined and single effect of Marjoram essential oil, ascorbic acid, and chitosan on fresh-cut lettuce preservation. Foods 10:575

Yadava RN, Khare MK (1995) A triterpenoid from *Majorana hortensis*. Fitoterapia 66(2)

Yashoda Bai S, Gayatri MC, Leelavathi D (2011) *In vitro* propagation of *Oreganum majorana* using axillary bud explants. J Cytol Genet 12:71–75

Yousefi M, Ghafarifarsani H, Hoseinifar SH et al (2021) Effects of dietary marjoram, *Origanum majorana* extract on growth performance, hematological, antioxidant, humoral and mucosal immune responses, and resistance of common carp, *Cyprinus carpio* against *Aeromonas hydrophila*. Fish Shellfish Immunol 108:127–133

Zargary A (2012) Medicinal plants, 7th edn. Tehran University Press, Tehran

Zawi'slak G (2008) Dependence on harvest date and yielding of Marjoram (*Origanum majorana* L.) cv. Miraz' cultivated from a seedling. Acta Sci Pol Hortorum Cultus 7:7

Zawi'slak G, Dzida K (2010) Yield and quality of sweet marjoram herb depending on harvest time. Acta Sci Pol Hortorum Cultus 9:65–67

Zgheib R, Chaillou S, Ouaini N et al (2016) Chemometric tools to highlight the variability of the chemical composition and yield of Lebanese *Origanum syriacum* L. essential oil. Chem Biodivers 13:1326–1347

Chapter 4
Breeding Strategies for Moringa (*Moringa oleifera* Lam.)

N. Manikanda Boopathi

Abstract Development of novel cultivars in Moringa (*Moringa oleifera* Lam.,) is the prime and actively growing research field owing to their commercial and nutritional importance. Though traditional breeding strategies such as introduction, pure line selection and hybridization have shown to be successful in developing new moringa cultivars, it is off late opined that such strategies demand huge time, effort and cost. Alternatively, leveraging the advances in the genomics and other omics field for the genetic improvement of Moringa for various traits may fasten and simplify the process of releasing novel cultivars in Moringa. Particularly, genetic improvement of its resilience, yield, nutritional quality, and ability to respond to a variety of agroecological environments would help to fully realize the promise of moringa as a nutrient-dense, sustainable crop for global food and nutrition security. Enhancing characteristics including seed size, pod shape, leaf size, and plant architecture can improve the agronomic performance and marketability of moringa. The nutritional and therapeutic value of moringa can be increased by breeding for higher concentrations of bioactive substances found in moringa (including phytochemicals, anti-inflammatory agents, and antioxidants). This chapter describes the principles and applications of different next generation plant breeding strategies in Moringa breeding, besides providing details on troubleshooting measures that can speed-up the cultivar releasing process.

4.1 Introduction

One of the most serious global problems is inability to access minimum nutrition requirement and meeting the demands of daily calories intake. Thus, undernourishment distresses globally 809.9 million people and creates several health issues, which, may sometimes lead to fatality (FAO, IFAD, UNICEF, WFP and WHO 2019). To end

N. Manikanda Boopathi (✉)
Department of Plant Biotechnology, Centre for Plant Molecular Biology and Biotechnology, Tamil Nadu Agricultural University, Coimbatore, India
e-mail: nmboopathi@tnau.ac.in

J. M. Al-Khayri et al. (eds.), *Biodiversity and Genetic Improvement of Medicinal and Aromatic Plants II*, Advances in Plant Breeding Strategies 10,
https://doi.org/10.1007/978-3-031-81857-8_4

this issue, UN's Sustainable Development Goals (SDGs) fix the target of achieving food and nutritional security by 2030 by gearing up the processes that lead to increase the agricultural productivity of nutritionally improved crops and income of small-scale food producers. Crisis due to novel Coronavirus (COVID-19) pandemic also have a unique worldwide effect on food accessibility, health and nutrition and the livelihoods of people. Further, if the current trends in climate change continues, it would be difficult to achieve these targets in major field crops by 2030. Thus, there is an urgent need to focus on alternative crops, but at the same time such crops can provide better yield even under changing climatic conditions and helps to ensure the nutritional security with affordable efforts.

Moringa (*Moringa oleifera* Lam.,) is one such alternative crop, which can thrive not only in hardy climate but also provide enough nutrients, minerals and vitamins that are associated the human health. Moringa is also referred as drumstick tree or horseradish tree or ben oil tree. Moringa was originated from north eastern foothills of Indian Himalayan region and belongs to the family Moringaceae. There are 13 species in the genus *Moringa* and due to their food, nutritional and medicinal values, they have been widely cultivated throughout Asia and Africa. Among the 13 species, *M. oliefera* is widely cultivated in tropical and sub-tropical regions including Asia and Africa and it was first botanically described by the French naturalist Jean-Baptiste Lamarck in 1785. It grows quickly and can reach a hight of 5 m within a year even under water limited environments. The trunk of the *M. oleifera* may develop up to 45 cm and it produces young shoots which may be eighter purplish or greenish-white with hairy bark. Usually the bark is in whitish-grey colour and has a thick cork. The tree may attain an open crown of drooping. Stem branches are fragile and have tripinnate leaves that lead to a feathery foliage (Fig. 4.1).

Moringa is generally propagated through seeds or by limb cuttings. It starts to flower 6 months after planting and it produces fragrant and hermaphrodite flowers. Flowers are tiny (1.0–1.5 cm long and 2.0 cm broad) and enclosed by five unequal, thinly veined, yellowish-white petals (Fig. 4.1). Flowers are produced on a slender and hairy stalk (which may have 10–25 cm length) as a spreading or drooping clusters. Under an unchanged climatic condition, Moringa flowers twice in a year or sometime throughout the year, if the climate is conducive. Moringa fruits are colloquially called as drumstick, which is actually a hanging, three-sided brown capsule that has 20–45 cm length with 3–5 cm diameter. The fruits are usually bearing dark brown globular seeds, which may have a diameter of up to 1 cm. Seeds are naturally dispersed by wind as well as water as it has three whitish papery wings. Moringa tree is annually pruned to 1–2 m height and allowed to grow to a reasonable hight in order to facilitate the easy harvest of the fruits.

4.1.1 Health Benefits of Moringa

Historically, all parts of the Moringa (leaves, flowers, pods, roots, stem and root barks, and exudates from the stem) have been employed for food, medication and other

Fig. 4.1 *Moringa oliefera* var. PKM1 grown in natural environment in Tamil Nadu, India. Morphology of leaves and flowers are also shown in other panels. Photo credit: N. Manikanda Boopathi

industrial and domestic purposes and hence it is considered as an industrially and therapeutically valuable plant species (Kashyap et al. 2022). Due to its outstanding health, nutritional and environmental importance, Moringa is often referred as "Miracle Tree" or "Tree of life". Moringa leaves were shown to contain 19–30% proteins and 16–24% fibres (Teixeira et al. 2014) and thus intake of 100 g of fresh moringa leaves can provide 17.5% of the daily requirement of protein. Importantly, Moringa proteins are rich in 19 amino acids including ten essential amino acids such as lysine, leucine, isoleucine, histidine, phenylalanine, methionine, tryptophan, threonine, tyrosine and valine (Falowo et al. 2018). As it has low calorific value, it has long been prescribed to obese persons. Moringa leaves also contain polysaccharide including MOP-2 which is an immunoregulatory agent in various functional foods (Dong et al. 2018). Similarly, Moringa possesses significant quantity of mono- and poly- unsaturated fatty acid (including α-Linolenic acid, omega-3 and omega-6) and less amount of saturated fatty acids (Tshabalala et al. 2020).

Moringa has widely been used to combat malnutrition and health-issues found in the pregnant women owing to its high content of nutrients. For example, Moringa leaves can supply four times more calcium and two times more digestible protein than milk. Hence it is proposed as affordable and good source of calcium and protein as an alternative to costly food supplements. Similarly, higher amounts of minerals such as potassium, zinc, magnesium, iron and copper have also been found in Moringa leaves when compared with other regular vegetables (Kasolo et al. 2010). Likewise, vitamins (including Fat-soluble vitamins (vitamin-A, D and E) and water-soluble vitamins (B complexes such as folic acid, pyridoxine and nicotinic acid and vitamin-C)), are also found in larger quantity than the daily required levels in Moringa (Mbikay 2012).

Besides, significant amounts of bioaccessible and bioavailable phytochemicals (which are known for their anti-cancerous properties) such as phenolic acids, flavonoids, tannins, saponins and alkaloids have also been found to be present in Moringa (Berkovich et al. 2013), which are the key molecules responsible for its nutritional, therapeutic and water purifying properties. The major flavonoids present in Moringa are quercetin, kaempferol, apigenin, luteolin, and myricetin glycosides, which are nowadays considered as key compounds in functional foods as they provide several health advantages including protection from cellular damage, ageing, and chronic illnesses.

Owing to its rich nutrients, vitamins, minerals and antioxidants, Moringa offers multiple health advantages such as antioxidant activity, anti-microbial activity, anti-cancerous activity, anti-inflammatory action, antidiabetic properties, central nervous system activity, cardiovascular activity and many more (recently reviewed in (Kashyap et al. 2022). Moringa flowers were found to be rich reservoirs of bioactive compounds. Use of crude flower extracts have demonstrated to possess antibacterial, antifungal, anti-larval, antioxidant, anti-inflammatory and anticancer properties. Despite of their wide applicability in food, nutritional and medicinal industries, Kalappurayil and Joseph (2017) opined that further studies on identification and characterization of the active compounds that are involved in many therapeutic potential of the Moringa flower extracts have huge benefits to those industries.

Mazumder et al. (1999) studied the hematological and hepatorenal activities of a methanolic extract of Moringa oleifera roots. The effects of different doses of the crude extract (CE) on liver and kidney functions as well as hematological markers in mice were investigated. The effects of crude extract (CE) doses upon liver and kidney function as well as hematological parameters in mice were investigated. There was no change in hematological or biochemical properties when a low and moderate dosage level of extract was administered daily, and a low dose level of extract was administered weekly. However, the extract altered serum aminotransferase and plasma cholesterol levels considerably in a weekly therapy at a modest dosage level. A high dosage altered total bilirubin, nonprotein nitrogen, blood urea, and plasma protein, in addition to the foregoing measures. The use of a high dose of CE on a daily basis, as well as a moderate and high dose of CE on a weekly basis, led to a dramatic increase in WBC count and a substantial reduction in clotting time.

With the support of the above evidences, Moringa has strongly been recommended as natural dietary supplements because of their high nutritional value and low anti-nutritional factors; moreover, herbal preparations have long been used as safe and without any adverse effects. Moringa based formulations and preparations have been globally used as food and food supplements and so far, no ill effects have been reported. In a study, it has been shown that feeding of 70 g of moringa leaf extract per day to the rats was found to be safe (Adedapo et al. 2009). Thus, Moringa has arose as a potential component for developing food products that are nutritionally rich and therapeutically active and global market demand for the Moringa based formulations is in increased trend due to its health benefits.

4.2 Conventional Breeding Strategies in Moringa

Currently, the methods that are used for plant genetic improvement Moringa are hybridization, mutagenesis, and molecular breeding. Alternatively, Moringa productivity can also be increased by following appropriate agronomic practices. For example, Mabapa et al. (2017) investigated the impact of plant density and cutting interval on biomass output by planting on four different densities and revealed that increasing plant density resulted in a higher biomass production, ranging from 527 to 2867 kg/ha. Similarly, Kagambèga et al. (2019) concluded that differences in location, soil type, and culture time influenced the morphophysiological characteristics and leaf yield of Moringa in the Sahelian environment. They also noted that the warm season (March/April) seems to be the best time to establish the two *M. oleifera* types for growth and biomass production.

The management practices do not need the use of any sophisticated biochemical or molecular techniques. However, inherent limitations of those Moringa cultivars greatly hampers the realization of full potential of Moringa leaf production. Therefore, breeding for improved Moringa cultivars with enhanced leaf biomass (as well as pod) production is considered as simple and affordable way. Breeders have effectively utilized desirable morphological characters (such as leaf biomass traits (including leaf area and dry weight, in this case) in breeding programs for genetic improvement of leaf biomass in variety of crops (Nadeem et al. 2018). However, the main drawbacks of morphological markers are that (i) they are restricted in number, (ii) are impacted by plant development phases, and (iii) are affected by different environmental variables (Eagles et al. 2001).

Because India has a climate that has been broadly classified as tropical wet and dry, subtropical humid climate, and temperate (mountain climate) by the Koeppen system of climate classification, breeding crops based on morphological characteristics is not best suited (Chaitra et al. 2020). Owing to its cross-pollination nature, the best conventional methods suitable for the development of desirable cultivar in Moringa are mass selection, recurrent selection, family selection, hybridization and synthetics. Among them, selection is the frequently used strategy in Moringa genetic improvement and used to release several cultivars around the world. Majority of the cases, selected in exercised by selecting the progenies that are preferred by farmers and other end users and obviously those lines that can provide higher economic yield (such as number and weight of the pods). It is also should be noted that there is nil report on the cultivars that were selected based on the superior nutritional or other biochemical traits.

The oldest technique of Moringa breeding programs, selection, has been used effectively by identifying the genetic diversity observed in the targeted traits. Though selection has been used in Moringa genetic improvement for a long time, it is mostly used to choose progenies that are aesthetically appealing and have a greater economic yield (Balakumbahan and Boopathi 2021). PKM 1, which is a pureline selection from Eppodumvendran, Kudumianmalai 1, Konkan Ruchira, Rohit, Bhaghya, are the notable cultivar developed through selection in moringa. The potential lines that are present in one region can be transferred to other region for breeding purpose. Such type of breeding method is known as Introduction. Jaffna variety of moringa is such a variety that was introduced from Sri Lanka (Sekhar et al. 2017).

Hybridization has also been tried in Moringa genetic improvement by making controlled crossing of compatible parents. Reports are available only with simple hybridization strategies and there is no other strategies such as diallele crosses has been tried in Moringa. Hybridization is the technique of crossing two genetically diverse species to result in a third person with a new, typically desirable, set of qualities. Plants of the same species cross readily and generate viable offspring. Wide crossings are difficult to accomplish and often yield sterile offspring because of chromosome-pairing issues during meiosis (Hadley and Openshaw 1980). PKM 2 is a hybrid variety of MP31 (Eppodumvendran Local) and MP28 (Arasaradi Local).

Various asexual propagation methods are followed for breeding in Moringa. Limb cuttings, shield budding and air layering are some of the vegetative propagation methods that have followed by the farmers (Boopathi et al. 2021a, b). Since these methods were suitable for small scale farming, in vitro culture was proposed as an upgrade of the traditional methods. Micropropagation means the large-scale aseptic and in vitro production of true-to-type complete plants by using a small part of the explant (mother plant) (Naik et al. 2020). Boopathi et al. (2021a) has optimized a protocol for the micropropagation of PKM1 variety of moringa by nodal culture in MS medium, supplemented with various concentrations of BAP, NAA and kinetin.

Mutation breeding is regarded as an effective way of enhancing the breeding efficiency of evolving novel moringa leafy types since this provides an easy method for inducing novel and heritable genetic variation in a short period of time in tree crops, easily creating new variations that are not found in nature, generating novel alleles in orders of magnitude faster than that occur spontaneously (Prasanth et al. 2020) and retaining such novel alleles through asexual propagation as there is an issue with linkage drag of undesirable traits through seed propagation in Moringa.

Mutagenesis is accomplished by chemical (such as ethyl methane sulfonate; EMS) or physical treatments followed by selection for heritable changes of specific genotypes, and this approach has been used successfully in the genetic improvement of crop plants (Badigannavar et al. 2022). The primary goal of mutation breeding is to broaden the spectrum and frequency of viable mutants via which desired plant characteristics may be chosen for direct or indirect application in plant breeding programs. Morphological mutants with high practical utility are immediately introduced as crop varieties. The mutagenic efficiency and efficacy of any mutagen influence the success of mutation breeding in various crops.

Therefore, it is critical to establish an appropriate mutagen dosage for any mutagenesis. Lethal dose is the percent of test individuals killed by a particular dosage of physical or chemical agents; at Lethal Dose 50 (LD50), half of the organisms will die. To achieve the greatest number of mutations, the mutagen dosage administered should be adequate to kill about half of the seed and induce considerable heritable changes. The LD50 was utilized by the majority of researchers to estimate the fatal dosage of mutagens. An effort to introduce genetic variation using chemical induced

Fig. 4.2 Induced mutagenesis using 0.25% Ethyl Methane Sulphate (EMS) created desirable leafy mutants in Moringa Photo credit: N. Manikanda Boopathi

mutagenesis in moringa has been initiated at the author's laboratory and preliminary studies have shown that EMS can induce sufficient variation in leaf biomass in Moringa and Fig. 4.2 illustrates the morphological feature of control (wildtype: PKM1) and a mutant line 14.

4.3 Scope for Next-Gen Plant Breeding Strategies in Moringa

In order to meet the increasing global demand on Moringa product, it is imperative to improve the rate of genetic gain. It certainly depends on accelerated crop breeding programs that enhance the genetic gain per unit time and thus ultimately the breeding cycle time. Since the inception of plant breeding in early 1940s, several strategies such as single-seed descent and shuttle breeding have been employed. Of late, employment of controlled-environment growth conditions has also been tried to reduce the plant lifecycle time and such techniques are commonly called as speed breeding. Several speed breeding methods such as accelerated single-seed descent (rapid development of homozygous lines), rapid generation cycling (more breeding cycles per year using DNA marker technology), fast generation cycling (more generations per year using stressed conditions and in vitro culture of immature embryos), and rapid generation turnover (increasing number of generations per year using immature seed harvest and photoperiod response) has been tried in several Poaceae, Fabaceae, and Brassicaceae crop species. However, such strategies are still in its toddler stage in Moringa. Most of the above strategies rely on molecular markers in one or other way.

Rapid developments in molecular markers have shown their potentials in fast and accurate recovery of the target traits in several field crops and same can also be expected in Moringa crop improvement. When new advances in deoxyribose nucleic acids (DNA) and ribo nucleic acid (RNA) develops, evolving next generation molecular markers (such as SNPs and InDels) become popular in several field crops and they were rapidly replacing previous marker classes. However, Moringa has witnessed the progress so far only with previous set of low-throughput markers such as RAPD and SSRs. These multi-locus markers have been used to capture the genetic variation among the investigated Moringa accessions and invariably established that these marker classes were rapid and can be used as cost-effective fingerprinting techniques (Boopathi et al. 2021a, b). Though such marker classes were found to be low-throughput, significant amount of genetic variation were found among the Moringa accessions and those genetically dissimilar and inimitable accessions would be a treasure for future efforts that focus on genetic improvement of new traits such as improved nutritional content.

Realising the importance of developing huge numbers of simple sequence repeat (SSR) markers in Moringa, this laboratory has made an attempt to identify 60,000 + SSRs using the genomic and genic sequences of Moringa cultivar PKM1 (Fig. 4.3; Boopathi et al. 2021b). After identifying these SSRs, they were validated in Moringa ecotypes and demonstrated their usefulness in Moringa research such as varietal protection, germplasm management, trait mapping, gene isolation and selections and biotic and/or abiotic stress resistance improvement.

Despite of these efforts, there is no large-scale strategy employed in Moringa that captured the SNPs and InDels and development of Chip- or Sequence-based approaches that facilitate high-throughput genotyping of Moringa accessions. Such facility/infrastructure development is extremely essential to begin molecular breeding approaches in Moringa.

4.4 Progress in Molecular Breeding Strategies in Moringa

Though Moringa has been shown to possess multi-various medicinal applications and has recognized in the international market as a high valued commodity and having several agronomic traits (such as minimizing the antagonistic effects of drought and can improve the soil of arid regions), programs that focus on genetic improvement of Moringa using molecular breeding are minimal compared to other crops (Boopathi et al. 2021b). Few studies have reported use of molecular markers for characterizing the available Moringa germplasm (Lakshmidevamma et al. 2021) and the utility of DNA barcoding in Moringa (Boopathi et al. 2021a). However, thousands of SSR markers (Boopathi et al. 2021b), genome sequence data (Tian et al. 2015; Lin et al. 2019; Shyamli et al. 2021) and other potential biotechnological tools (Jain et al. 2021) are available as genomic resources in Moringa. Such genomics resources in Moringa are historic accomplishment and it provided starting point for its improvement by engaging trendy biotechnological tools.

Different omics strategies have also shown their potential in generating genomics resources in Moringa and which are considered as a valuable resource and can provide new perspectives in breeding new Moringa cultivars. For example, RNA-Seq have been employed to generate transcriptomic data which have led to the identification

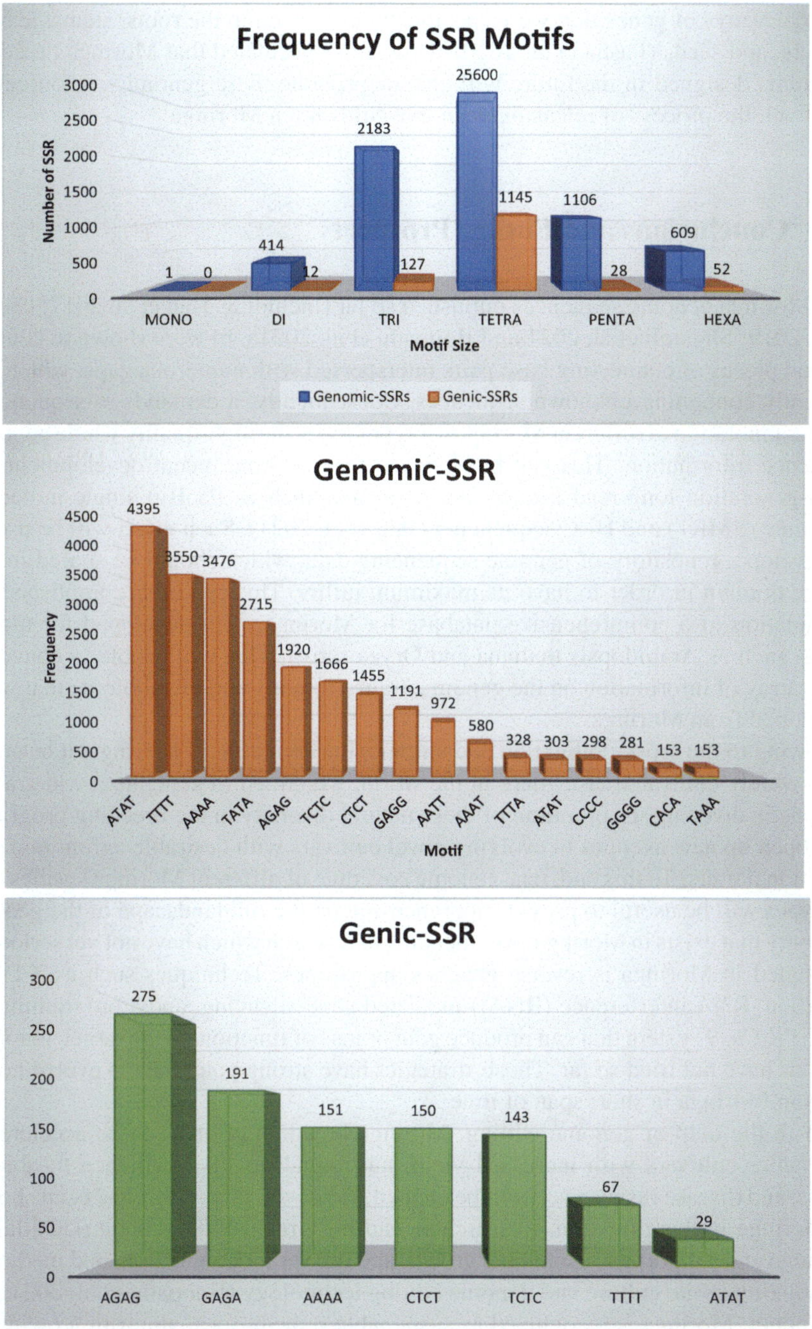

Fig. 4.3 Details of SSRs identified at the author's laboratory along with the motif details

of large arrays of genes that were specifically expressed in the roots, stems, leaves, flowers, and seeds (Pasha et al. 2020). Thus, it is speculated that Moringa breeding programs designed in the future will also incorporate these genomics resources to fast-track the process of releasing desirable cultivars in Moringa.

4.5 Conclusions and Future Prospects

The Moringa genome sequences published so far (including Tian et al. 2017; Chang et al. 2019; Shyamli et al. 2021 and Boopathi et al. 2021a, b) were shown to contain several pieces of connecting base pairs interspersed with numerous gaps, which are currently containing unknown sequences. Consequently, it demands re-sequencing and re-annotation of different Moringa ecotypes to get the high-quality whole genome sequence information. This can be achieved by exploring recent developments in third-generation long-read sequencing platforms such as PacBio single-molecule real-time (SMRT) and Hi-C sequencing (Jain et al. 2021). Such effort will be useful to develop a repository of genome sequencing data, which should be shared in the public domain in order to have its maximum utility. This can be accomplished by formulation of a comprehensive database for Moringa, as it developed for model plants such as Arabidopsis thaliana and Oryza sativa. This will enable to converge large array of information on the genome, transcriptome, and metabolome that were developed from Moringa.

Owing to the cross-pollination and entomophilous nature of Moringa, it has now been widely cultivated elsewhere in the world, which led to generate a wide range of genetic diversity. Exploration of these natural diversity in the breeding programs will open up new avenues in evolving novel cultivars with desirable agronomic and nutritional traits. To this end, pan-genome coverage of different Moringa species and ecotypes will be useful to provide apprehension on the full landscape of the genetic diversity that exists in Moringa. Another area of research which have not yet seriously attempted in Moringa is reverse genetics approaches. Techniques such as T-DNA insertion, RNA interference (RNAi)-mediated gene-silencing, and a programmable *CRISPR/Cas9* system that can produce gain or loss of function on the target genomic regions have not tried so far. Those strategies have strong potentials to evolve novel traits in Moringa in short span of time.

With the help of genome editing, Moringa breeding efforts can be accelerated and better cultivars with increased yield, nutritional quality, resilience to abiotic stress, and disease resistance could be created. However, little effort has been shown in Moringa in this direction. Researchers can fully realize the genetic potential of moringa and support its sustainable growth and use for food, nutrition, and medicine by utilizing tissue culture and genome editing technology (Boopathi et al. 2021a).

Though Moringa is recognized as sustainable remedy for malnutrition owing to its dense phytochemicals, minerals, vitamins and antioxidants, intensive Moringa breeding efforts have been focused only on few pockets of the world. Perhaps intensive efforts in breeding Moringa cultivars as well as releasing significant number of cultivars with improved acclimatization to different ecosystems and resistance to pest, disease and drought stresses have been achieved in India during the last five decades (Boopathi et al. 2021a, b). Especially, the cultivar PKM1 released from this author's University needs a special attention. It has been released nearly three decades ago but still widely cultivated in all over the world due to its adaptability to several types of growing environments, yield and nutritional quality traits. Any special trait (such as enhanced seed oil production and improved leaf biomass) introgression into the PKM1 through molecular breeding methods outlined above would be a boon to the industrial and medicinal sectors. Hence, it is strongly believed that molecular breeding in Moringa would have great potential in not only improving the rural economy but also ensuring nutritional security.

References

Adedapo AA, Mogbojuri OM, Emikpe BO (2009) Safety evaluations of the aqueous extract of the leaves of Moringa oleifera in rats. J Med Plants Res 3(8):586–591

Badigannavar AM, Mondal S, Suprasanna P (2022) Role of induced mutagenesis and improved crop varieties in food security: impact in the Indian context. CABI Rev. https://doi.org/10.1079/cabireviews202217051

Balakumbahan R, Boopathi NM (2021) Classical genetics and traditional breeding in Moringa. In: The Moringa genome. Springer, Cham, pp 41–49

Berkovich L, Earon G, Ron I, Rimmon A, Vexler A, Lev-Ari S (2013) Moringa Oleifera aqueous leaf extract down-regulates nuclear factor-kappaB and increases cytotoxic effect of chemotherapy in pancreatic cancer cells. BMC Complement Altern Med 13(1):1–7

Boopathi NM, Raveendran M (2021) Molecular markers and DNA barcoding in Moringa. In: The Moringa genome. Springer, Cham, pp 85–100

Boopathi NM, Williams M, Veera Ranjani R, Paul AE, Jayakanthan M, Saranya N, Raveendran M (2021b) Development of novel SSR markers derived from genomic and transcriptomic data of Moringa oleifera L. var. PKM1 and their applicability. J Hortic Sci Biotechnol 1–9

Boopathi NM, Harshitha J, Santhanakrishnan VP, Raveendran M (2021a) Tissue culture and genetic engineering in Moringa. In: The Moringa genome. Springer, pp 67–83

Chaitra KC, Sarvamangala C, Manikanta DS, Chaitra PA, Fakrudin B (2020) Insights into genetic diversity and population structure of Indian carrot (Daucus carota L.) accessions. J Appl Genet 61(3):303–312

Chang Y, Liu H, Liu M, Liao X, Sahu SK, Fu Y, Song B, Cheng S, Kariba R, Muthemba S, Hendre PS (2019) The draft genomes of five agriculturally important African orphan crops. GigaScience 8(3):giy152

Dong Z, Li C, Huang Q, Zhang B, Fu X, Liu RH (2018) Characterization of a novel polysaccharide from the leaves of Moringa oleifera and its immunostimulatory activity. J Funct Foods 49:391–400

Eagles HA, Bariana HS, Ogbonnaya FC, Rebetzke GJ, Hollamby GJ, Henry RJ, Henschke PH, Carter M (2001) Implementation of markers in Australian wheat breeding. Aust J Agric Res 52(12):1349–1356

Falowo AB, Mukumbo FE, Idamokoro EM, Lorenzo JM, Afolayan AJ, Muchenje V (2018) Multifunctional application of Moringa oleifera Lam. in nutrition and animal food products: a review. Food Res Int 106:317–334

FAO, IFAD, UNICEF, WFP and WHO (Food and Agriculture Organization of the United Nations, International Fund for Agricultural Development, United Nations Children's Fund, World Food

Programme and World Health Organization) (2019) The State of Food Insecurity in the World 2019. Strengthening the Enabling Environment for Food Security and Nutrition. FAO, Rome

Goll MG, Bestor TH (2002) Histone modification and replacement in chromatin activation. Genes Dev 16(14):1739–1742

Hadley H, Openshaw S (1980) Interspecific and intergeneric hybridization. Hybrid Crop Plants 133–159

Jain A, Poli Y, Goud MD, Ravi RD, Bhaskaran S, Wang X, Das SS, Gupta S, Jain R, Kachhwaha S, Sharma P (2021) A comprehensive review on the biotechnological intervention for deciphering the pharmacological and other multifarious properties of miracle tree Moringa oleifera. Ind Crop Prod 170:113807

Kagambèga FW, Bagagnan AR, Ayantunde A, Sawadogo L (2019) Seasonal effect on growth performance and leaf biomass production of two Moringa oleifera Lam. Varieties grown under sahelian condition in Burkina Faso. J Exp Agric Int 1:12

Kalappurayil TM, Joseph BP (2017) A review of pharmacognostical studies on Moringa oleifera lam. flowers. Pharmacogn J 9(1)

Kashyap P, Kumar S, Riar CS, Jindal N, Baniwal P, Guiné RPF, Correia PMR, Mehra R, Kumar H (2022) Recent advances in drumstick (Moringa oleifera) leaves bioactive compounds: composition, health benefits, bioaccessibility, and dietary applications. Antioxidants 11:402

Kasolo JN, Bimenya GS, Ojok L, Ochieng J, Ogwal-Okeng JW (2010) Phytochemicals and uses of Moringa oleifera leaves in Ugandan rural communities. J Med Plants Res 4(9):753–757

Lakshmidevamma TN, Ugalat J, Apoorva KA, Suresh SG, Doddamani M, Kadam S, Nayana RS, Jagadeesha RC, Fakrudin B (2021) Genetic diversity of Moringa (Moringa Oleifera Lam.). In: The Moringa genome. Springer, Cham, pp 57–65

Lin W, Dai S, Chen Y, Zhou Y, Liu X (2019) The complete chloroplast genome sequence of Moringa oleifera Lam.(Moringaceae). Mitochondrial DNA Part B, 4(2):4094–4095

Mabapa MP, Ayisi KK, Mariga IK (2017) Effect of planting density and harvest interval on the leaf yield and quality of moringa (Moringa oleifera) under diverse agroecological conditions of Northern South Africa. Int J Agron 2017

Mazumder UK, Gupta M, Chakrabarti S, Pal D (1999) Evaluation of hematological and hepatorenal functions of methanolic extract of Moringa oleifera Lam. root treated mice. Indian J Exp Biol 37:612–614

Mbikay M (2012) Therapeutic potential of Moringa oleifera leaves in chronic hyperglycemia and dyslipidemia: a review. Front Pharmacol 3:24

Nadeem MA, Nawaz MA, Shahid MQ, Doğan Y, Comertpay G, Yıldız M, Hatipoğlu R, Ahmad F, Alsaleh A, Labhane N, Özkan H (2018) DNA molecular markers in plant breeding: current status and recent advancements in genomic selection and genome editing. Biotechnol Biotechnol Equip 32(2):261–285

Naik R, Bhushan A, Gupta RK, Walia A, Gaur A (2020) Low cost tissue culture technologies in vegetables: a review. Int J Biochem Res Rev 29(9):66–78

Pasha SN, Shafi KM, Joshi AG, Meenakshi I, Harini K, Mahita J, Sajeevan RS, Karpe SD, Ghosh P, Nitish S, Gandhimathi A (2020) The transcriptome enables the identification of candidate genes behind medicinal value of Drumstick tree (Moringa oleifera). Genomics 112(1):621–628

Prasanth CS, Pulical AJ, Govindaswamy M, Vanjinathan SS (2020) Genetic improvement through chemical and physical mutagenesis for enhanced leaf production in moringa (Moringa oleifera Lam.). Veg Sci 47(1):34–38

Sekhar C, Venkatesan N, Vidhyavathi A, Murugananthi M (2017) Post-harvest processing of Moringa and socio-economic appraisal of Moringa orchards in Tamil Nadu. Int J Hortic 7

Shyamli PS, Pradhan S, Panda M, Parida A (2021) De novo whole-genome assembly of moringa oleifera helps identify genes regulating drought stress tolerance. Front Plant Sci 12

Teixeira EMB, Carvalho MRB, Neves VA, Silva MA, Arantes-Pereira L (2014) Chemical characteristics and fractionation of proteins from Moringa oleifera Lam. leaves. Food Chem 147:51–54

Tian Y, Zeng Y, Zhang J, Yang C, Yan L, Wang X, Shi C, Xie J, Dai T, Peng L, Zeng Huan Y (2015) High quality reference genome of drumstick tree (Moringa oleifera Lam.), a potential perennial crop. Sci China Life Sci 58(7):627–638

Tshabalala T, Ndhlala AR, Ncube B, Abdelgadir HA, Van Staden J (2020) Potential substitution of the root with the leaf in the use of Moringa oleifera for antimicrobial, antidiabetic and antioxidant properties. S Afr J Bot 129:106–112

Chapter 5
Exploring Neem (*Azadirachta indica*): A Comprehensive Review of Its Uses and Advanced Breeding Techniques

Barka Peter Mshelmbula, Efeota Bright Odozi, Jasini Alexander Wahedi, Beckley Ikhajiagbe, and Geoffrey Obinna Anoliefo

Abstract Neem (*Azadirachta indica*), indigenous to the subcontinent part of Indian, is esteemed for its remarkable versatility and extensive health benefits. The tree contents abundant natural antioxidants and possesses significant antimicrobial and anti-inflammatory properties. This review aims to thoroughly investigate and evaluate the diverse applications of neem, alongside examining breeding strategies for its enhancement. These strategies encompass traditional approaches like cell and tissue culture studies, as well as advanced techniques such as genetic engineering and innovative methods like in vitro culture and genetic transformation. Additionally, we emphasize the importance of collaborative efforts among molecular biologists, breeders, and bioinformaticians to accelerate the development of enhanced breeding programs. Such collaboration is crucial for promoting sustainable agriculture and discovering new bioactive compounds for pharmaceutical and industrial use.

B. P. Mshelmbula
Department of Plant Science and Biotechnology, Federal University of Lafia, Lafia, Nigeria
e-mail: barka.mshelmbula@science.fulafia.edu.ng

E. B. Odozi
Department of Medical Laboratory Science, College of Medical Sciences, University of Benin, Benin City, Nigeria

J. A. Wahedi
Department of Biological Sciences, Federal University of Health Sciences, Benue State, Otukpo, Nigeria
e-mail: jasini.wahedi@fuhso.edu.ng

B. Ikhajiagbe (✉) · G. O. Anoliefo
Department of Plant Biology and Biotechnology, University of Benin, Benin, Nigeria
e-mail: beckley.ikhajiagbe@uniben.edu

G. O. Anoliefo
e-mail: obidimbuanoliefo@uniben.edu

J. M. Al-Khayri et al. (eds.), *Biodiversity and Genetic Improvement of Medicinal and Aromatic Plants II*, Advances in Plant Breeding Strategies 10,
https://doi.org/10.1007/978-3-031-81857-8_5

Keywords Neem · Advanced breeding · Cell and tissue culture · Genetic engineering

5.1 Introduction

Neem (*Azadirachta indica*), indigenous to the Indian subcontinent, is renowned for its remarkable versatility and extensive health benefits. This acclaim is primarily attributed to its rich array of bioactive compounds, including azadirachtin, nimbin, and nimbidin, which exhibit antioxidant, anti-inflammatory, antimicrobial, and anti-cancer properties (Awadhiya 2021; Zeeshan et al. 2024). Traditionally, Ayurvedic medicine has harnessed neem for treating various health conditions, such as eye disorders, liver problems, and skin diseases (Gautam and Mittal 2021). Furthermore, neem's bioactive components are now widely known in its application in dermatology and cosmetics because of its antimicrobial and anti-inflammatory characteristics (Baby et al. 2022). Recent studies have underscored neem's diverse applications across numerous industries, including medicine, environmental solutions, veterinary care, cosmetics, disinfectants, rubber, and textiles (de Alba et al. 2023). In agriculture, neem leaf extract enhances drought stress tolerance in maize, and its medicinal properties are widely recognized for their beneficial effects (Hussain et al. 2023). Additionally, neem products have shown promise as pest antifeedants, improving yield production and influencing pollinators in agro-biodiversity (Habarurema et al. 2021).

To enhance neem's potential further, research has explored various breeding methodologies, such as clonal propagation and genetic variability studies (Pal 1995; Swati et al. 2023). Techniques like somatic embryogenesis and other cell and tissue protocols is now showing promise and making breakthroughs in genetic improvement and large-scale propagation of neem (Srivastava and Chaturvedi 2018). However, there is an increasing emphasis on advanced breeding methods, including genetic engineering, to optimize neem's genetic potential (Swati et al. 2023). Vegetative propagation has also been identified as an effective strategy for mass multiplication of superior neem trees (Palanisamy and Kumar 2001). Researchers highlight the ongoing need for genetic improvement research, particularly targeting genes with significant utility and economic value (Swati et al. 2023). Other researches further examined the utilization of the in vitro culture of oil found in neem and explored the antibacterial properties of neem extracts, which could benefit livestock and poultry breeding (Xu et al. 2021; Regni et al. 2023). Additionally, research has assessed neem's genetic diversity across different regions and discussed the broader implications of plant tissue culture in crop improvement (Chadipiralla et at. 2021; Mshelmbula et al. 2022). These collective efforts underscore the potential of advanced techniques to augment neem's traits and contribute to its improvement. In this review, we aim to comprehensively explore and assess the various uses of neem, along with examining breeding approaches for its improvement, encompassing traditional methods like cell and tissue culture studies, as well as advanced techniques such as genetic engineering and other innovative breeding methods like in vitro culture and genetic transformation.

5.2 Traditional Uses of Neem

5.2.1 Historical Significance and Cultural Uses

Neem, scientifically known as *Azadirachta indica*, holds substantial historical and cultural importance globally, particularly in India (Winterbottom 2021). The medicinal properties as well as its utilization for the purpose of the ecology is well documented, this in turn makes it a cornerstone of Ayurvedic medicine where it is employed to treat various skin conditions, enhance immune function, and aid digestion (Gautam and Mittal 2021). There is no part of the tree that is not integral to traditional healing practices which includes its leaves, bark, seeds, and oil. Additionally, neem is applicable in agriculture as a natural pesticide in organic farming, protecting crops while preserving beneficial insects (Srirangarayan et al. 2020). Culturally, neem is known as the "village pharmacy" in India, revered in Hinduism, and incorporated into rituals and festivals. Its presence near homes and temples symbolizes protection and purification. Beyond its traditional uses, neem is a focal point in intellectual property debates and scientific research, reflecting its extensive cultural and historical significance. Research highlights its diverse medicinal properties, including antimicrobial and anti-inflammatory effects, and its potential in combating diseases such as malaria (Zeeshan et al. 2024; Devi and Sharma 2023; Osazee and Eribe 2023).

5.2.2 Traditional Medicinal Applications

Neem has a longstanding history in crude medical application, including Ayurveda, Unani, and Homeopathy, and is esteemed for its multifaceted health benefits (Reddy and Neelima 2022). Various neem parts, such as its leaves, flowers, seeds, fruits, roots, and bark, are utilized by practitioners to treat different ailments, including inflammation, infections, fever, skin diseases, and dental disorders (Sarah et al 2019). The medicinal properties of neem are attributed to its abundance of biologically active compounds.

Neem's use in traditional medicine spans centuries globally, with a particularly prominent role in Ayurvedic practices in the Indian subcontinent (Rajaiah et al. 2022). Scientific research has confirmed neem's effectiveness in the treatment of skin diseases such as acne, eczema, and psoriasis because of the antibacterial, antifungal, and anti-inflammatory properties (Farzana et al. 2020). Additionally, neem has been shown to address digestive issues, with studies indicating its efficacy in combating ulcers and alleviating gastric distress symptoms (Obiajulu and Chukwuemeka 2020).

The versatility of neem as a medicinal plant is further emphasized by its bioactive compounds, such as nimbolide and azadirachtin, which possess antioxidant, anti-inflammatory, and immunomodulatory properties (Sarkar et al 2021). Its anti-inflammatory, antibacterial, and antioxidant attributes make neem a promising therapy for skin disorders (Singh et al 2021a, b). Furthermore, research suggests neem's potential in cancer control, with its compounds showing therapeutic promise against various cancer types (Azam et al 2023). These findings collectively underscore neem's diverse bioactivity and its significance in medicine (Zeeshan et al. 2024).

5.2.3 Neem in Pest Control and Agriculture

Neem is a well-researched tree with diverse applications in pest control and agriculture. Its bioactive compounds, including azadirachtin, nimbin, and nimbidin, provide potent insecticidal and pesticidal properties, offering a natural alternative to synthetic pesticides (Agbo et al. 2019; Reddy and Neelima 2022). Neem acts primarily as a biopesticide in agriculture, effectively targeting pests such as aphids, mites, whiteflies, and thrips, while posing minimal risk to beneficial insects (Wahab et al. 2023). Additionally, neem functions as a plant growth regulator, promoting crop development and enhancing resilience to environmental stresses (Sohail et al 2022; Razzaq et al. 2022). It also improves soil fertility by providing essential nutrients like nitrogen, phosphorus, potassium, and micronutrients, thereby addressing nutrient deficiencies in crops (Sharma et al 2023). Research supports neem's efficacy against a wide range of pests and its potential as a natural fungicide in managing crop diseases (Adusei and Azupio 2022). Recent studies have emphasized neem's effectiveness in pest management and agriculture, highlighting its broad spectrum of action, low toxicity, and potential for enhancing pesticide formulations through nano-encapsulation (, Akama et al 2023; Fenin et al. 2022; Charpe 2022; Gusan et al. 2022). Neem offers a sustainable solution in agriculture by managing pests effectively while minimizing environmental harm. As the demand for sustainable farming practices grows, neem-based products have the potential to contribute significantly to ecologically sound agricultural systems. However, further research is required to fully harness neem's potential and optimize its application methods.

5.3 Breeding Approaches for Neem Improvement

5.3.1 Conventional Breeding Methods

Conventional breeding methods for improving neem involve selecting and crossing trees with desirable traits (Parthiban et al 2020). These methods include clonal propagation, mass selection based on phenotypic traits, and progeny testing to ensure the presence of desired traits before releasing improved varieties. Additionally, techniques such as grafting and budding are used to maintain genetic identity. Swati has emphasized the necessity for further research on neem genetic improvement, while Regni has demonstrated the potential of neem oil in enhancing shoot production in pear and olive tree cultivars. Furthermore, Ewansiha has highlighted the practical benefits of neem leaf residue in managing the fall armyworm pest in maize, resulting in improved plant growth and yields (Swati et al. 2023; Regni et al. 2023; Ewansiha et al 2023).

5.3.2 Challenges in Neem Breeding

Neem breeding faces several challenges, including issues with seedling growth, frost susceptibility, the need for genetic enhancement, and resource exploration (Sharma et al 2020; Swati et al. 2023). Economic considerations, such as restrictions on seed collection and sales, underscore the importance of neem seeds in efforts to improve farmers' income in India (Baburaj et al 2021). Recent studies in Nigeria have highlighted challenges such as the economic infeasibility of neem production in certain areas and the need for alternative pest control measures (Muhammad et al 2020). Genetic diversity, cross-pollination, and propagation difficulties further complicate neem breeding efforts, along with the arduous task of screening for disease and pest resistance (Izahudin et al 2022). Additional obstacles include the scarcity of molecular markers, environmental sensitivity, and limited resource allocation (Isewon et al 2022). Despite these challenges, advancements in biotechnological tools offer promising solutions for improving neem breeding programs, facilitating more effective trait selection and propagation.

5.4 Cell and Tissue Culture in Plants

5.4.1 Cell and Tissue Culture in Neem

Like other plants, the basics of cell and tissue culture in neem involve growing either the cell, tissues, or organs of plants in a controlled, aseptic conditions on a nutrient culture medium in laboratory conditions (Anuruddi et al 2023). Key steps and components include selecting an explant, such as leaves, stems, roots, or seeds, sterilizing it with chemical agents like ethanol or sodium hypochlorite, and placing it on a growth media containing abundant nutrients necessary for its growth (Fernández-Da et al 2021). The medium typically consists of carbon sources, inorganic salts, vitamins, amino acids, and plant hormones (auxins and cytokinins). Cultured tissues are maintained under controlled environmental conditions, with regular subculturing to ensure ongoing growth or manipulate development (Regni et al 2023). By adjusting plant hormone concentrations, neem tissue cultures can be directed to differentiate into shoots and roots, ultimately regenerating whole plants (Padilla et al 2021). Before transplantation into soil, the developed plantlets must first acclimatize to natural conditions. This tissue culture method, called micropropagation, allows for the mass production of neem trees while preserving parent plant traits, facilitating the rapid multiplication of desired cultivars, conserving genetic resources, and supporting research efforts such as genetic engineering and breeding programs (Tripathi et al 2021). Research indicates the presence of autotoxicity in neem secretory cells, which can affect cell health and signaling reactions (Canaveze et al 2021). Successful in vitro production of azadirachtin, a key compound in neem, has been achieved using tissue culture techniques (Fernández-Da et al 2021). Neem oil has been found to enhance the in vitro rapid growth of an old cuitivar of pear tree (Regni et al. 2023), while acetylsalicylic acid has been shown to improve the antioxidative nature as well as the storing it in a cold state viability of encapsulated nodal segments of neem (Padilla et al 2021).

5.4.2 Applications in Plant Propagation and Disease Management

Neem cell and tissue culture techniques are highly valuable for both plant propagation and disease management. In plant propagation, these techniques allow for the mass production of neem trees with desired traits while preserving genetic fidelity (Gosal et al. 2018). Micropropagation enables the rapid multiplication of disease-free plants, ensuring uniformity in characteristics such as growth rate and resistance to pests and diseases (Suman 2017). Moreover, tissue culture facilitates the propagation of elite neem cultivars, aiding in the widespread dissemination of superior genetic material (Salgotra et al. 2019). In disease management, neem cell and tissue culture play a crucial role in producing secondary metabolites with antimicrobial properties, such as azadirachtin, which can be formulated into eco-friendly pesticides or fungicides (Ozyigit et al 2023). Tissue culture techniques also facilitate the examination of plant-pathogen interactions at the cellular and molecular levels, shedding light on the underlying mechanisms of disease resistance. Neem tissue culture has applications in rapid propagation, disease-free stock production, pest and disease resistance enhancement, genetic resource conservation, research, and secondary metabolite production, highlighting its potential in horticulture (Devanakonda 2021; Kundu et al. 2021).

5.5 Genetic Engineering

Genetic engineering is a multifaceted toolkit that offers a broad range of techniques for manipulating an organism's genetic makeup, with profound implications across various domains (Clouse and Wagner 2021). Recombinant DNA technology, a cornerstone in this field, combines DNA fragments from different sources to generate novel genetic sequences that are subsequently integrated into host organisms for expression (Lanigan et al 2020). Another pivotal technique, gene cloning, meticulously isolates and amplifies specific genes, facilitating their insertion into vectors like plasmids for further manipulation (Bridgen et al 2018). CRISPR-Cas9 is now seen as a revolutionary approach, enabling accurate genome editing by inducing targeted double-strand breaks, allowing for meticulous modifications at desired loci (Aljabali et al 2024). Transfection and transformation techniques are instrumental in introducing exogenous DNA into cells, achieved through methods such as electroporation or viral vectors, thereby broadening the scope of genetic manipulation (Bridgen et al 2018). Gene silencing technologies, exemplified by RNA interference, provide an avenue for modulating gene expression by homing in on specific messenger RNA molecules (Zhou and Li 2023). Genetic modification via T-DNA harnesses the possibility of DNA to be transferred naturally using Agrobacterium tumefaciens, exploiting it to ferry tailored gene sequences into plant genomes, thereby enabling precise genetic engineering in plant systems (Bridgen et al 2018).. Recent advances in neem genetic improvement highlight the application of cutting-edge biotechnological tools such as nicking endonuclease-mediated DNA assembly (NEMDA) for gene functional research, furthering our understanding of neem genetics. In vitro propagation techniques, as exemplified by recent research focusing on other medicinal plants like Phlogacanthus thyrsiformis Nees, emphasize the importance of maintaining genetic stability in propagated materials (Baro and Das 2022). Moreover, ongoing investigations explore the potential of utilizing rootstocks such as Melia azedarach to enhance neem plant growth, illustrating the diverse avenues through which genetic engineering techniques can be harnessed to improve neem plants (Sharma et al 2020).

Genetic modification has to do with the conscious change of a plant's genetic makeup to introduce beneficial character or improve the one that is already there, with diverse applications across agriculture and biotechnology (Bailey-Serres et al

2019). This approach seeks to address key challenges in plant improvement, spanning increased yield, pest and disease resistance, herbicide tolerance, enhanced nutritional content, improved abiotic stress tolerance, quality traits enhancement, and pharmaceutical and biofuel production (Ninama et al 2024). These techniques undergo rigorous safety assessments to ensure their suitability for consumption and minimal environmental impact, with regulations varying globally (Bhowmik et al. 2023). Recent studies have explored methods to enhance traits in the neem plant, including genetic improvement, utilizing potential rootstocks, and assessing genetic diversity, laying the groundwork for further research into maximizing neem's beneficial properties (Sharma et al 2020). Leveraging advanced genome editing technologies like CRISPR/Cas9 holds great potential for introducing value-added traits in neem plants (Kiran and Abdin 2020). Therefore, genetic modification offers a potent avenue for unlocking neem's full potential as a sustainable and multifunctional crop with applications spanning agriculture, medicine, and beyond.

5.6 In Vitro Culture and Genetic Transformation

5.6.1 Techniques for in Vitro Culture

In vitro culture procedures for neem involve a series of steps similar to general plant tissue culture methods, including explant preparation, initiation, callus formation, shoot induction, root induction, multiplication, elongation and hardening, and acclimatization (Bello et al. 2022). The first step is to choose a suitable part of the plant, such as leaves, stems, or seeds, and meticulously sterilize it to eliminate surface contaminants. The sterilized explant is then placed on a nutrient-rich culture medium containing mineral salts, vitamins, carbon sources, and plant growth regulators to encourage cell differentiation and proliferation. Subsequent steps involve inducing callus formation, promoting shoot and root induction, multiplying shoots by subculturing, and preparing plantlets for transfer to soil by enhancing tissue strength and acclimatization to external conditions (Bhowmik et al. 2023; Bokri et al. 2023). These techniques facilitate rapid propagation of selected genotypes with desirable traits and aid in conserving genetic material and producing disease-free and uniform plant material for research or agricultural purposes.

Various studies have explored in vitro culture of neem through different techniques. Fernández-Da et al. (2021) optimized azadirachtin production using specific callus types and chemical agents. Regni et al. (2023) demonstrated neem oil's potential in enhancing the proliferation of ancient pear tree cultivars. Akshaya et al. (2020) highlighted neem extract's antimicrobial properties against pathogens like Staphylococcus aureus and Clostridium perfringens. Padilla et al. (2021) proposed the use of acetylsalicylic acid to improve the antioxidative status and cold storage of encapsulated nodal segments of neem. These studies illustrate the great possibilities of in vitro culture techniques that will promote neem's properties and traits.

5.6.2 Genetic Transformation

Plant biotechnology has significantly advanced with the development of genetic transformation methods, which have facilitated the creation of genetically modified organisms (GMOs) with novel traits and improved agronomic properties (Singh et al 2021a, b). Among the most prevalent techniques for genetic transformation are Agrobacterium-mediated transformation, which uses Agrobacterium tumefaciens to transfer foreign genes into the cells of plants, and biolistic or particle bombardment, which involves direct delivery of DNA-coated particles into plant cells (Singh et al 2021a, b; Gheysen et al 2022). Recent innovations, such as CRISPR/Cas9-mediated genome editing, offer precise gene manipulation without DNA insertion, broadening possibilities for trait development (Jacinto et al 2020). Studies have highlighted these methods' potential in enhancing crop productivity, nutrient content, and resistance to pests and diseases (Ye et al. 2022; Malakhova et al 2021). Shrestha et al. (2020) provided a comprehensive overview of these techniques, reflecting the dynamic landscape of plant genetic transformation methods. Overall, these advancements hold promise for agricultural sustainability and curbing global challenges, such as climate change and food insecurity.

5.7 Advanced Breeding Methods

5.7.1 Marker-Assisted Selection in Neem Breeding

The use of Marker-Assisted Selection (MAS) has transformed neem breeding by enabling targeted improvement of desired characters with enhanced precision and efficiency. Integrating molecular markers linked to specific traits into conventional breeding programs, MAS facilitates the early selection of superior genotypes Shrestha et al. (2020), Boopathi and Boopathi (2020). This technique shows promise in accelerating neem breeding, especially for traits with complex inheritance patterns or those influenced by environmental factors (Feng et al 2020; Martinez et al 2021). By identifying molecular markers associated with traits like pest resistance, high oil content, or boosting medicinal properties, breeders can streamline the selection process, significantly wasting of time as well as all that is needed for traditional phenotypic evaluations (De Mori and Cipriani 2023; Vargas et al. 2023). Furthermore, MAS allows for the introgression of favorable alleles from wild or exotic germplasm into elite neem cultivars, enhancing genetic diversity and broadening the genetic base for future breeding efforts (Demircioglu 2024). However, successful MAS implementation in neem breeding requires well-characterized molecular markers, robust genetic linkage maps, and access to high-throughput genotyping technologies Collaboration among molecular biologists, breeders, and bioinformaticians is essential to fully leverage MAS and enhance neem improvement programs. As research in genomics and molecular biology progresses, MAS is showing great potential in developing improved neem varieties for agriculture, forestry, and the pharmaceutical industry.

5.7.2 Genome Editing Technologies

Genome editing technologies have revolutionized molecular biology, providing precise tools for modifying genetic material with significant implications across various fields (Bhattacharyya and Mukherjee 2020). The CRISPR-Cas9 system, in particular, has gained widespread attention for its simplicity, efficiency, and versatility in different organisms (Javaid et al 2022). This technology enables targeted DNA modifications with unprecedented precision and ease (Zhang et al 2020). Additionally, TALENs and zinc finger nucleases (ZFNs) offer alternative methods for precise DNA manipulation (Shamshirgaran et al 2022). These advanced tools have facilitated groundbreaking discoveries in gene function, disease modeling, and therapeutic development. In agriculture, genome editing holds immense promise for crop improvement, enhancing yield, nutritional content, and resistance to pests and diseases (Kumar et al 2022). Despite ethical and societal concerns, particularly regarding human germline modification, ongoing advancements in genome editing continue to drive innovation, addressing pressing challenges in human health, agriculture, and beyond. Notably, in 2020, significant advancements in plant DNA manipulation were made using genome editing technologies (Nadakuduti et al. 2021). The CRISPR/Cas9 system has been a game-changer, enabling the creation of crops resistant to disease, pests, and abiotic stresses (Komisarenko and Romaniuk 2021). This system, along with other genome editing tools such as ZnF and TALE nucleases, RNA interference, and Cre/loxP, has opened new possibilities in biotechnology, medicine, and agriculture (Schulze and Lammers 2020).

5.7.3 Integration of Omics Technologies in Neem Breeding

The integration of omics technologies in neem breeding represents a sophisticated approach that utilizes high-throughput molecular profiling techniques to optimize breeding programs with greater efficiency and precision. Omics technologies, including genomics, transcriptomics, proteomics, and metabolomics, allow for a comprehensive examination of the genetic, transcriptomic, proteomic, and metabolic profiles of neem plants, providing deeper insights into their biological processes and traits (Yang et al 2021; Nadakuduti et al. 2021). Genomics aids in understanding neem's genetic makeup, identifying genes related to desirable traits such as pest resistance and high oil content (Chao et al 2024). Transcriptomics reveals gene expression patterns across various tissues and environmental conditions, uncovering regulatory mechanisms and candidate genes for targeted manipulation (Zenda et al

2021). Proteomics examines protein composition and function, elucidating biochemical pathways crucial for trait expression (Kaya 2024). Metabolomics complements these approaches by profiling small molecule metabolites, providing a holistic view of neem's biochemical phenotype and responses to various stresses (Kaya 2024). Integrating omics data through systems biology approaches identifies key genes, proteins, and metabolites associated with complex traits, facilitating marker-assisted selection and genomic breeding strategies (Chao et al 2024). Additionally, omics technologies aid in developing predictive models for trait prediction, accelerating breeding efforts and improving neem's resilience and productivity across diverse agroecological contexts (Zenda et al 2021). With advancements in omics technologies, bioinformatics, and computational tools, neem breeding programs can be significantly enhanced, contributing to sustainable agriculture and uncovering novel bioactive compounds for pharmaceutical and industrial applications.

5.8 Challenges and Future Directions

The adoption of advanced breeding techniques in plant breeding has emerged as a powerful tool to address various agricultural challenges and revolutionize crop improvement strategies (Rönspies et al. 2021). These techniques include marker-assisted selection (MAS), genomic selection, genome editing, and the integration of omics technologies. By identifying genetic markers associated with desired traits, MAS and genomic selection enhance crop resilience to stresses such as diseases, droughts, and changing environmental conditions (Gosal et al 2020). Additionally, CRISPR/Cas9 genome editing enables precise modifications in plant genomes, resulting in crops with improved agronomic traits and resistance to pests and diseases (Delhove et al 2020). Omics technologies provide insights into genetic and molecular processes, facilitating the identification of key genes and pathways for crop improvement, thus accelerating breeding cycles and developing new varieties more efficiently (Dwivedi et al 2020). These methods also allow for tailoring crops' nutritional profiles and quality attributes to meet consumer preferences and dietary needs, supporting sustainable agriculture by creating high-yield, resource-efficient crops while conserving natural resources (Bharti and Chimata 2019).

The potential applications of advanced breeding techniques in plant breeding are vast and promising, particularly in addressing emerging challenges such as climate change and global warming (Gosal et al 2020). CRISPR/Cas-mediated chromosome engineering has the potential to transform plant breeding by altering genetic linkages between traits, which is crucial for developing climate-resilient cultivars (Rönspies et al. 2021). Integrating structural genomics into breeding and eugenics analysis has proven essential in identifying rare genes and alleles crucial to significant agronomic traits, further enhancing genetic gains in plant breeding (Johar and Salgotra 2023). Furthermore, the use of advanced analytics, phenomics, and biotechnology approaches can improve crop plants' resilience and resource-use efficiency while enhancing their nutritional content (Dwivedi et al 2020)..

However, the challenges of advanced breeding techniques in plant breeding are multifaceted, including the complexity and cost of implementation, regulatory frameworks, potential unintended effects, and equitable access for smallholder farmers in developing countries [109, 113, 114]. Addressing these challenges requires interdisciplinary collaboration, stakeholder engagement, and transparent communication to foster responsible innovation in agricultural biotechnology.

5.9 Conclusion and Prospects

Neem (*Azadirachta indica*) stands as a botanical treasure with extensive applications ranging from traditional medicine to modern biotechnology. Its rich cultural heritage, along with its numerous health benefits and ecological contributions, underscores its importance in global agriculture and healthcare. Advanced breeding techniques, including genetic engineering, marker-assisted selection, and omics integration, offer promising avenues for enhancing neem's traits and addressing agricultural challenges as researchers continue to unlock its genetic potential. However, the successful implementation of these advanced breeding methods requires navigating complex regulatory frameworks and ensuring equitable access. By fostering interdisciplinary collaboration, transparent communication, and responsible innovation, neem breeding programs can utilize these advanced tools to develop resilient, sustainable crop varieties that meet the evolving needs of agriculture and society, while promoting environmental stewardship and global health.

References

Adusei S, Azupio S (2022) Neem: a novel biocide for pest and disease control of plants. J Chem 8:45–53

Agbo B, Nta A, Ajaba M (2019) Bio-pesticidal properties of Neem (*Azadirachta indica).* Adv Trends Agric Sci 1:17–26

Akama CK, Amiteye S, Appiah AS, Kpentey PB, Appiah R, Kutufam JT, Dzimega DA, Amenorpe G (2023) Assessment of different rates of neem extracts as bio-pesticide for the control of insect vectors and associated viral diseases in okra. Afr J Agric Res 19(4):364–373

Akshaya S, Bhuvaneshwari G, Grace JL, Vennila R (2020) In vitro efficacy testing of neem *(Azadirachta indica)* extract against aerobic and anaerobic bacteria. Int J Curr Pharm Res 119–121

Aljabali AA, El-Tanani M, Tambuwala MM (2024) Principles of CRISPR-Cas9 technology: advancements in genome editing and emerging trends in drug delivery. J Drug Deliv Sci Technol 105338

Anuruddi H, Nakandalage N, Fonseka DK (2023) New insights for the production of medicinal plant materials: ex vitro and in vitro propagation. In: Biosynthesis of bioactive compounds in medicinal and aromatic plants: manipulation by conventional and biotechnological approaches. Springer, pp 181–212

Awadhiya N (2021) Neem: *Azadirachta Indica* and its medicinal applications medicinal plants, pp 23–38

Azam H, Khan MT, Lodhi MS, Sadiqa A, Yasin T (2023) Therapeutic role of neem *(Azadirachta indica Adr. Juss.)* in different types of cancer: a systematic review. Int J Appl Exp Biol 2(2):97–113

Baburaj K, Venkata Pirabu J, Suresh Kumar D, Vidhyavathi A, Pushpavalli R (2021) Livelihood status of neem seed pickers and constraints in seed collection and procurement in Tamil Nadu, pp 4–17

Baby AR, Freire TB, Marques G.d.A, Rijo PF, Lima V, Carvalho JCM, Rojas d, J, Magalhães WV, Velasco MVR, Morocho-Jácome AL (2022) *Azadirachta indica* (Neem) as a potential natural active for dermocosmetic and topical products: a narrative review. Cosmetics 9(3):58

Bailey-Serres J, Parker JE, Ainsworth EA, Oldroyd GE, Schroeder JI (2019) Genetic strategies for improving crop yields. Nature 575(7781):109–118

Baro T, Das S (2022) In vitro propagation and somaclonal variation study of Phlogacanthus thyrsiformis Nees an ethnic medicinal plant. Plant Science Today 5(5):45–61

Bello AM, Kamali Aliabad K, Saravi AT, Sodaei Zade H (2022) Determination of the best culture medium and plant growth regulators for micropropagation of neem tree *(Azadirachta indica A. Juss)*. Int J Hortic Sci Technol 9(2):237–245

Bharti G, Chimata MK (2019) Review on new plant breeding techniques. Int J Sci Res 8(4):723–730

Bhattacharyya S, Mukherjee A (2020) CRISPR: The revolutionary gene editing tool with Far-reaching applications. Biotechnol Bus-Concept Deliv 47–56

Bhowmik P, Konkin D, Polowick P, Hodgins C.L, Subedi M, Xiang D, Yu B, Patterson N, Bokri H, Kahlaoui S, Hcini K, Dhoueibi M, Harzallah-Skhiri F, Stambouli-Essassi S (2023) Induction of organogenesis and callogenesis in Limbarda crithmoïdes L.(Asteraceae) explants cultured on MS media supplemented with various concentrations of Na+ and K+. NotE Bot Horti Agrobot Cluj-Napoca 51(2):13147–13147

Boopathi NM, Boopathi NM (2020) Marker-assisted selection (MAS), genetic mapping marker assisted selection: Basics, practice benefits 343–388.

Bridgen MP, Van Houtven W, Eeckhaut T (2018) Plant tissue culture techniques for breeding. Ornam Crop 127–144

Canaveze Y, Scudeler EL, Rodrigues MS (2021) Neem secretory cells: developmental cytology and indications of cell autotoxicity. Protoplasma 258:415–429

Chadipiralla K, Gayathri PK, Rajani V, Reddy PB (2021) Plant tissue culture and crop improvement 43

Chao H, Zhang S, Hu Y, Ni Q, Xin S, Zhao L, Ivanisenko VA, Orlov YL, Chen M (2024) Integrating omics databases for enhanced crop breeding. J Integr Bioinform 20(4):20230012

Charpe AM (2022) Nano-encapsulation of neem based pesticide formulations. J Pl Dis Sci 17(2):82–92

Clouse KM, Wagner MR (2021) Plant genetics as a tool for manipulating crop microbiomes: opportunities and challenges. Front Bioeng Biotechnol Bus-Concept Deliv 9:567548

de Alba SL, de García-González SL, Coronado Ortega MA, Ayala Bautista JR, Alpírez GM, Montes DG (2023) Extraction methods and applications of bioactive compounds from neem (*Azadirachta indica*): a mini-review. Mini-Rev Org Chem 20(7):644–654

De Mori G, Cipriani G (2023) Marker-assisted selection in breeding for fruit trait improvement: a review. Int J Mol Sci 24(10):8984

Delhove J, Osenk I, Prichard I, Donnelley M (2020) Public acceptability of gene therapy and gene editing for human use: a systematic review. Hum Gene Ther 31(1–2):20–46

Demircioglu HB (2024) Genomic selection in animal breeding. In: BIO web of conferences, EDP sciences, p 01069

Devanakonda RR (2021) Plant tissue culture applications in horticultural crops. Int J Creat Sci Thoughts 9(5):156–160

Devi J, Sharma RB (2023) Medicinal importance of *Azadirachta indica*: an overview. J Drug Deliv Ther 13(6):159–165

Dwivedi SL, Goldman I, Ceccarelli S, Ortiz R (2020) Advanced analytics, phenomics and biotechnology approaches to enhance genetic gains in plant breeding. Adv Agron 162:89–142

Ewansiha S, Enogieru L, Aigbe U, Kamara A. (2023) Managing fall armyworm (Spodoptera frugiperda JE Smith) using leaf residue of neem *(Azadirachta indica A. Juss)* on maize (Zea mays L.) under natural infestation. Biol Agric Hortic 39(4):305–315

Farzana S, Saha SP, Sultana N, Khan M (2020) Gastroprotective effect of *Azadirachta indica* leaves (neem) extract on aspirin induced gastric ulcer in rats. Delta Med CollE J 7(2):61–66

Feng J-L, Jiang Y, Yang Z-J, Chen S-P, El-Kassaby CH (2020) Marker-assisted selection in hybrid population. Silvae Genetica 69(1):63–72

Fenin M, Meena DM, Meena DOP (2022) Lumpy skin disease: an emerging contagious threat. Justagriculture 2(12):2582–8223

Fernández-Da SR, Salomón J, Villarroel A, Montilla V (2021) Evaluación de la producción in vitro de azadiractina en Neem según el tipo de callo y los agentes químicos. Revista Ingeniería UC 28(2):282–293

Gautam P, Mittal A (2021) Review on Traditional Plant *Azadirachta indica*: natural source for disease curability and health promotion. J Pharm Res Int 33(56A):1–7

Gheysen G, Angenon G, Van Montagu M. (2022) Agrobacterium-mediated plant transformation: a scientifically intriguing story with significant applications. In: Transgenic plant research. Routledge, pp 1–34

Gosal SS, Wani SH (2018) Cell and tissue culture approaches in relation to crop improvement. Biotechnol Crop Improv Vol 1 Cell Approaches 1–55

Gosal SS, Pathak D, Wani SH, Vij S, Pathak M (2020) Accelerated breeding of plants: methods and applications. Accel Plant Breed Vol 1 CeR Crop 1–29

Gusan A, Popa A, Tretiacova T, Todiras V, Savranschii D, Gladcaia A (2022) Biological efficacy of the neem oilfor the control of aphidae populations. Int J Adv Res 10(09):848–852

Habarurema G, Habinshuti J, Mukarunyana B, Uwamariya C, Safari J, Ndayambaje JB (2021) Activity assay of neem products as pest antifeedant in agro-biodiversity. Int J Trop Insect Sci 42:109–115

Hussain R, Shehzadi I, Sadiq M, Ahmad A, Akram NA, Garcia-Caparros P (2023) Leaf extract of neem (*Azadirachta indica* Adr. Juss.) modulates morphological and biochemical characteristics in maize (*Zea mays* L.) under water deficit stress. Int J Appl Exp Biol 3(2):1–10

Isewon I, Apata O, Oluwamuyiwa F, Aromolaran O, Oyelade J (2022) Machine learning algorithms: their applications in plant omics and agronomic traits' improvement. Research 11:1256

Izahudin MH, Rosmi MH, Mohd Fauzi NA (2022) Neem oil as biopesticides. Chemical process and sustainability in agricultural biotechnology, pp13–28

Jacinto FV, Link W, Ferreira BI (2020) CRISPR/Cas9-mediated genome editing: from basic research to translational medicine. J Cell Mol Med 24(7):3766–3778

Javaid D, Ganie SY, Hajam YA, Reshi MS (2022) CRISPR/Cas9 system: a reliable and facile genome editing tool in modern biology. Mol Biol Rep 49(12):12133–12150

Johar P, Salgotra R (2023) Elementary and advanced mechanisms for genetic engineering in crops. Plant Cell Biotechnol Mol Biol Rep 24(7–8):33–43

Kaya (2024) Microbial modulation of hormone signaling, proteomic dynamics, and metabolomics in plant drought adaptation. Food Energy Secur 13(1):e513

Kiran U, Abdin MZ (2020) Genome editing technologies for value-added traits in plants. In: Transgenic technology based value addition in plant biotechnology. Elsevier, pp 51–67

Komisarenko S, Romaniuk S (2021) Prospects of genome editing using CRISPR/CAS or how to master genetic scissors. Nobel Prize in Chemistry 2020. Ukr Biochem J 93(1):113–128

Kumar D, Yadav A, Ahmad R, Dwivedi UN, Yadav K (2022) CRISPR-based genome editing for nutrient enrichment in crops: a promising approach toward global food security. Front Genet 13:932859

Kundu S, Sindhusha P, Kumar A (2021) Micro-propagation techniques in horticultural crops and various factors affecting it: a review. Mod Phytomorphol 15:137–141

Lanigan TM, Kopera HC, Saunders TL (2020) Principles of genetic engineering. Genes 11(3):291

Malakhova N, Skiba Y, Iskakova G, Naizabayeva D, Tezekbaeva B, Ismagulova G, Maltseva E (2021) A positive experience in applying the biolistic approach to potato varieties Aksor and Nevskiy. Vavilov J Genet Breed 25(2):157

Martinez I, Michael VN, Fu Y, Shrestha S, Meru GJA .J.o.P.S (2021) DNA extraction from a single seed for marker-assisted selection in squash. Am J Plant Sci 12(12):1912–1925

Mshelmbula BP, Anoliefo GO, Ikhajiagbe B, Edegbai BO (2022) Genetic diversity assessment of neem (*Azadirachta indica* A. Juss) in northern Nigeria. J Mater Environ Sci 13(04):547–560

Muhammad Y, Ilu K, Jibo A, Salami K, Ahmed B (2020) Economic analysis of Neem *(Azadirachta indica)* production in Yobe state, Nigeria. J Res For Wildl Environ 12(2):249–256

Nadakuduti SS, Enciso-Rodríguez F (2021) Advances in genome editing with CRISPR systems and transformation echnologies for plant DNA manipulation. Front Plant Sci 11:637159

Ninama N, Gangal L, Khayum A, SB H, HM S, Singh A.J.J.o.A.i.B (2024) Biotechnology, post-harvest biotechnology or genetic engineering solutions: extending shelf life and reducing food waste. J Adv Biol Biotechnol 27(4):1–26

Obiajulu CE, Chukwuemeka EF (2020) Antidiabetic and modulatory effect of ethanol extract of neem leaf on some essential biochemical parameters of streptozotocin-induced diabetic rats. Int J Biochem Res Rev 28(4):1–11

Osazee O, Eribe M (2023) Therapeutic potentials of neem against malaria parasite: a review. J Pure Appl Sci 9:180–189

Ozyigit II, Dogan I, Hocaoglu-Ozyigit A, Yalcin B, Erdogan A, Yalcin IE, Cabi E, Kaya Y (2023) Production of secondary metabolites using tissue culture-based biotechnological applications. Front Plant Sci 14:1132555

Padilla I, Barba-Espín G, Hernández J, Piqueras A, Burgos L (2021) Acetylsalicylic acid improved antioxidative status and cold storage of encapsulated nodal segments of neem (Azadirachta indica A. Juss.). Plant Cell, Tissue Organ Cult 144:261–270

Pal M (1995) Clonal approaches for yield improvement in neem: strategies and protocols for selective use of genetic diversity. Indian For 121:1033–1039

Palanisamy K, Kumar P (2001) Vegetative propagation and genetic improvement of neem. Indian Forester 127(3):347–350

Parthiban KT, Krishnakumar N, Devanand P (2020) Tree breeding and improvement theory and technology. Scientific Publishers, p 391

Rajaiah Yogesh H, Gajjar T, Patel N, Kumawat R (2022) Clinical study to assess efficacy and safety of Purifying Neem Face Wash in prevention and reduction of acne in healthy adults. J Cosmet Dermatol 21(7):2849–2858

Razzaq M.K, Hina A, Abbasi A, Karikari B, Ashraf HJ, Mohiuddin M, Maqsood S, Maqsood A, Reddy I, Neelima P (2022) Neem *(Azadirachta indica)*: a review on medicinal Kalpavriksha. Int J Econ Plants 9(1):59–63

Regni L, Facchin SL, da Silva DF, Proietti P, Silvestri C, Micheli M (2023) Optimization of the in vitro proliferation of an ancient pear tree cultivar ('Decana d'inverno') through the use of neem oil. Plants 12

Rönspies M, Schindele P, Puchta H (2021) CRISPR/Cas-mediated chromosome engineering: opening up a new avenue for plant breeding. J Exp Bot 72(2):177–183

Salgotra R, Sharma M, Pandotra P (2019) Biotechnological interventions for sustainable conservation of plant genetic resources in the scenario of climate change. Nat Resour Conserv Res 2(1):1–13

Sarah R, Tabassum B, Idrees N, Hussain (2019) Applications, Bio-active compounds isolated from neem tree and their applications. M.K.J.N.B.-a.C.V.P.509-528

Sarkar S, Singh RP, Bhattacharya GJB (2021) Exploring the role of *Azadirachta indica* (neem) and its active compounds in the regulation of biological pathways: an update on molecular approach. 3 Biotech 11(4):178

Schulze S, Lammers M (2020) The development of genome editing tools as powerful techniques with versatile applications in biotechnology and medicine: CRISPR/Cas9, ZnF and TALE Nucleases, RNA Interference, and Cre/loxP. ChemTexts 7(1):3

Shamshirgaran Y, Liu J, Sumer H, Verma PJ, Taheri-Ghahfarokhi A (2022) Tools for efficient genome editing; ZFN, TALEN, and CRISPR. Appl Genome Modul Ed 29–46

Sharma B, Tiwari S, Kumawat KC, Cardinale M (2023) Nano-biofertilizers as bio-emerging strategies for sustainable agriculture development: Potentiality and their limitations. Sci Total Environ 8(60):160476

Sharma K, Sharma SD, Kumar S, Darek (2020) (*Melia azedarach)-*A potential root stock for raising vigorous Neem *(Azadirachta indica)* plants. Indian J Agric Sci 90(3):659–661

Shrestha S, Subedi S, Jiban J (2020) Marker-assisted selection: a smart biotechnological strategy for modern plant breeding. Peruvian J Agron 4(3):104–120

Singh V, Roy M, Garg N, Kumar A, Arora S, Malik DS (2021b) An insight into the dermatological applications of neem: a review on traditional and modern aspect, recent advances in anti-infective drug discovery formerly. Recent Pat Anti-Infect Drug Discov 16(2):94–121

Singh A, Rajput V, Singh AK, Sengar RS, Singh RK, Minkina T. (2021) Transformation techniques and their role in crop improvements: a global scenario of GM crops. In: Policy issues in genetically modified crops. Elsevier, pp 515–542

Sohail M, Pirzada T, Opperman CH, Khan SA (2022) Recent advances in seed coating technologies: transitioning toward sustainable agriculture. Green Chem 24(16):6052–6085

Srirangarayan RS, Ramasamy M, Paulin R, Thangam, (2020) Hidden elixir in indian sub continent–*Azadirachta Indica Linn*

Srivastava V, Chaturvedi R (2018) Somatic Embryogenesis in Neem 22:87–102

Suman S (2017) Plant tissue culture: a promising tool of quality material production with special reference to micropropagation of banana. Biochem Cell Arch 17(1):23–26

Swati S, Dhaka RK, Ayate (2023) Genetic improvement in neem-a potential multipurpose tree: a review. Agric Rev 67–72

Tripathi MK, Tiwari S, Tripathi N, Tiwari G, Bhatt D, Vibhute M, Gupta N, Mishra N, Parihar P, Singh P (2021) Plant tissue culture techniques for conservation of biodiversity of some plants appropriate propagation in degraded and temperate areas. Current Topics in Agricultural Sciences BP International Publisher, Bhanjipur, pp 656–658

Vargas AM, Fernández-Pastor M, Castro FJ, Martínez MA, Gómez-Cifuentes A, Espinosa-Wahab A, Batool F, Muhammad M, Zaman W, Mikhlef RM (2023) Neem, current knowledge, research progress, and future prospects of phyto-synthesized nanoparticles interactions with food crops under induced drought stress. Sustainability 15(20):14792

Winterbottom A (2021) Becoming "Traditional" a transnational history of neem and biopiracy discourse. Osiris 36(1):262–283

Xu ML, Li C, Luan YP, Zheng Y(2021) Study on antibacterial effect of the neem extracts. Jounal Sci 2:56–68

Yang Y, Saand MA, Huang L, Abdelaal WB, Zhang J, Wu Y, Li J, Sirohi MH, Wang F (2021) Applications of multi-omics technologies for crop improvement. Front Plant Sci 12:563953

Ye X, Shrawat A, Williams E, Rivlin A, Vaghchhipawala Z, Moeller L, Kumpf J, Subbarao S, Martinell B, Armstrong C (2022) Commercial scale genetic transformation of mature seed embryo explants in maize. Front Plant Sci 13:1056190

Zeeshan A, Amjad W, Masood M, Akram W, Yameen I, Mansoor M, Hassan H, Majeed K (2024) Neem's bioactive marvels: a therapeutic review. J Health Rehabil Res 4(1):185–195

Zenda T, Liu S, Dong A, Li J, Wang Y, Liu X, Wang N, Duan H (2021) Omics-facilitated crop improvement for climate resilience and superior nutritive value. Front Plant Sci 12:774994

Zhang D, Hussain A, Manghwar H, Xie K, Xie S, Zhao S, Larkin RM, Qing P, Jin S, Ding F (2020) Genome editing with the CRISPR-Cas system: an art, ethics and global regulatory perspective. Plant Biotechnol J 18(8):1651–1669

Zhou J, Li QQ (2023) Stress responses of plants through transcriptome plasticity by mRNA alternative polyadenylation. Mol Hortic 3(1):19

Chapter 6
Opium Poppy (*Papaver somniferum* L.)

Jozef Fejér

Abstract Opium poppy (*Papaver somniferum* L.) is a multifunctional crop with many uses. It is grown legally in many countries around the world. However, it is also abused for the illegal production of drugs. The aims and methods of breeding this crop depend on the purpose of legal use. The most important breeding objectives are the yield and quality of the poppy seeds, the yield of the poppy straw and the alkaloid content. Also, marginally for ornamental purposes. This chapter explains the basic breeding objectives of the opium poppy depending on the cultivation purpose. Conventional breeding methods have brought many successes in the creation of new varieties. They are used today and will be used in the future, even if they have certain limitations. With the new findings of genetic engineering, molecular breeding methods, in vitro methods and genetic engineering are promising for the future.

Keywords Alkaloids · Breeding · Capsules · Opium poppy · Papaver · Poppy seeds · Poppy straw

6.1 Introduction

Opium poppy (*Papaver somniferum* L.) is a multifunctional plant used for the production of poppy seeds and poppy straw. It is also used as an ornamental plant (Bernáth (ed) 1998). However, its history and present are associated with both exploitation and abuse. Legal cultivation focuses on the production of poppy seeds for direct consumption and oil production. Poppy straw for the extraction of alkaloids, especially morphine, codeine, thebaine, papaverine, narcotine and others. In some Asian countries (India), opium is produced legally for pharmaceutical processing. Poppy is grown illegally to produce opium, which is used to make the dangerous drug heroin.

J. Fejér (✉)
Department of Ecology, Faculty of Humanities and Natural Sciences, University of Prešov, Prešov, Slovakia
e-mail: jozef.fejer@unipo.sk

J. M. Al-Khayri et al. (eds.), *Biodiversity and Genetic Improvement of Medicinal and Aromatic Plants II*, Advances in Plant Breeding Strategies 10,
https://doi.org/10.1007/978-3-031-81857-8_6

Poppy seeds are therefore an important functional food. In many countries of the world it is used in the form of various types of bread for direct consumption (Fig. 6.1a–d). The seeds contain a high amount of proteins 12.0–13.5 (Özcan and Atalay 2006). Mishra et al. (2013) reported up to 24% protein in poppy seeds. The oil content depends on the place of origin and color of the seeds. The poppy from India states a content of 41.4–49.1% (Singh et al. 1990), in Pakistani poppy from 47.0 to 53.0% (Azcan et al. 2004) and the Turkish poppy from 44.0 to 57.0% (Demirbas 2008; in Azcan et al. 2004). Regarding seed color Azcan et al. (2004) reported an oil content of 49.2% in yellow-seeded varieties, 36.8% in white-seeded varieties and 33.6% in blue-seeded varieties. In contrast, Hlinková et al. (2012) reported an oil content of 49.9% in the Slovakian blue-seeded variety Opal. The oil is dominated by linoleic acid with a content of 68.0–74.0% (Rahimi et al. 2015), which decrease blood cholesterol levels and has a positive effect on the treatment of cardiovascular diseases. Lančaričova et al. (2016) reported a linoleic acid content of 63.1–74.3%. The lowest content was found in the Hungarian morphine variety Buddha, the highest Czech varieties Redy and Racek with an ochre color, as well as the Slovakian white-seeded variety Albín. The difference in the content of this fatty acid between genotypes and between localities was observed. The second dominant oil acid is oleic acid. Its amount is negatively correlated with linoleic acid. Most oleic acid was found in the seeds of morphine variety Budha 19.4–23.4% and the lowest in varieties with ocher seeds and white-seeded variety 13.3–16.3% (Lančaričová et al. 2016). In connection with the consumption of poppy seeds, the risk of alkaloid content on the seed as a consequence of insect damage to the capsules or contamination with poppy dust particles at harvest is stated. The European Food Safety Authority reported the acute reference dose (ARfD) of 10 μgmorphine/kg body weight (Knutsen 2018). Carlin et al. (2020) evaluated the content of alkaloids on poppy seeds from different countries of the world. In the conclusion of this study was repored that the recommended acute reference dose is not exceeded in harvested poppy seeds and heat-treated seeds, when used in normal amounts. Heat treatment reduces the amounts of alkaloids. In some Central European countries, poppies are traditionally grown in gardens; for example, in the Slovak Republic it is possible to grow poppies without a permit on an area of 100 m^2 for the production of poppy seeds for own use (Fejér and Šalamon 2014). The matured capsules are collected by hand. The poppy seeds are poured out of the capsules by hand. In this way, minimal contamination with dust particles is achieved In this way, a high-quality and safe soft seed is obtained for consumption. Empty capsules must be disposed of.

Apart from the food industry, poppy seeds can also be used for the production of biodiesel. Demirbas (2008) mentions poppy seeds as one of the sources for biodiesel production. The oil from 2nd generation poppy seeds (*Papaver somniferum*) can be considered as one of the feedstocks that is cost effective and does not compete with food crops for biodiesel production (Bhuiya et al. 2020).

The pharmaceutical industry relies on the extraction of raw materials from opium poppy for the extraction of alkaloids. For this purpose, dry capsules with a short stem (approx. 150 mm) are usually used after the plants have matured. It is a concentrate made from poppy straw (CPS). Poppy cultivation for the production of CPS was

Fig. 6.1 **a–c** Different types of cakes with poppy seeds, **d** Round yeast cake with ground poppy seeds and sugar—a typical Christmas dish in Slovakia named "opekance" (Fig. 6.1a–d author). **e** Poppy seeds harvested with poppy straw crushing on the ground, **f** harvest of poppy seeds and poppy straw (capsules + 150 mm stems) with a special adapter, **g** a mixture of poppy seeds and poppy straw, **h** storage of poppy seed and poppy straw until separation (Fig. 6.1e–f author). Different types of flowers of ornamental varieties and capsules (Fig. 6.1i–l author)

adopted worldwide after the invention of the Hungarian pharmacist János Kabay in 1928 (Bernáth (ed) 1998). He developed a new method for extracting morphine from dry capsules with a short stem. Compared to the complexity of opium production, this method is cost-efficient.

Raw opium, which is legally obtained in some countries, as already mentioned, e.g. in India, is a source for alkaloid extraction. It is obtained from green immature capsules by making an incision of 1–2 mm in the epidermal wall of the capsule (Mishra et al. 2013). Sharma et al. (2002) reported that India is the only country where opium is legally produced for alkaloid extraction. In other countries where poppy is cultivated, the alkaloids are extracted from the CPS after the seeds are removed.

Ornamental varieties play a subordinate role in the importance of this plant. These are varieties intended for ornamental flower beds and have different colors and shapes of inflorescences (Fig. 6.1i–l). Dry capsules with stems in various shapes and sizes (Fig. 6.1k–l) are used for arranging bouquets.

6.1.1 Botanical Classification and Distribution

The genus Poppy (*Papaver* L.) has more than 100 described species, which are divided into several categories—sections Vašák et al. (2010). The basis of the division is formed by morphological features and the area of distribution. The new taxonomic division classifies opium poppy into the order *Ranunculales.*

The Taxonomic classification of opium poppy according to the Integrated Taxonomic Information System is as follows (www.itis.gov):

Kingdom	*Plantae*—plantes, Planta, Vegetal, plants
Subkingdom	*Viridiplantae*—green plants
Infrakingdom	*Streptophyta*—land plants
Superdivision	*Embryophyta*
Division	*Tracheophyta*—vascular plants, tracheophytes
Subdivision	*Spermatophytina*—spermatophytes, seed plants, phanérogames
Class	*Magnoliopsida*
Superorder	*Ranunculanae*
Order	*Ranunculales*
Family	*Papaveraceae*—poppies
Genus	*Papaver* L.—poppy
Species	*Papaver somniferum* L.—opium poppy

According to Vašák (2010), the genus poppy (*Papaver* L.) is divided into the following sections: *Rhoeadium* section, *Argemonidium* section, *Papaver* section, *Glauca* section, *Meconidium* section, *Pilosa* section, *Macrantha* section, *Meconella* section, *Californicum* section, *Carinatae* section, *Horrida* section. Opium poppy belongs to section Papaver. The literature mentions the existence of two subspecies of opium poppy. Ellis et al. (1985) described two subspecies within the species *Papaver somniferum* L.

- Oilseed Poppy *Papaver somniferum* L. subsp. *Hortense,*
- Opium Poppy *Papaver somniferum* L. subsp. *Somniferum.*

However, the fact that these are two subspecies has not been precisely proven and described. More detailed research is needed to confirm the existence of poppy subtypes or ecotypes.

The poppy is a self-pollinating plant or, optionally, an out-pollinating plant. Maturation of the stamens and pollen and subsequent pollination occurs is in the closed bud at the time of its erection. After erection, the scar is pollinated, but after flowering it is able to take pollen. Thanks to pollinators, especially bees, there is also foreign-pollination, which is reported to varying degrees. Self-pollination is the most important pollination and the proportion of out-pollination is estimated at 10–30% (Schreier and Zájeda 1994; Bechyně et al. 2001). Under favorable weather conditions there is 15–35% share of foreign-pollination, while the share of colonies ranging from 15 to 25% (Voškeruša (ed) 1965). The wind contribution on pollination is only 0–10%. Little is known about the biology of poppy pollination, but by the study of poppy pollination, wind pollination is negligible (Miller et al. 2005). Foreign - pollination is mainly due to insects. Thus, self-pollination of poppies predominates. The degree of foreign-pollination is mainly influenced by the conditions at flowering time—humidity, temperature and the presence of pollinators.

6.1.2 Economic Importance

The economic importance of poppy stems mainly from cultivation for the production of alkaloids, poppy seeds for direct consumption and oil production. To a lesser extent for decorative purposes. The dominant poppy alkaloids, especially morphine, are used in the manufacture of drugs. They are made from poppy straw (or opium) and cannot be produced synthetically (Pienkny et al. 2009).

Poppy seed is a popular delicacy, especially among the Slavic population and the population of neighboring Central European countries, such as Austria, Romania, Hungary and Germany. In Central and Eastern Europe, as far as we know, poppy mixed with sugar and/or another ingredient is used as a filling for pastries, strudels, cakes, etc. In other European countries and North America, poppy is used mainly in the bakery industry as a pastry, mainly for decoration (Ministry of Agriculture of the Czech Republic). Vašák (2010) reported the use of poppy seeds as a delicacy in other countries such as Turkey, Pakistan, Afghanistan, Iraq, Australia, New Zealand, South Africa, Togo, India, Malaysia. Greagh (2018) states that the global poppy market is set to grow over the next 10 years. In 2016, 111,000 tons of poppy were produced. In terms of value, the global poppy market had total wholesale revenues of 183.4 million dollars' in 2016 (Greagh 2018). The largest consumer of poppy seeds is India. It is mainly white poppy seeds. Annual consumption is estimated at around 25,000 tons. However, per capita consumption in India is only 20 g. This corresponds to around five poppy capsules. Central and Eastern European countries probably have the highest poppy consumption. In the Czech Republic, for example, poppy consumption is up to 400 g per person per year. In addition, the Czech Republic is also the largest grower of poppy seeds for food purposes. (Ministry of Agriculture of the Czech Republic).

6.1.3 Domestication, Selection and Early Improvements

Poppy seeds are a plant already known in the Neolithic by the inhabitants of wheeled buildings in Switzerland (Robenhausen), Italy (Lagozza) and in the Bronze Age in France (Bourget). It is not entirely clear whether this is a poppy. There could be used a weed species. The poppy seeds found in Robenhausen were medium in size. They can be classified as a weed and cultural poppy species in size. Cultural poppy has a five- to hexagonal structure on the surface of the seeds. This is not arranged in exact rows. The findings from wheeled buildings in Switzerland are similar to cultural poppies, according to this structure. They have larger angles than the weed poppy (*Papaver rhoeas*), five to hexagonal, with well-defined squares. Their size is the same as for *Papaver somniferum*. However, the seeds are smaller and there are fewer fields in the network on the surface. These clues, together with the proximity to the eight lobes, show that the poppy from the excavations is not entirely identical to today's cultural poppy. It was apparently a poppy integrating characters of several species. It is already extinct and is probably the ancestors of *Papaver somniferum* and *Papaver setigerum* (Sinskaja 1973). Hajnalová and Hajnalova (2001) also anticipate the arrival of poppy (*P. somniferum*) from southwestern Europe, whose grandparent *P. setigerum* is known from archaeobotanical finds of the Middle and Small Neolithic from the western part of the Mediterranean coast, Germany and Switzerland. Recent studies on the domestication and early diffusion of opium poppy using small botanical remains from archaeological sites have been conducted by Salavert et al. (2020). By using radiocarbon chronology have been shown that the age of these remains from the central and western Mediterranean, northwestern temperate Europe and the western Alps ranges from 5622 to 4050 BC. The results show that opium poppy is present from at least the middle of the sixth millennium in the Mediterranean, where it possibly grew naturally and was cultivated by pioneer Neolithic communities. Its dispersal outside of its native area was early, being found west of the Rhine in 5300–5200 BCE. It was introduced to the western Alps around 5000–4800 BCE, becoming widespread from the second half of the fifth millennium. The probability of domestication of poppy seeds in the western Mediterranean is also reported by Jesus et al. (2021), where his supposed wild ancestor occurs, from where it then spread to Central and Northern Europe. In a case study with a Neolithic poppy seed set from a locality in Switzerland (Zurich-Parkhaus Opéra, ca. 3170 BC), he indicates the presence of mixed populations of domestic and wild seeds belonging to the *P. somniferum* group, suggesting that the plant was already late at the end of the 4th millennium BC in the process of domestication. They used elliptic Fourier transforms, with other morphometric descriptors, to describe and identify of the seeds *Papaver setigerum, Papaver somniferum*, and other *Papaver* taxa (Fig. 6.2a).

Makovnyka (2020) concludes that future research, specifically anthropological field research grounded in historical and sociopolitical contexts, can offer important insights into the lived experiences of individuals and cultures that produce, distribute, and consume the poppy as food and medicine. Future research may offer critical insight into the relationship between the cultural constructs of food and medicine and the effects of narcotic substance consumption and also offer insight into the possible restructuring of cultural meanings and economies in order to mitigate the harmful effects of narcotic substances within foods.

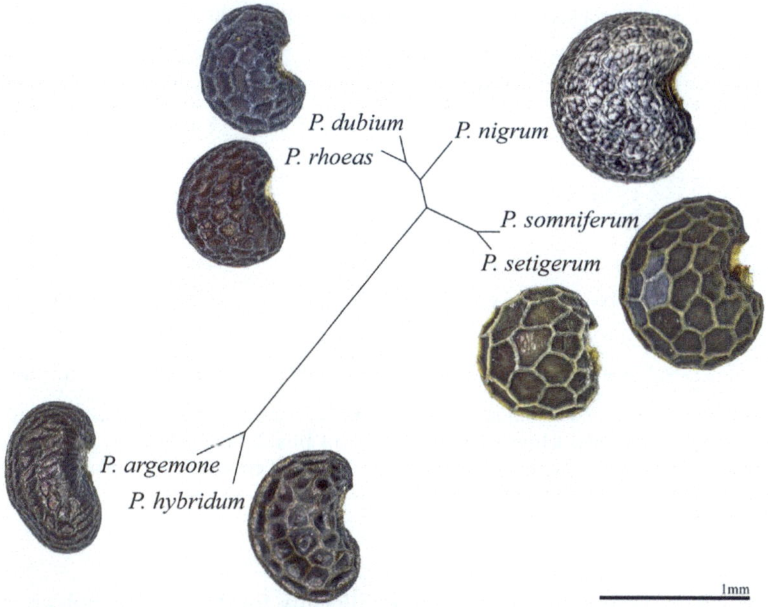

Fig. 6.2 The unrooted tree obtained with hierarchical clustering on the Euclidean distance matrix between Fourier coefficients averaged per taxa. *Source* Jesus et al. (2021)

6.2 Current Cultivation Practices and Challenges

6.2.1 Current Cultivation Practices

Opium poppy (*Papaver somniferum* L.) are grown as a cultivated plant almost all over the world. In various parts of Europe, North Africa, Australia, Japan, South and North America (Kapoor 1995). Poppy cultivated areas for legal alkaloid production are given in Table 6.1 (INCB 2020). Agriculture Organization of the United Nations (FAO) has listed 15 poppy seed growing countries (Austria, Bulgaria, Croatia, Czech Republic, France, Germany, Hungary, Netherlands, North Macedonia, Palestine, Romania, Serbia, Slovakia, Spain, Turkey).

Poppies are also grown in other countries that are listed in the FAO statistics. These include, for example, China, Great Britain, Ukraine and Poland. Information on the production of poppy seeds in these countries is not generally available. This is a by-product of the production of poppy straw for pharmaceutical purposes. It is reported (FAOSTAT) that the average area under cultivation in the last ten years was 115 thousand hectares. In individual years, it fluctuated between 72,000 ha and almost 150,000 ha. These statistics only show the legal cultivation areas. Illegal areas for the production of narcotics are not included here (FAOSTAT 2020; in MACZ 2020). The

Table 6.1 Total actual area harvested with opium poppy 2016–2020

Country	Total actual area harvested (ha)				
	2016	2017	2018	2019	2020[a]
Australia*	14 053	8 620	12 455	7 450	14 059
France*	9 475	8 054	6 359	7 541	n/a
Hungary**	3 540	2 023	514	3 100	n/a
India***	557	8 721	4 710	6 107	4 308
Slovakia***	1 928	1 790	1 604	3 500	3 023
Spain*	9 505	5 655	5 685	9 452	10 921
Turkey***	29 921	23 731	45 123	56 511	35 249
Total	68 979	58 594	76 450	93 661	67 560

A Figures for area sown and actual area harvested in 2020 are based on advance data submitted by Governments to the International Narcotics Control Board; "n/a" indicates that data are not yet available; "*" Opiumpoppy rich in morphine, thebaineand codeine; "**" Opium poppy rich in morphine and thebaine; "***" Opium poppy rich in morphine
Source INCB (2020)

countries where the opium poppy is grown illegally for drug production are shown in Fig. 6.3.

Different cultivation methods are used in the cultivation of poppy seeds. It is grown on small areas and large-scale cultivation technologies are also developed. The Czech Republic played an important role in the development of large-scale cultivation technology. All agro-technical operations from sowing to harvest are fully mechanized. Control of weeds, diseases and pests is with the use of registered pesticides (Vašák 2010; Fejér 2014).

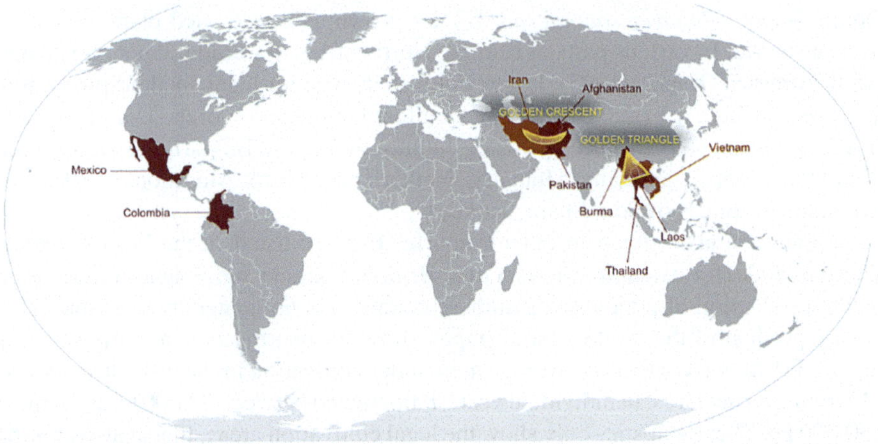

Fig. 6.3 Areas of illicit opium poppy cultivation and opium production. *Source* https://en.wikipedia.org/wiki/Golden_Crescent

6.2.2 Current Cultivation Challenges

In recent years, there has been a problem with the optimal establishment of stands in the spring due to the drought (in the middle Europe countries). Growers react to this fact by modifying the technology. They are looking for technologies that significantly save soil moisture. The land is leveled already in the fall, and in the spring only sowing is done with minimal soil movement. As an alternative, spring poppy varieties are sown in autumn. It is carried out in countries and localities there are a mild winter, but despite the choice of localities, the stands often freeze. The second alternative can be the sowing of winter varieties of the poppy, which have high winter resistance. However, the problem with winter poppies grown in Europe is their inferior taste properties of poppy seeds compared to spring poppies (food varieties). Cultivation technologies are being developed that will ensure good soil coverage and thereby improve moisture management. It is mainly about growing covering crops together with poppy. Luhmer et al. (2021) explains that poppy as an intercrop is a suitable strategy to minimize crop losses.

The method of harvesting is important in poppy growing technology. For industrial varieties, it is important to collect capsules with a stem up to 150 mm (Fig. 6.1f). Even when harvesting food varieties, it is advisable to collect whole pods. This reduces harvesting losses and damage to the seed. The mass can be dried as needed and later separated (Fig. 6.1g, h). For this reason, it is necessary to improve the harvesting technique.

6.3 Germplasm Biodiversity and Conservation

6.3.1 Germplasm Diversity

Khanna and Singh (1975; in Mishra et al. 2013) evaluated 190 poppy genotypes from Russia, Hungary, Poland, the United Kingdom and other temperate countries. They concluded that most cultivars belonging to European countries required a long photoperiod and therefore were unsuitable for Indian climatic conditions. Evaluation of the genetic source of poppy originating in India have been found that this genotype had a short photoperiod (Fejér 2015). This genotype had a short period of vegetative growth after emergence and quickly entered the generative phase. The plants hardly formed side branches at all. There were presented one, at most two, capsules. The stigmatic disc had a small number of stigmatic rays. It germinated 10–14 days earlier compared to Central European poppies. While the Indian genotype was in the green maturity phase, the Central European genotypes were in the pre-flowering phase (Fig. 6.4a). The theory of the existence of two ecotypes within the poppy species arises, which differ from each other in the length of the photoperiodic reaction.

Sequence analyzes can be used to assess genetic diversity—genotyping. Such genotyping using sequencing was performed by Hong et al. (2022). He evaluated 91

Fig. 6.4 Front genetic resource from India, back genotypes from Central Europe (Slovakia), (Fig. 6.4 Author)

poppy germplasm samples from 30 countries from Europe, North Africa, America and Asia. He identified five genetically distinct subpopulations using discriminant principal component analysis and STRUCTURE analysis.

6.3.2 Cultivars Characterization and Phylogeny

In the first half of the twentieth century, traditional regional populations were mainly cultivated. Poppy varieties were scarcely known (Bernáth and Németh 2009). The science-based breeding began e.g. in Hungary in 1930 (Bernáth and Németh 2009). Fejér (2014) reported the beginning of poppy breeding in the Slovak Republic (part of the former Czechoslovakia) since 1948. Some varieties emerged from the collected regional gene pool, such as Dunajský, Galantský, Váhovecký, which were not registered. The first officially registered variety was the Blankyt in 1967. Bernáth and Németh (2009) states the varietal basis according to growing areas:

- Areas with highly developed industrial production, such as Western Europe, Australia, use patented strains without variety registration by the variety authority. Detailed information is lacking if these strains or varieties were bred directly by pharmaceutical companies. The varieties are homogeneous and bred by different methods.
- In other Central European countries, new varieties are registered by national testing authorities. Candidate varieties are evaluated by these authorities according to the current DUS guideline, which assesses distinctness, uniformity and stability. In the case of opium poppy, the evaluation of overloading is carried out according to the TG 166/4 test guideline (www.upov.int). Basic information is available on

most varieties. In the case of industrial varieties, the propagating material is practically distributed by the processing company to their growers. The best varieties can be patented.

In the Common European Catalog of varieties of agricultural plant species 2020 (www.uksup.sk) there are 65 registered poppy varieties. These are varieties intended for industrial and culinary use. Németh and Bernáth (2007) stated that in Hungary the limit value for morphine in dry capsules has been 0.20% since 2010. Varieties above this limit are considered industrial. Varieties with a morphine content below 0.20% are used for food purposes. A maximum morphine value of 0.01% in dry capsules is permitted in Germany. In the Czech Republic, the morphine content limit is 0.80%. Varieties of Czech and Slovak provenance are grown, which are able to accumulate 0.30–0.70% morphine in dry capsules (eagri.cz). These types of varieties provide a high yield of quality poppy seed for consumption. In low morphine varieties as well as varieties with a high alkaloid content, the seed yield decreases and its quality is also worse. This was confirmed by research in 2008–2010 by Vlk et al. (2011), which evaluated 30 low, medium and high morphine varieties, registered in the Common European Catalog (Fig. 6.5).

The yield potential of poppy seeds of the Slovak varieties ranges from 1.77 to 1.90 t ha^{-1} (Fejér et al. 2021).

Registration of poppy varieties is also carried out in Turkey. Güçlü et al. (2014) conducted research on the genetic characterization of commercial poppy varieties. He states that in the work were used 17 registered Turkish poppy varieties used for commercial alkaloid production.

In India, there have been developed new varieties in the National Botanical Research Institute, Lucknow and the Central Institute of Medicinal and Aromatic Plants, Lucknow (Mishra et al. 2013). There are several specific varieties intended

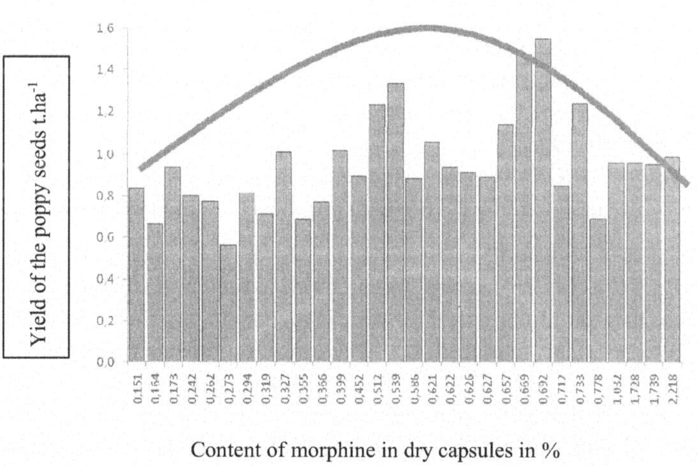

Fig. 6.5 Dependence of the poppy seeds yield from morphine content in dry capsules. *Source* Vlk et al. (2011)

for the production of opium, poppy seed or combined use. Thebaine alkaloid accumulating lines are also introduced. Opium less, alkaloid less and non-narcotic variety Sujata are also mentioned. Several other universities and research institutes in India have bred disease-resistant, frost-resistant or lodging-resistant varieties. A major research effort has been made at CSIR CIMAP, Lucknow, in this direction and many cultivars of *P. somniferum* have been developed. Various breeding approaches like mass selection ('Shweta', 'Shyama', 'Sampada', 'Shubhra'), half-sib selection ('Sanchita'), induced mutation breeding ('Vivek', 'Sujata') and selection in intra-specific hybrids for diverse resistance ('Rakshit') have been used for this (Patra and Kumar 2005). Morphine-rich CPS cultivar 'CIMAP Ajay' has been released that was developed through recombination breeding approach (Lal et al. 2011).

6.3.3 Genetic Resources Conservation Approaches

The study of plant genetic resources is essential for the preservation and use of existing variability, which is one of the means to increase the genetic potential of new varieties (Benediková et al. 2009). Morphological markers are an essential tool for characterizing genetic resources. They have been used in breeding for a long time. It is an evaluation of visible and measurable features. In essence, these markers generally represent genetic polymorphisms which are easily identified and manipulated. Some of these markers are linked with other agronomic traits and thus can be used as indirect selection criteria in practical breeding (Jiang 2013). Different types of descriptor list are used to characterize poppy genetic resources. They represent a list of descriptors (characters). For each descriptor, there are levels of expression to capture the variability of the genetic resource being evaluated. Faberová (2000) divided descriptor list as follows:

- *Descriptors list* used to describe genetic resource (GR) collections. These are used to characterize GR within the collection, to record the diversity of GR—they include all categories of descriptors. It distinguishes two levels—international and national descriptor list,
- *Test guidelines* UPOV. They are used to confirm the distinctness, uniformity and stability of the new varieties. The main emphasis is on characterization.

The use of these types of descriptor list is different. Although they evaluate the same trait, they use different scales and methodologies in the evaluation.

Descriptor lists should be a tool for standardized characterization of plant genetic resources. International standards should be considered when developing descriptor lists (Bioversity International 2007). Bioversity International is responsible for the development of databases for plant genetic resources. It plays an important role in the exchange of data and in the development of cooperation between institutions involved in the conservation of genetic resources. The prerequisite for the use of these resources is the quality of the information about them. Their description of economically significant traits is the basis for their utilization (Gotor and Cherfas 2010).

In the former Czechoslovakia there was developed the poppy descriptors list by Martínek (1971). In the Czech Republic there was developed the national descriptors list by Havel et al. (2008), which contains 40 descriptors. Dittbrenner et al. (2008) reported the use of a 34-descriptor lists that was developed to standardize characterization and also includes data such as date of sowing, flowering etc., which are essentially phenological data. In the Slovak Republic a manual for the evaluation of genetic resources of poppy was created by Fejér (2015), which has 59 descriptors. Lal et al. (2014) stated that descriptors list should be comprehensive for descriptors that contain and should be used when useful to the curator for managing and maintaining the collection and/or for users of poppy genetic resources. The poppy descriptor list proposed by Lal et al. (2014) includes, in addition to descriptive descriptors, passport data, soil and climatic characteristics of the experimental site where the assessment is performed together with agro-technical measures. Abiotic and biotic stresses or biochemical characteristics are considered too.

Opium poppy are conserved ex situ in gene banks in seed collections. When storing the seeds of the genetic resources, passport and descriptive databases are created, which characterize individual genotypes.

6.3.4 Cytogenetics

Papaver somniferum is diploid (n = 22) while *Papaver setigerum* is tetraploid (n = 44). No other species of *Papaver* is known to have the number n = 11 or multiple of 11. This indicates that *P. setigerum* cannot be the ancestoral species of *P. somniferum* since a diploid would not be expected to be derived from a tetraploid (Farnilo et al. 1953). Morphine content has been confirmed in these two species (Farmilo et al. 1953; Garnock-Jones and Scholes 1990). Malik et al. (1979) demonstrated that hybridization and subsequent genes change between fertile triploids is possible between these two subspecies. A study of the karyological characteristics of *P. somniferum* and *P. bracteatum* was performed by Rezaei et al. (2014). All studied populations were diploid and the basic number of chromosomes was n = 7 for *P. bracteatum* and n = 11 for *P. somniferum*. Component Analysis (PCA) showed that *P. somniferum* populations separated from *P. bracteatum* populations.

6.4 Traditional Breeding

Traditional breeding procedures can include the use of classical methods such as selection, crossing and subsequent evaluation of offspring with the selection of desired genotypes. Genetic diversity is needed for plant breeding as an essential resource for improving plant varieties (Gepts and Papa 2002). The basis of success is the provision of wide variability, which can be ensured by obtaining germplasm from different geographical areas. It is the introduction of different plant genotypes, with diverse genetic bases, which represent a source for creating variability. It follows that the study of genetic diversity is an important prerequisite for the success of conventional breeding. Obtaining available information on morphological, biological, economic and biochemical characteristics is used to select the desired genotypes in breeding programs. The collected collections of poppy genetic resources in gene banks, research institutions or universities are mainly used for this purpose. The study and obtaining of genetic resources for breeding work, especially with regard to specific varieties—industrial varieties with a high content of alkaloids, accumulating in addition to morphine other alkaloids (thebaine and others) is difficult. Such varieties are bred directly by pharmaceutical companies and in most cases are legally protected and inaccessible. Kapoor (1995) stated that in Eastern European countries, success has been achieved using conventional breeding methods. Success has been achieved with increased seed yield, morphine yields as well as another alkaloid composition.

6.4.1 Improvement Strategies and Goals

The breeding strategy is based on the use of poppy, which is described in the introduction. Bernáth and Németh (1999) based on the UN Convention on Narcotics described three different types in the creation and registration of poppy varieties:

1. Varieties accumulating high morphine content, or specially varieties with another alkaloid composition, which accumulate thebaine, codeine, narcotine, noscapine, papaverine and others,
2. Varieties for the poppy seed and oil production with a low or no alkaloid accumulation,
3. Ornamental types with a special flower or capsule shape with limited alkaloid accumulation capacity.

In the case of industrial varieties intended for the production of alkaloids, the breeding of varieties for the production of CPS as well as varieties with a high content of alkaloids in opium is actual. Breeding of poppy seed varieties with a medium morphine content in dry capsules (up to 0.80%) is successful in Central Europe (Slovakia, Czech Republic, Hungary, etc.). These types of varieties provide a high seed yield with very good quality.

6.4.1.1 Current Goals of Breeding

The current poppy breeding objectives are based on a specific type of variety.

Increasing the poppy seed yield is a basic goal in the breeding of food types of varieties. The seed yield is given by the number of plants per unit area, the number of capsules per plant and the weight of seeds per plant. The number of plants is regulated by agro-technical measures. The number of capsules is determined by the genotype. Martínek et al. (1975) stated the heritability coefficient for the number of capsules per plant of 40–80%. This feature is greatly influenced by stand density and environmental conditions. The weight of the seed of the plant is determined by the number of capsules. The weight of the seed in one capsule is determined by the number of seeds and the size of the seeds. The number of seeds in the capsule ranges from 2000 to 12,000 (Voškeruša (ed) 1965) and depends on the number and size resp. areas of fruiting bodies in the capsule. The weight of the seed is positively correlated with the area of the fruit leafs, as well as with the number of rays in the stigma. The number of rays basically indicates the number of fruits leaves in the capsule. Hlaváčková (1962) founded a positive correlation between the number of stigmatic rays and the weight of the seeds per plant. A positive correlation was also founded between seed yield per plant and plant height, number of the leaves, number and weight of the capsules per plant (Yadav and Singh 2011). The heritability coefficient of 30–80% is given for the weight of the seeds from one plant (Hlaváčková 1978). Seed size is characterized by a measurable parameter weight of thousand seeds (WTS). It is an inherited trait and largely influenced by environmental conditions. Poppy WTS ranges from 0.3 to 0.7 g. Kara (2017) stated the difference in WTS depending on the date of sowing. During the sowing in the autumn, he found poppy WTS 0.38–0.40 g. During the spring sowing, the poppies had 0.27–0.34 g WTS. The heritability coefficient for this trait is 30–40% (Martínek et al. 1975). The size and shape of the capsule determine the economical use of the capsule space and good seed placement. The most suitable shape is one which allows the sun's rays to fall on as large a surface area of the capsule as possible. The optimal shape is considered to be spherical or one that approaches a round shape. The variety has a statistically significant effect on the seed weight of the capsule and WTS (Matyášová 2010).

Seed color breeding. The color of the seed is a characteristic varietal feature. It is to some extent influenced by the ecological conditions of the environment as well as the degree of ripeness of the seeds. The most common and desired color is blue. In addition to blue, the varieties are white-seeded, with gray and brown colors. There are several color variations and shades. There are also a black, pink or purple color of the seeds. Matyášová (2010) presented a statistically significant influence of variety on seed color. Poppy seed color breeding is therefore necessary to focus on the selection of the desired color, which is best evaluated on a black background.

Breeding for poppy seed quality. According to Popovec (1999), the oil content in the seed is a hereditary property based on polygen and influenced by ecological environmental conditions. The influence of environmental conditions on the oil content has been confirmed by several studies (Lančaričová et al. 2016). It is known that the content of oil and the composition of its fatty acids affect the taste properties of poppy seeds. The total oil content and its composition in relation to the effect on taste properties have been proven by many partial works and research (Fejér 2007; Lančaričová et al. 2016). The white and ocher color varieties had a higher oil content than the blue varieties. In addition to the oil content, the content of the dominant linoleic acid and oleic acid, for which a negative correlation has been found, probably also has an effect on the taste properties (Lančaričová et al. 2016).

Capsules yield is one of the goals in breeding industrial varieties for poppy straw production for alkaloid extraction. As with seeds yield, the yield of the capsules depends on the number of plants per unit area and the number of capsules per plant. The third indicator is the weight of the capsules from one plant. As stated at seed yield, the number of capsules is determined by genotype. The heritability coefficient of this trait is reported to be 40–80%. In contrast, at a capsule weight per plant, a low heritability of 15–32% was found (Martínek et al. 1975). It is significantly influenced by environmental conditions and agricultural technology. A larger number of stigmatic rays as well as a larger surface area are needed for maximum pollination and higher poppy straw yield (Kumar and Patra 2010).

Breeding for high content of alkaloids is the main goal of breeding industrial types of varieties for pharmaceutical purposes. There are basically two types of varieties—varieties for poppy straw production (Europe, Australia) and varieties for opium production (Asian countries). Since the 1950s, attention has focused mainly on increasing the content of the dominant alkaloid morphine in poppy straw. Its content was gradually increased to 1.0% and later to 2.0% and more (Bernáth and Németh 2009). The Tasmanian alkaloids (2004) lists the top 10 poppy growers who have achieved a morphine content from 2.95 to 3.46%, which represents a yield of 30.15–31.81 kg ha^{-1}. The content of thebaine ranged from 2.49 to 3.10%, which represented a yield from 30.99 to 34.38 kg ha^{-1}. Raw opium is also a raw material for the pharmaceutical industry. Mishra et al. (2013) stated that, based on demand, scientists and breeders have accepted the call for the creation of varieties with high opium yields. As with poppy yield, the number of stigmatic rays and their area affect the alkaloid content (Kumar and Patra 2010). Khanna a Singh (1975; in Mishra et al. 2013) investigated the germplasm of the genetic resources of the Indian poppy. The correlation analysis confirmed that morphine is synthesized from codeine and less from oripavine. The absence of narcotine led to a decrease in total alkaloid content. The accumulation of morphine was reduced by the influence of codeinone and morphinone.

The study by Pal et al. (2021) notes that knowledge of the effect of epistasis on different traits is important for setting breeding goals in opium poppy. Greater additive genetic variation (D) compared to dominance (H) for all traits confirmed that this practice, when selected for, represents a significant step towards increasing opium poppy yield. Considerable emphasis should be placed on traits such as plant height, number of days to maturity, number of branches per plant, number of capsules per plant, opium yield per plant, morphine content, narcotine content and papaverine content in opium poppy pre-selection. In addition, the heterosis could be effective to achieve maximum yield potential (Pal et al. 2021). This study could be a milestone for the development of high-yield latex varieties rich in specific alkaloids.

Biparental mating followed by recurrent selection involving the desired recombinants can be used to improve the components of the quantitative traits. This is stated by Kumar and Patra (2010), who analyzed the gene interactions involved in the inheritance of quantitative traits, namely plant height, number of branches per plant, number of capsules per plant, stem length, capsule index, number of stigmatic rays, yield of poppy straw per plant and content of morphine.

The trend in breeding of poppy varieties for seeds production is the creation of varieties with a low content of alkaloids or without alkaloids, which are described as safe. Low alkaloid content is also required in ornamental poppy varieties. Meos et al. (2017) evaluated in experiments 22 samples from the United Kingdom using an internet page (http://chilternseeds.co.uk/) and 12 samples of ornamental poppy from the mall Prisma in Tartu, Estonia. He found a relatively high morphine content in the range of 152–676 mg. 100^{-1} g in dry capsules, which is a remarkably high concentration allowing the potential illegal use of ornamental poppies.

Breeding for Disease Resistant. During the growing season, opium poppy is attacked by several diseases that cause a reduction in yield and affect the accumulation of alkaloids. Downy Mildew (*Peronospora arborescens* (Berkeley) de Bary) can be included among the most serious poppy diseases. In Australia, the disease *Peronospora somniferi* was identified based on morphological and molecular analyzes (Thangavel et al. 2017). Another dangerous disease is Leaf Spots (*Helminthosporium papaveris* (Hennig)), which also attacks capsules and seeds. Powdery Mildew (*Erysiphe communis* (Wallr.) Link., syn. *Erysiphe polygoni*), Rot disease (*Rhizoctonia solani* Kühn). Seed borne diseases—Leaf Blight (causal agent—*Pleospora calvescens*), Seedling Blight (causal agents—*Phytium ultimum* and *Phytium mammimatum*), Wilt & Root Rot (causal agent—*Fusarium semitectum*). Diseases caused by bacteria and diseases caused by viruses are also dangerous. When breeding for disease resistance, the selection of resistant plant strains is primarily used. Increased resistance to Downy Mildew by 60–70% was found during in vitro mutagenesis (see Sect. 3.3).

Breeding for growing conditions. Here we could include plant height, resistance to lodging and abiotic stress as basic parameters of varieties. Poppies are not specially bred at plant height. Kumar and Patra (2010) stated that lower plants are suitable for Indian conditions. Bajpai et al. (2000) found out by research on Indian germplasm of poppy that plants with lower high achieved high morphine yields, had larger capsules, not ribbed with a small amount of the seeds. A specific category is the resistance of poppy plants to pesticides, especially herbicides. This can be achieved by applying permitted poppy herbicides from early breeding generations, i.e. from the F1 generation with subsequent selection of resistant lines. This ensures a resistant material suitable for large-scale production growing conditions (Vašák 2010; Fejér 2014).

Resistance to undesirable opening of capsules—prevent creating of openings on corona under stigma from which seed can drop out during the maturation stage. It is ensured by negative selection—excluding plants with this undesirable feature from the breeding process.

6.4.2 Traditional Breeding Methodologies

6.4.2.1 Crossing Method

Crossbreeding is one of the oldest conventional methods used in plant breeding. Bernáth (ed) (1998) stated that hybridization had been used in poppy breeding since the 1940s. This method requires information on the available gene pool—varieties, landraces populations, lines, which are the basis for the selection of parental pairs to cross according to the breeding goals. We can say that this is one of the basic methods of extending the variability of breeding material. The crossing can be divided into four separate steps (Rod 1982):

1. castration of mother plants,
2. isolation of castrated plants,
3. preparation of pollen of paternal plants,
4. own pollination.

The castration of the mother plants must be done at the crocheting stage (Fig. 6.6a). It is performed using small shears. At the base of the bud, a cut is made around the stem. This cut separates calyx leaves, crown flakes and stamens. The flower shells together with the stamens are removed from the testicle—the future capsule (Fig. 6.6a and c). After castration, the plant is isolated so that no insects—pollinators—get to the testicle. Various cellophane or paper bags with holes are used (Fig. 6.6d).

Fig. 6.6 **a** Castration start phase. **b** Cutting of calyx flakes, crown flakes and stamens. **c** Pull of flower shells and stamens from the testicle. **d** Isolation of castrated plant (Fig. 6.6a–d author). **e** The bud suitable for preparing pollen from the paternal plant, **f** the ripening stamens on the paper, **g** pollination with paternal pollen using a brush, **h** marking of the crossed plant (Fig. 6.6e–h author)

Castration of mother plants should be done in the morning. Buds are collected from the paternal plants at the stage when they are upright (afternoon) (Fig. 6.6e). As with castration (Fig. 6.6a, c), the flower wrappers are removed and the flower sticks are poured on the paper (Fig. 6.6f). They let them mature until the next day. The pollen is then poured into a bottle and the mother plants are pollinated with a brush (Fig. 6.6g). After pollination, the capsules are re-isolated. This process will be repeated the next day. After the scar has dried, when it is no longer able to absorb pollen, the isolator is removed and the crossed plant is marked (Fig. 6.6h).

Cleavage of the desired genotype with the desired gene recombination is often only random. This deficiency can be remedied to some extent by backcrossing. The fundamentals of the backcrossing lie in the fact that the F1 generation crosses with one of the parents who is the donor of the desired trait, with a parent who has good economic characteristics (usually it is a transfer of genes such as resistance to diseases, etc.). It follows from the laws of genetics that a heterozygous hybrid in one trait forms all gametes in a 1: 1 ratio. The dominant allele is 50%, similar to the 50% recessive allele. If the hybrid of the F1 generation crosses with the recessive parent, a 1: 1 splitting ratio is achieved in the generation, and thus a certain trait that was suppressed during the first crossing can be recovered. The backcross can be one-time or repeated. Generally, 5–6 backcrosses are used, which is usually sufficient to transfer the character of the recurrent parent (Rod 1982; Boháč et al. 1990).

6.4.2.2 Selection Method

Selection is the basic breeding method. Deliberate selection is most often used, which is a targeted selection of the desired plants (according to breeding goals). It can be one-time or recurring selection. Natural selection represents the influence of external environmental factors on plants. This can also be used in breeding work. The selection can be positive, i.e. the best genotypes (lines) are selected or negative, where inappropriate genotypes (lines) are removed. Individual selection is applied, i.e. selection of individual plants for subsequent establishment of new progeny from selected plants of so-called stem mothers (SM). Mass selection, in which a larger number of plants (individuals) is selected from a given population and the establishment of a new generation from a seed mixture of these plants (Chloupek 2008; Rod 1982). According to Acquaah (2007), the genetic progress achieved by selection depends on three factors:

- the total variation (phenotypic) in the population in which selection will be conducted,
- heritability of the target character,
- the selection pressure to be imposed by the plant breeder.

In breeding work, the individual selection of plants is preferred. It is important to evaluate the progeny of selected individual plants and to ensure self-pollination. For this reason, it is necessary to isolate the flowers in order to exclude allogamous pollination (Bernáth and Németh 2009).

The pedigree selection method is most often used. According to Levy and Milo (1998), to increase the yield of capsules, opium and seeds, the pedigree method is effective in combining and stabilizing these traits. Selection for the desired traits and characteristics of the genotypes with their subsequent stabilization takes 3–4 generations. Voškeruša (ed) (1965) reported achieves stabilization of hybrid generations after crossbreeding by pedigree selection in 4–6 generation. In these generations, any heterogeneous effect will be eliminated. It will be possible to verify the actual and stabilized performance of the individual plants in the performance tests. In the first three to five years, it is necessary to ensure the isolation of individual plants for homozygotisation. It is best to insulate the main bud of the plant. Isolation is recommended on 5–8 selected progeny plants. Thus, after isolation in the F_1–F_3 generation, according to Voškeruša (ed) (1965) decomposes the starting material into individual morphologically and genetically different lines. This is followed by the selection of individual plants and the evaluation of their progeny according to the set breeding goals. The basis of the future variety may be a single strain or a mixture of strains (4–10) with identical morphological and economic characteristics.

When evaluating individual plants for quantitative traits such as poppy seed yield, basic mathematical statistical quantities are used—arithmetic mean and standard deviation.

$$\overline{x} = \frac{SX}{n}$$

Arithmetic mean ($\overline{x}$): SX = sum of measured values; n = number of evaluated plants

$$s = \pm\sqrt{\frac{Sx^2}{n}}$$

Standard deviation (s): Sx^2 = mean error of the mean; n = number of evaluated plants.

Parental identity zones include selected material identical to the better parent (P^+) or worse parent (P^-) in the selected trait. This is true for selection in hybrid generations of the poppy breeding process, but this procedure is also applicable to the evaluation of a set of selected plants when assessing their yield. Identity zones are calculated as follows:

$$Z1 = \overline{x} + \delta;\ Z2 = \overline{x} + 2\delta;\ Z3 = \overline{x} + 3\delta \text{ or } Z1 = \overline{x} + s;\ Z2 = \overline{x} + 2s;\ Z3 = \overline{x} + 3s \tag{6.1}$$

Thus, after adding the standard deviation to the average one, two or three times, plants with a higher yield compared to a better parent are selected, or a comparison can be made to the average of the evaluated group (Fejér et al. 2018). This fact must be verified in the progeny of these plants.

In selection, morphological markers are commonly used as an auxiliary tool to select plants with desired traits. Their use in plant breeding started a long time ago. The markers used include mainly visible traits such as leaf shape, flower color, pod color, seed color, seed shape and so on. These morphological markers generally represent genetic polymorphisms that are easy to identify and manipulate. Some of these markers are also associated with some agronomic traits and can therefore be used as indirect selection criteria in practical breeding.

However, available morphological markers are limited and many of these markers are not associated with important economic traits (Jiang 2013). In order to evaluate morphological traits, several descriptors list or test guideline are used, which are described in Sect. 3.3.

Sowing of the trials and poppy breeding trials at the rosette stage are shown in Fig. 6.7a and b. In the picture Fig. 6.7c are shows poppy breeding trial plots at CIMAP India (obtained during the First International Symposium on *Papaver* in the year 2011).

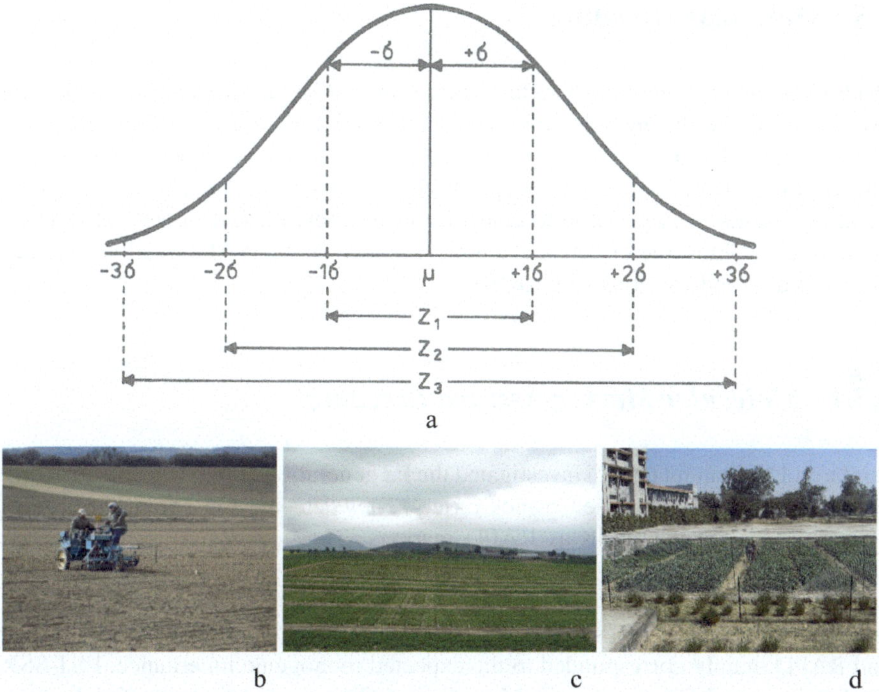

Fig. 6.7 **a** Zonal analyses—derivation of zones (Boháč et al. 1990). **b** Sowing of the poppy trials with small plot seed machine, **c** poppy breeding trials at the rosette stage (Malý Šariš, Slovakia), **d** poppy breeding trials (CIMAP, India), (Fig. 6.7a–c author)

6.4.3 *Traditional Breeding Limitations*

Several successes have been achieved in the breeding of some medicinal plants. However, these species are characterized by certain peculiarities, so that many questions remain unanswered when it comes to defining breeding objectives. At the same time, it must be borne in mind that there are certain limitations to the use of traditional and modern methods when breeding these plants (Wang et al. 2020). Traditional poppy breeding practices have their limits, which stem from the genetic basis and the limited possibilities of conventional methods. These are mainly limited possibilities in terms of changing the alkaloid profile of the poppy plant. For example, Nyman and Hall (1976) obtained by selection a spontaneous mutant that contained a high amount of thebaine. After several years of efforts to stabilize and create the thebaine variety, the attempts failed. The populations were characterized by chromosomal instability, and a large proportion of each new population returned to the morphine type.

6.5 Molecular Breeding

Traditional or conventional breeding uses morphological characteristics that are externally visible during selection. These are an expression of genetic diversity. Modern methods can use molecular methods based on changes in the DNA to characterize their effects on the phenotype. With the help of markers, which are given by allelic forms, it is possible to characterize an individual, a tissue, cells, a cell nucleus, a chromosome or a gene. Genetic markers can be used to facilitate studies of inheritance and variation (XU 2010).

6.5.1 *Molecular Marker-Assisted Breeding*

Straka and Nothangel (2002) investigated the F2 generation of poppies after crossing a low-morphine line with the Cosmos variety. They detected 125 polymorphic fragments, 77 as AFLP and 48 as RAPD markers. They evaluated six morphological markers (plant height, number of stigma rays, petal color, peduncle hairiness, capsule shape, and dehiscence of capsules). Most of the traits could be assign to two phenotypic classes. The observed segregation ratios for morphological traits, AFLP and RAPD mainly corresponded to the expected monogenic inheritance. EST-SSR markers are used for diversity and breeding studies in opium poppy (Selale et al. 2013). The evaluation of regional varieties and cultivars of Turkish poppy using SSR and AFPL analyzes was addressed by Celik et al. (2016). It was found that one SSR and three AFLP loci were responsible for morphine content. Other markers (six SSR and 14 AFLP loci) were significantly associated with economically important traits such as plant height, number of stigma rays, capsule index, seed and straw yield. These results are useful for the selection process in poppy breeding.

Verma et al. (2016) also report on the use of AFLP markers in poppy breeding for the creation of industrial varieties. Using the AFLP method, he studied the genetic diversity of opium poppy germplasm originating from different regions of India. Based on this method, he selected lines suitable for breeding a new variety.

Short Tandem Repeat (STR) markers are also used to characterize opium poppy genotypes. They are derived from Expressed Sequence Tags (ESTs), which are obtained from DNA. Their use for commercial purposes is described by Mičianová et al. (2017). It is possible to use them to create varieties for food purposes for seed production and also varieties for pharmaceutical processing of alkaloids. These markers are suitable for forensic PCR analysis.

The RAPD method was used to evaluate the genetic diversity of the hybrid lines and their parents. Polymorphism was detected by 12 primers at 84.8%. The detected genetic differences between parents and hybrids can be used in combining the desired genes to increase the production parameters of poppy (Khatik et al. 2017).

Molecular methods are also used to evaluate the genetic differentiation of varieties and the origin of varieties. Svoboda et al. (2020) reported the use of SSR length polymorphism to distinguish blue seeded and non-blue seeded varieties. In this way, the 'Czech blue poppy' varieties were distinguished from the blue seeded varieties of foreign origin. With these markers it is possible to verify the origin of genotypes when checking goods on the market. Distinguish high-quality blue poppies from inferior ones.

The use of STR markers with potential application in forensic plant medicine is described by Vašek et al. (2021). The Sanger method was used for sequencing. It was used to propose a standardized nomenclature and create three allelic ladders. With this system, pharmaceutical varieties can be distinguished from food, dual or ornamental poppy varieties.

The methodology for the description of the genetic diversity of poppy genetic resources using microsatellite analysis and the IRAP method was developed by Jozová et al. (2020). These methods not only enable the description and characterization of genetic resources, but also find their application in plant breeding, seed production (evaluation of seed purity and authenticity) and food industry (verification of the authenticity of "Czech poppy").

6.5.2 Functional Genomics and Bioinformatics

The DNA code of the opium poppy genome is currently known. Key steps have been identified in how the plant has evolved to produce pharmaceutical compounds used to make vital medicines. Firstly, reported the reference of opium poppy genome Guo et al (2018), which has a large complex genome with over 70% of repetitive sequences. The final assembled genome size is 2.72 Gb, with annotation of 51,213 protein-coding genes and 9494 non-coding RNAs (Hu et al. 2018).

Pei et al. (2021) reported the de novo assembly and genomic analysis of *P. somniferum* traditional landrace 'Chinese Herbal Medicine'. In order to better understand the evolutionary history that led to a massive genome expansion, he compared the genome of the traditional opium poppy CHM with that of other members of the *Papaveraceae* family. He identified specific genetic variations (SNPs and InDels). The knowledge gained forms the basis for the identification of allelic variations and genes. This can contribute to the introgression and improvement of the genetic base of commercial poppy varieties.

6.6 Tissue Culture Applications

Thanks to the totipotency of the cell, its regeneration and the formation of the entire organism can be initiated from a living plant cell isolated from the tissue. During the regeneration of a cell into a complete organism, certain changes occur at the genetic level and genetic variability arises. The cultivation and regeneration of somatic embryos takes place under stress conditions, which also contribute to variability (Xu 2010).

6.6.1 Micropropagation Approaches

The procedure for induction of somatic embryos in opium poppy is described by Nessler (1982). A callus was obtained from the hypocotyl of the seedling plants. The callus was cultivated in two steps. First, it was cultivated on a solid medium. The meristems were then differentiated in the liquid medium, from which the growth regulators were removed. Cellular origin of shoot buds, cell morphogenesis and differentiation during direct and indirect shoot regeneration of *Papaver somniferum* L. in vitro were studied by Ovečka et al. (2000). In immature somatic embryos, it induced shoot formation by direct organogenesis. In the subepidermal and epidermal layers of the hypocotyl, cell division and the formation of protomeristems occurred. Indirect regeneration was initiated from the callus culture. Auxins and cytokinins were used in this process. These supported the formation of compact globular meristemoids. Tisserat and Berhow (2009) states that poppy shoot cultures (from *P. bracteatum* Lindl., and *P. somniferum* L.) readily manifest pharmaceutical alkaloids in vitro. Alkaloid production was greatly influenced by temperature with optimum alkaloid production occurring at 18.5 and 20 °C. Thebaine was the dominant alkaloid found in shoot cultures for all poppy cvs. grown in vitro. In vitro poppy cultivation methodologies are currently available that are useful in poppy breeding. Klíma et al. (2014) developed a methodology for in vitro cultivation of opium poppy in order to obtain homozygous materials for the needs of breeding work. It describes the process of creating dihaploid poppy lines of the winter and spring type from the initial F1 hybrids, from the preparation of donor plants, the establishment of wormwood cultures, regeneration of whole plants to the transfer of regenerants to non-sterile conditions and their multiplication.

6.7 Genetic Engineering

Belny et al. (1997) reported the development of a transformation of opium poppy (*Papaver somniferum*) with *Agrobacterium tumefaciens*. For the cultivation, a cell suspension culture was used together with *A. tumefaciens* GV3101(pMP90), which is a binary vector carrier. *Papaver somniferum* calluses transformed with the sam-1 gene from *Arabidopsis thaliana*, which encodes a SAM-synthetase, were subcultured by Nabha et al. (1999).

A protocol for the transformation of poppy seeds for pharmaceutical purposes was developed by Chitty et al. (2003). This protocol allows, for the first time, the transformation of this plant to obtain highly productive varieties that can be used commercially. The method is based on the infection of explants obtained from the hypocotyl of plants with *A. tumefaciens*. An embryogenic callus is formed on the infected explants. This is resistant to antibiotics and herbicides. In the next phase, the formation of somatic embryos and plants from the callus was induced. The resulting transgenic plants were tested under field conditions to estimate the frequency of hybridization. The transgenic plants were sown at different distances with non-transgenic and related *P. somniferum spp. setigerum* and *P. dubium*. The frequency of crossing decreased significantly with distance. At a distance of 0.2 m, hybridization was 2.6%. At a distance of 2.5–5.0 m, the hybridization frequency was zero. The use of *Agrobacterium tumefaciens* for poppy transformation is reported by Fachini et al. (2008). The pCAMBIA3301 vector was used. This vector carries the phosphinothricin acetyltransferase (pat) gene. The gene transformation protocol developed makes it possible to produce fertile plants that are resistant to herbicides.

The genetic transformation was carried out by Larkin et al. (2007). It was performed with a constitutively expressed cDNA of codeinone reductase (PsCor1.1). This step is the penultimate step in the biosynthesis of morphine. In experiments (both in the greenhouse and in the field), the transgenic lines obtained yielded a higher amount of alkaloids in the capsules. The increase in morphinan alkaloids was 15–30% compared to the non-transgenic genotypes.

6.8 Mutation Breeding

Mutation induction is one of the ways to spread variability and thus obtain basic materials for breeding work. Mutant lines can be homogenized and stabilized by selection in the desired direction or used in a crossing process. Advances in mutational breeding, irradiation techniques, the use of chemo-mutagenesis in seed-propagated and vegetative propagated plants are elaborated in detail in the Manual on Mutation Breeding. This manual provides a comprehensive overview and procedures for phenotypic and genotypic screening methods. These are currently available and are used to detect specific and valuable mutant traits. They also provide techniques for increasing breeding efficiency through mutagenesis (FAO/IAEA 2018).

6.8.1 Conventional Mutagenesis, Enhanced Traits and Improved Cultivars

The use of poppy mutagenesis is considered by Bernáth (ed) (1998) in making changes to the alkaloid profile of plants. Desgagne—Penix et al. (2009) stated that mutagenesis is a powerful approach for the creation of new traits and the identification of new genes. In the case of opium poppy, mutants generated using chemicals or ionizing radiation can be screened for altered alkaloid profiles using high-throughput analytical methods. Alternative measures, such as transcript profiling techniques must be used to link a specific gene with an observed phenotype.

Based on the research of Nyman and Hall (1976), who found instability of spontaneous mutants with higher thebain content, in Tasmanian Alkaloids in Tasmania, they chose mutation breeding in the creation of a high thebain poppy variety. The underlying hypothesis for the breeding of such a variety was that the biosynthetic pathway would be interrupted, resulting in the accumulation of thebaine, rather than the biosynthesis continuing to produce morphine. EMS (Ethyl Methane Sulfonate) was used as the mutagen. The work resulted in the selection of a plant in the M2 generation which was named 'Norman' (Fist 2001). The strategy was used to isolate the top1 mutant ('Norman') with a high thebaine/oripavine content and a low morphine/codeine content from a chemically (EMS) mutagenized population (Millgat et al. 2004). The mutagenesis procedure is patented in the USA. The success of the use of radiomutagenesis (gamma rays), chemomutagenesis (EMS) and their combinations thereof is reported by Chatterjee et al. (2012a). He selected mutant lines with a high thebaine content. Another study found the effect of different doses of gamma radiation in combination with EMS on the content of morphine, thebaine, codeine and narcotine. A dose of kR10 + 0.4% EMS produced positive mutations for high thebaine and codeine and low morphine. At a dose of kR40 + 0.6% EMS, a high narcotine content and a low morphine content was found by Chatterjee et al. (2012b).

Mutation breeding has received much attention in India. Chopra (2005) reported that several universities and research institutes are involved in mutagenesis of agricultural crops in India. By 1990, the number of mutant varieties developed in India was 205. By 2004, the number had grown to 313. Mutation Breeding Newsletter (2003) listed several mutant varieties of poppy—'BC-28/9/4—Vivek', 'Shweta', 'Shyama', 'Sanchita'. The Mutant Variety Database (https://nucleus.iaea.org) lists one mutant poppy variety 'BC-28/9/4-Vivek' registered in 1992.

Lal et al. (2020) summarizes the success of mutational breeding of medicinal and aromatic crops, including poppy seeds at the Central Institute of Medicinal and Aromatic Plants Lucknow India. He lists the 'Sujata' poppy variety as opium less, very low alkaloids, nonnarcotic oil-seed crop. Sharma et al. (1999) described the induction of a mutation performed by gamma rays (^{60}Co) at a dose of 200 Gy in combination with a 0.4% aqueous solution of EMS (Ethyl Methane Sulfonate) by treating poppy seeds. The best non-narcotic mutant line LL-34, named Sujata, was selected. It is characterized by a high potential of poppy seed and oil yield. The variety is patented in the United States.

Naseem and Kumar (2013) demonstrated the use of EMS in a mutation breeding program in poppy to obtain paternal sterile lines.

6.8.2 In Vitro Mutagenesis and Selection

In a study by Mathur et al (1993), a mutation was induced in vitro under two varieties, Shweta and Shyama, using gamma radiation and EMS. Of the various mutagenic doses, the combined dose of 50 Gy + 0.4% EMS was most effective in creating viable, visible and micro mutations. Evaluation of these mutants over 4–5 generations in terms of yield attributes and response to *Peronospora arborescens* disease revealed that BC and M26/100/2 micromutant were the best in morphine yield, while PCL, DM and M8/102/4 micromutant experienced a 60–70% increased level of resistance to downy mildew. By combining radiomutagenesis and chemomutagenesis using 10kR (kilorad) + 0.4% EMS, a high thebaine poppy plant was selected. Evaluation of M2 generation in vitro conditions revealed a high content of thebaine in the stem, followed by leaf callus, stem callus and cotyledons (Chaterjee et al. 2010). Various germplasms have been produced using different breeding methods and other methods, including molecular methods. This is a change in the modulated biosynthesis and accumulation of alkaloids other than the usual ones. Nowadays, there is a need to modulate the biosynthetic pathways of alkaloids using transgenic methods (Misra and Saema 2016). In opium poppy, induction of embryogenic callus formation from root explants in the presence of 2,4-D and kinetin is reported. This initiated somatic embryogenesis. It was only possible from the root callus. The amount of alkaloids was determined both in the callus and in the plants grown under field conditions (Pathak et al. 2012).

6.9 Hybridization

6.9.1 Conventional Hybridization

The conventional cross-hybridization procedure to extend variability is described in Sect. 4.2.1. This procedure can be used in hybrid poppy breeding in the production of hybrid poppy varieties. However, it is very time consuming and labor intensive. It is more appropriate to use pollen sterility of paternal plants, which can be achieved, for example, by chemo-mutagenesis (Nasem and Kumar 2013). Charles Darwin had already conducted experiments in 1877 to show that crosses of related strains did not achieve heterosis, i.e. increase the yield of F1 generation compared to parents. By crossing two unrelated cultivars within one species, it is possible to achieve a heterosis effect of either 50% (Xu 2010). A heterogeneous effect was observed in the poppy seeds yield as well as in the alkaloid content. Németh-Zámbori et al. (2011) evaluated the alkaloid profiles of F1–F3 progenies. The effects of genetic regulation were demonstrated by these crossing experiments. These are enzymatic processes at three levels involved in the biosynthesis of alkaloids. Experiments have shown that selection can already be effective in the F2 generation to fix a low narcotine content. Selection before the F3 generation is not suitable for the morphine content. Yazici and Yilmaz (2021) report an increase in the basic alkaloids morphine, thebaine, noscapine, codeine, oripavine and papaverine in hybrids compared to parents. In a study performed by Yazici and Yilmaz (2020), 36 hybrids, offspring of 9 parents, were evaluated. He found a significant heterosis effect in the F1 generation in alkaloid content compared to the middle parent. These crosses can be evaluated as suitable for breeding studies. Dogramaci and Arslan (2019) studied the heterosis effect in opium poppy. He found heterosis effect of capsule yield in terms from − 56.76 to 95.81%, seed yield in terms from − 57.40 to 89.29%, morphine content in terms from − 13.21 to 15.70%, morphine yield in terms from − 58.21 to 95.53% showed variation. General and specific combining ability and heterosis were studied by Singh and Pandey (2011) for morphine content, opium yield plant^{-1}, seed yield plant^{-1}, capsule weight plant^{-1}, capsule size, capsule plant^{-1}, plant height and days to 50% flowering. Genetic variability plays an important role in the inheritance of all evaluated traits (additive and non-additive). In view of this fact, the system of selective diallel mating may prove unsuitable to produce an improved variety due to the resulting variability. Selection for certain desirable traits will not be effective. Non-additive genetic variance, which was greater for almost all traits.

Hybridization of cultivars and novel genotypes with specific characteristics can propitiates higher heterotic values in the segregating population (Shukla et al. 2016).

6.9.2 Hybrid Cultivars

In Slovakia the result of long-term breeding effort was also a distant hybrid MK—131 = *Papaver caucasicum* x MK 11-7-52 (regional Slovak genotype). It stood out in extreme earliness, small size and suitability for mechanized harvest. However, it reached low yield of seed, so its growing did not expand. combination crossbreeding of the varieties Hanácky modrý x Dětěnický bělosemenný resulted in creation of Czechoslovak Hybrid HD. It reached about 15–20% higher seed yield than line varieties. Its growing did not expand due to high costs of seed production (Fejér 2014). Lal et al. (2011) developed Hybrid CIMAP Ajay as part of a hybridization program at CIMAP (Central Institute of Medicinal and Aromatic Plants, Lucknow India). This hybrid offers to enrich the alternative providing an economically most viable high morphine in concentrate (obtained from poppy straw) variety of opium poppy. The top highest yielder hybrid has maintained its superiority over check for high seed yield, with high alkaloids contents. It also offers the initial base (parental material) for next poppy breeding.

6.10 Conclusion and Future Prospects

Poppy (*Papaver somniferum* L.) is an economically important plant. Its economic importance results from the production of the raw material (concentrate from poppy straw or opium) for the production of alkaloids, which cannot be produced synthetically. In addition, the poppy is a source of poppy seeds used in the food industry for direct consumption and oil production. The seeds are a promising raw material for the production of biodiesel. The opium poppy is therefore a promising crop for the future.

Global climate change is affecting the cultivation of all types of crops, including poppies. In the course of these changes, the poppy is attacked by various abiotic and biotic stress factors. This places increased demands on breeding work to increase resistance to these stress factors. This is related to the need to breed winter poppy varieties, which is currently the case in Central Europe. In addition to breeding suitable varieties, it is important for breeders to research cultivation technologies that focus on water management.

Conventional poppy breeding methods and techniques have led to success in the form of new varieties in many countries around the world. In the future, it can be assumed that molecular methods, tissue culture techniques and genetic engineering will be increasingly used in poppy breeding. This will primarily involve the breeding of stress resistance, the modification of alkaloid profiles or the fatty acid content of the oil. The creation of hybrid varieties could be used to a greater extent in the future in breeding for seed yield using the heterozygosity effect.

References

Acquaah G (2007) Principles of Plant Genetics and Breeding. Blackwell Publishing Ltd. 569

Azcan N, Kalender BO, Kara M (2004) Investigation of Turkish poppy seeds and seed oils. Chem Nat Compd 40(4):370–372

Bhuiya MMK, Rasula MG, Khana MMK, Ashwath N (2020) Biodiesel production and characterisation of poppy (*Papaver somniferum* L.) seed oil methyl ester as a source of 2nd generation biodiesel feedstock. Indus Crops Prod 152:112493. https://doi.org/10.1016/j.indcrop.2020.112493

Bajpai S, Gupta AP, Gupta MM, Sharma S, Govil CM, Kumar S (2000) Inter-relation between descriptors and morphine yield in Asian germplasm of opium poppy *Papaver somniferum*. Genetic Resources and Evolution 47:315–322

Bioversity International (2007) Guidelines for the development of crop descriptor lists. Bioversity Technical Bulletin Series No. 13. Bioversity International, Rome, Italy

Bechyně M (ed) (2001) Mák [Poppy]. Agrospoj Praha [Agrospoj Prague], [In the Czech language]

Belny M, Herouart D, Thomasset B, David H, Jacquin-Dubreuil A, David A (1997) Transformation of Papaver somniferum cell suspension cultures with saml from *A. thaliana* results in cell lines of different S- adenosyl-L-methionine synthetase activity. Physiologia Plantarium 99:233–240

Benediková D, Benková M, Drobná J, Čičová I, Hauptvogel P, Hauptvogel R, Mendel L, Pastirčák M, Beluský J (2009) Ochrana a využitie agrobiodiverzity pre výživu a poľnohospodárstvo – princípy, legislatíva, dokumentácia [Protection and use of agrobiodiversity for food and agriculture—principles, legislation, documentation] Piešťany, november 2009, [In the Slovak language]

Bernáth J (ed) (1998) The Genus Papaver. Harwood academic publishers, Amsterdam

Bernáth J, Németh E (1999) New trends in selection of poppy (*Papaver somniferum* L.). Int J Hort Sci 5:69–75

Bernáth J (2005) Possibilities and limitation of poppy cultivation in Middle-European countries in accordance with UN Convention on narcotics. In: 11th International Congress of Polish Herbal Committie. Poznan, June 24–25, 2005

Bernáth J, Németh E (2009) Poppy. In: Vollman J, Rajcan I (eds) Hanbook of Plant Breeding. Springer, Oil Crops, pp 459–478

Boháč J, Andonov I, Čermín Ľ, Vlk J (1990) Šľachtenie rastlín [Plant Breeding]. Príroda Bratislava [Príroda Publishing House Bratislava], [In the Slovak and Czech language]

Carlin MG, Dean JR, Ames JM (2020) Opium Alkaloids in Harvested and Thermally Processed Poppy Seeds. Front Chem 8:737. https://doi.org/10.3389/fchem.2020.00737

Celik I, Camci H, Kose A, Kosar FC, Doganlar S, Frary A (2016) Molecular genetic diversity and association mapping of morphine content and agronomic traits in Turkish opium poppy (*Papaver somniferum*) germplasm. Mol Breeding 36:46. https://doi.org/10.1007/s11032-016-0469-8

Chaterjee A, Shukla S, Mishra P, Rastogi A, Singh SP (2010) Prospects of in vitro production of thebain in opium poppy (*Papaver somniferum* L.). Industrial Crops and Products 32: 668–670. https://doi.org/10.1016/j.indcrop.2010.04.007

Chaterjee A, Shukla S, Rastogi A, Mishra BK, Singh SP (2012a) Induction of Variability and Selection of Elite Lines for Specific Alkaloids from a Population Induced Mutagenesis in Opium Poppy (*Papaver somniferum* L.). Bioremediation, Biodiversity and Bioavailability 6: (Special Issue 1) 94–102

Chaterjee A, Shukla S, Mishra BK, Rastogi A, Singh SP (2012b) Induction of variability through mutagenesis in opium poppy (*Papaver somniferum* L.). Turk J Agric for 36:1–11. https://doi.org/10.3906/tar-1007-977

Chitty JA, Allen RS, Fist AJ, Larkin PL (2003) Genetic transformation in commercial Tasmanian cultivars of opium poppy, *Papaver somniferum*, and movement of transgenicpollen in the field. Functional Plant Biol 30:1045–1058

Chloupek O (2008) Genetická diverzita, šlechtení a semenářství [Genetic diversity, breeding and seed production] Academia Praha [Academia Prague], 308. [In the Czech language]

Chopra VL (2005) Mutagenesis: Investigating the process and processing the outcome for crop improvement. Curr Sci 89(2):353–359

Common Catalogue of Varieties of Agricultural Plant Species (2020) 351–352

Creagh B (2018) Global consumption of poppy seed set to continue increasing in next decade. Food&Beverade (https://www.foodmag.com.au/poppy-seed-global-consumption/)

Demibras A (2008) Biodiesel. A realistic fuel alternative for diesel engines. Springer, London. https://doi.org/10.1007/978-1-84628-995-8

Desgagne-Penix I, Hagel JM, Facchini PJ (2009) Mutagenesis as a functional genomics platform for pharmaceutical alkaloid biosynthetic gene discovery in opium poppy. In: Shu QY (ed) Induced plant mutations in the genomics era. Food and Agriculture Organization of the United Nations, Rome, pp 411–418

Dittbrenner A, Lohwasser U, Mock HP, Börner A (2008) Molecular and phytochemical studies of *Papaver somniferum* in the context of intraspecific classification. Acta Hort 799:81–88

Dogramaci S, Arslan N (2019) he Effect of Heterosis on Yield Components of Opium (*Papaver somniferum* L.) Cultivars and Their Reciprocal Crosses. Ekin J. 5(2):84–91

Knutsen HK (ed) (2018) Update of the scientific opinion on opium alkaloids in poppy seeds. EFSA J. 16(5): 5243 https://doi.org/10.2903/j.efsa.2018.5243

Ellis RH, Hong TD, Roberts EH (1985) Handbook of Seed Technology for Genebanks. Vol. II. International Board for Plant Genetic Resources, Rome 521–522.

Faberová I (2000) Klasifikátor jako základní nástroj hodnocení genetických zdrojů rostlin [The descriptor list as a basic tool for evaluating plant genetic resources]. In: Hodnotenie genetických zdrojov rastlín [Evaluation of plant genetic resources]. Zborník referátov a posterov zo seminára, Piešťany, 29.2–1.3. 2000 [Proceedings of papers and posters from the seminar, Piešťany]. [In the Czech language]. 48–59

Facchini PJ (2008) Natalia N and Blanche V (2008) Genetic transformation via somatic embryogenesis to establish herbicide-resistant opium poppy. Plant Cell Rep 27:719–727. https://doi.org/10.1007/s00299-007-0483-8

Farmilo CG, Rhodes HLJ, Hart HRL, Taylor H (1953) Detection of Morphine in *Papaver Setigerum* DC. Bull Narc 5(1):26–31

FAO/IAEA (2018). Manual on Mutation Breeding - Third edition. Spencer-Lopes M.M Forster BP, Jankuloski L (ed), Food and Agriculture Organization of the United Nations. Rome, Italy 301

Fejér J (2007) Nové trendy a perspektívy šľachtenia maku siateho (*Papaver somniferum* L.) [New trends and perspectives of poppy breeding (*Papaver somniferum* L.)] In: Osivo a sadba: VIII. odborný a vědecký seminář, sborník referátů. Praha, ČZU [In: Seed and seedling: VIII. professional and scientific seminar, Proceedings, Prague], [In the Slovak language] 122–128

Fejér J (2014) Breeding aspects of poppy in Slovakia. Acta Hort 1036:101–105

Fejér J, Šalamon I (2014) Poppy (*Papaver somniferum* L.) as a special crop in the Slovakian history and culture. Acta Hort 1036:119–121

Fejér J (2015) Morfologicko - biologická diverzita druhu mak siaty (*Papaver somniferum* L.) a jej hodnotenie [Morphological and biological diversity of opium poppy species (*Papaver somniferum* L.) and its evaluation] 1. vyd. Prešov, Prešovská univerzita v Prešove, Fakulta humanitných a prírodných vied [1st ed. Prešov, University of Prešov in Prešov, Faculty of Humanities and Natural Sciences], 115. [In the Slovak language]

Fejér J, Havrlentová M, Brezinová B (2018) Variabilita krajových populácii maku siateho (*Papaver somniferum* L.) a potenciál ich využitia pre rozšírenie genetickej diverzity pri tvorbe nových odrôd [Variability of regional poppy (*Papaver somniferum* L.) populations and the potential of their use for the expansion of genetic diversity in the creation of new varieties]. 107. [In the Slovak language]

Fejér J, Nastišin Ľ, Majdanová J (2021) Results of Oilseed Poppy Breeding and Production Potential of New Varieties. Biol. Life Sci. Forum 3:14. https://doi.org/10.3390/IECAG2021-09681

Fist AJ (2001) The Tasmanian poppy industry: a case study of the application of science and technology. 'Proceedings of the Australian Agronomy Conference 2001'. Australian Society of Agronomy. (http://www.regional.org.au/au/asa/2001/plenery/1/fist.htm)

Garnock PJ, Scholes P (1990) Alkaloid content of *Papaver somniferum* subsp. *setigerum* from New Zealand. NZ J Bot 28:367–369

Gepts P, Papa R (2002) Evolution during Domestication. Encyclopedia of Life Science 5.

Gotor E, Cherfas J (2010) The history and usefulness of Bioversity International's descriptor lists. Bioversity International Series of Impact Assessment Briefs, no. 3. Bioversity International, 4: 1235

Guo L, Winzer T, Yang X, Li Y, Ning Z, He Z, Teodor R, Lu Y, Bowser TA, Graham IA and Ye K (2018) The opium poppy genome and morphinan production. 55. (https://eprints.whiterose.ac.uk/). https://doi.org/10.1126/science.aat4096

Güçlü GB, Gürkök T, Koyuncu M, Arslan N, Parmaksız İ (2014) Genetic Characterization of Turkish Commercial Opium Poppy (*Papaver somniferum* L.) Cultivars Using ISSR and SSR Markers. Journal of New Results in Science 7:48–57

Hajnalová E, Hajnalová M (2001) Archeológia a história pestovania rastlín na Slovensku. [Archeology and history of plant cultivation in Slovakia] Genofond [Genepool], 5: 8–10. [In the Slovak language]

Havel J, Hájková M. Koprna R, Faberová I. Vrbovský V (2008) Klasifikátor Descriptor list *Papaver somniferum* L. Rada genetických zdrojů kulturních rostlin, „Genetické zdroje č. 96“, [Council of genetic resources of cultural plants, "Genetic resources no. 96"] VÚRV, v.v.i. Praha, OSEVA PRO s.r.o., o.z. VÚO Opava, 15. [In the Czech language]

Hlaváčková Z (1962) Korelace šlechtitelsky důležitých znaků u odrůd máku [Correlation of breeding important features in poppy varieties]. Rostlinná Výroba [crop Production] 8:379–388 [In the Czech language]

Hlaváčková Z (1978) Aplikace tří a šestiparametrového testu při genetické analyze hmotnosti semen na rostliny a výšky rostliny [Application of three and six-parameter test in genetic analysis of seed weight on plants and plant height]. Genetika a Šlechtění [genetics and Breeding] 14(2):153–160 [In the Czech language]

Hlinková A, Havrlentová M, Šupová J, Bednárová A (2012) Journal of Microbiology, Biotechnology and Food Sciences 1: (February Special issue) 908–922.

Hong UVT, Tamiru-Oli M, Hurgobin B, Okey CHR, Abreu AR Lewsey MG (2022) Insights into opium poppy (*Papaver* spp.) genetic diversity from genotyping-by-sequencing analysis. Scientific Reports (2022) 12:11. https://doi.org/10.1038/s41598-021-04056-3

Hu Y, Zhao R, Xu P, Jiao Y (2018) The Genome of Opium Poppy Reveals Evolutionary History of Morphinan Pathway. Genomics Proteomics Bioinformatics 16:460–462. https://doi.org/10.1016/j.gpb.2018.09.002

INCB (International Narcotics Control Board) (2020) Supply of opiate raw materials and demand for opiates for medical and scientific purposes. Narcotic Drugs 2020, 125–134. https://www.incb.org/documents/Narcotic-Drugs/Technical-Publications/2020/8_NAR_ 2020 Part_III_Supply_and_Demand_E.pdf

Jesus A, Bonhomme V, Evin A, Ivorra S, Soteras R, Salavert A (2021) Antolín F Bouby L (2021) A morphometric approach to track opium poppy domestication. Sci Rep 11:9778. https://doi.org/10.1038/s41598-021-88964-4

Jiang G-L (2013) Molecular Markers and Marker-Assisted Breeding in Plants, Chapter 3 In: Anderse, SB (ed.) (2013). Plant breeding from laboratories to fields. InTech. 45–83. https://doi.org/10.5772/3362

Jozová E Stará M, Horáček J, Ludvíková M Čurn V (2020) Metodika pro genotypizaci genetických zdrojů máku setého (*Papaver somniferum* L.) pomocí SSR a IRAP marker [Methodology for genotyping genetic resources of poppy (*Papaver somniferum* L.) using SSR and IRAP marker]. Jihočeská univerzita v Českých Budějovicích [University of South Bohemia in České Budějovice], 42. [In the Czech language]

Kapoor LD (1995) Opium poppy. Botany, Chemistry and Pharmacology. The Haworth Press, Inc. New York 326

Kara N (2017) The effects of autumn and spring sowing on yield, oil and morphine contents in the Turkish poppy (*Papaver somniferum* L.) cultivars. Turk J Field Crops 22(1):39–46. https://doi.org/10.17557/tjfc.301829

Khanna KR, Singh UP (1975) Genetic effects of irradiation in opium poppy. In: Proceeding 1st Indian congress of cytology and genetics, Chandigarh, India 274

Khatik CL, Sharma SP, Maloo SR, Dodiya NS, Joshi A. Jain RK (2017) Diversity analysis among opium poppy (Papaver somniferum L.) crosses and parents using RAPD. Electron J Plant Breed 8(1):342–346. https://doi.org/10.5958/0975-928X.2017.00051.5

Klíma M, Vyvadilová M and Kučera V (2014) Metodika *in vitro* kultivace máku setého a produkce homozygotních materiálů pro šlechtitelské využití [Methodology of *in vitro* cultivation of opium poppy and production of homozygous materials for breeding use], VÚRV, Praha-Ruzyně [Crop Production Research Institute, Prague-Ruzyně], 22. [In the Czech language]

Kumar B, Patra NK (2010) Genetic analysis of capsule and its associated economic traits in opium poppy (*Papaver somniferum* L.). J Hered 101(5):657–660. https://doi.org/10.1093/jhered/esq043

Lal RK, Chandra R, Misra HO, Shankar K, Singh AK, Dhawan OP, Gupta AK, Kumar B, Gupta MM, Verma RK, Alam AM, Singh SK, Kalra A, Singh HP, Chauhan HS, Singh HN (2011) Registration of a high yielding variety "CIMAP Ajay" morphine rich in concentrate of poppy straw of opium poppy (*Papaver somniferum*). J Med Aromat Plant Sci 33(2):204–208

Lal RK, Sarkar S, Singh S, Gupta P, Zaim M (2014) Inheritance pattern and conservative genetics study tool: descriptors on opium poppy (*Papaver somniferum* L.). Acta Hortic 1036:71–86. https://doi.org/10.17660/ActaHortic.2014.1036.7

Lal RK, Chanotiya ChS, Gupta P (2020) Induced mutation breeding for qualitative and quantitative traits and varietal development in medicinal and aromatic crops at CSIR-CIMAP, Lucknow (India): past and recent accomplishment. Int J Radiat Biol 96(12):1513–1527. https://doi.org/10.1080/09553002.2020.1834161

Lančaričová A, Havrlentová M, Muchová D, Bednárová A (2016) Oil content and fatty acids composition of poppy seeds cultivated in two localities of Slovakia. Agriculture 62(1):19–27. https://doi.org/10.1515/agri-2016-0003

Larkin PJ, Miller JAC, Allen RS, Chitty JA, Gerlach WL, Frick S, Kutchan TM, Fist AJ (2007) Increasing morphinan alkaloid production by over-expressing codeinone reductase in transgenic *Papaver somniferum*. Plant Biotechnol J 5:26–37. https://doi.org/10.1111/j.1467-7652.2006.00212.x

Levy A, Milo J (1998) Genetics and breeding of *Papaver somniferum*. In: Bernáth J (ed) Poppy—the genus Papaver. Harwood Academic Publishers, Amsterdam, pp 93–104

Luhmer K, Blum H, Kraska T, Döring, T, Pude R (2021) Poppy (*Papaver somniferum* L.) intercropping with spring barley and with white clover: benefits and competitive effects. Agronomy 11:948. https://doi.org/10.3390/agronomy11050948

Rezaei M, Naghavi MR, Hoseinzadeh AH, Abbasi A, Jahangiri B (2014) Study of Karyological characteristics in *Papaver bracteatum* and *Papaver somniferum*. Cytologia 79(2):187–194. https://doi.org/10.1508/cytologia.79.187

Makovnyka G (2020) Opium poppy agriculture and consumption: contextualizing its functions as food, medicine, and narcotic. Arbutus Rev 11(2). https://doi.org/10.18357/tar112202019609

Malik CP et al (1979) Cytogenetic studies on *P. somniferum* x *P. setigerum* hybrids and amphidiploids. Cytologia 44:59–69

Mathur AK, Patra NK, Chauhan SP (1993) In vivo and in vitro mutagenesis and research for downy mildew resistance in opium poppy (*Papaver somniferum* L.). In: Conference research co-ordination meeting on mutation breeding of oil seed crops, Vienna (Austria), 11–15 Jan 1993, p 179. https://www.osti.gov/etdeweb/biblio/31268

Martínek V (ed) (1971) Výzkum světového sortimentu kulturních rostlin - Klasifikátor máku - (*Papaver somniferum* L.) [Research of the world assortment of cultivated plants - Poppy descriptor list - (*Papaver somniferum* L.)]. Závěrečná zpráva [Final research report] ČAZ, VÚRV

Praha – Ruzyně [Czech Academy of Agriculture, Research Institute of Plant Production Prague - Ruzyně], [Unpublished in the Czech language] 54
Martínek V, Šíp V, Škorpík M (1975) Studium dedičnosti některých hospodářsky duležitých znaku máku [A study of the inheritance of some economically important poppy traits]. Záverečná správa [Final research report], VÚRV Praha-Ruzyně [Research Institute of Plant Production Prague - Ruzyně], 51. [In the Czech language, Unpublished]
Matyášová E (2010) Variabilita genových zdrojů máku [Variability of poppy gene resources] (*Papaver somniferum* L.). Disertační práce, Česká zemnědělská univerzita Praha, [Dissertation, Czech University of Life Sciences, Prague], 203. [In the Czech language]
Meos A, Saks L, Raal A (2017) Content of alkaloids in ornamental *Papaver somniferum* L. cultivars growing in Estonia. Proc Est Acad Sci 66(1):34–39. https://doi.org/10.3176/proc.2017.1.04
Mičianová V, Ondreičková K, Muchová D, Klčová L, Hudcovicová M, Havrlentová M, Mihálik D, Kraic J (2017) Forensic application of EST-derived STR markers in opium poppy. Biologia 72(6):587–594. https://doi.org/10.1515/biolog-2017-0076
Miller JAC, Henning L, Hazlewood VL, Larkin PJ, Chitty J, Allen R, Brown PH, Gerlach WL, Fist AJ (2005) Pollination biology of oilseed poppy, *Papaver somniferum* L. Aust J Agric Res 56:483–490
Millgat AG, Pogson BJ, Wilson IW, Kutchan TM, Zenk MH, Gerlach WL, Fist J, Larkin PJ (2004) Analgesia: morphine-pathway block in top1 poppies. Nature 431:413–414
MACZ (Ministry of Agriculture of the Czech Republic) (2020) The Czech Republic the largest producer of breadseed poppy. https://eagri.cz/public/web/file/668246/Cesky_mak___ENG.pdf
Mishra BK, Rastogi A, Siddiqui A, Srivastava M, Verma N, Pandey R, Sharma NC, Shukla S (2013) Opium poppy: genetic upgradation through intervention of plant breeding techniques. Plant Breed Lab Fields 209–238. https://doi.org/10.5772/53132
Misra P, Saema S (2016) Plant tissue culture for in vitro mutagenesis, large-scale propagation, and genetic transformation. In: Anis M, Ahmad N (eds) Plant tissue culture: propagation, conservation and crop improvement, pp 309–342. https://doi.org/10.1007/978-981-10-1917-3_14
Mutation Breeding Newsletter (2003) Joint FAO/IAEA, Vienna, Issue No. 46:54
Nabha A, Lamblin F, Gillet F, Laurin D, Fliniaux M, David A, Jacquin A (1999) Polyamine content and somatic embryogenesis in *Papaver somniferum* cells transformed with sam-1 Gene. J Plant Physiol 154:729–734
Naseem S, Kumar G (2013) Induced desynaptic variation in poppy (*Papaver somniferum* L.). Crop Breed Appl Biotechnol 13:363–366. https://doi.org/10.1590/S1984-70332013000400007
Németh É, Bernáth J (2007) Selection of poppy (*Papaver somniferum* L.) cultivars for culinary purposes. In: I International medicinal and aromatic plants conference on culinary herbs, 29 April–4 May 2007, Antalya, Turkey, Abstracts 1
Németh-Zámbori É, Jászberényi RP, Bernáth J (2011) Evaluation of alkaloid profiles in hybrid generations of different poppy (*Papaver somniferum* L.) genotypes. Ind Crops Prod 33:690–696. https://doi.org/10.1016/j.indcrop.2011.01.013
Nessler CL (1982) Somatic embryogenesis in the opium poppy, *Papaver somniferum*. Physiol Plant 55:453–458
Nyman U, Hall O (1976) Some varietes of *Papaver somniferum* L. with changed morphinane alkaloid content. Hereditas 84:69–76
Popovec M (1999) Program šľachtenia [Breeding program]. Mak siaty [Opium Poppy] (*Papaver somniferum* L.), VÚRV [Research Institute of Plant Production] Piešťany, VŠS [Research and Breeding Station] Malý Šariš 28. [In the Slovak language, unpublished]
Ovecka M, Bobak M, Sarna J (2000) A comparative structural analysis of direct and indirect shoot regeneration of *Papaver somniferum* L. *in vitro*. J Plant Physiol 157:281–289
Özcan MM, Atalay C (2006) Determination of seed and oil properties of some poppy (*Papaver somniferum* L.) varieties. Grasas Aceites 57:169–174. https://doi.org/10.3989/gya.2006.v57.i2.33

Pal PK, Maurya KN, Srivastava A, Shukla S (2021) Influence of epistatic effects on various traits in relation to the genetic improvement of opium poppy (*Papaver somniferum* L.). J Appl Res Med Aromat Plants 25:100322. https://doi.org/10.1016/j.jarmap.2021.100322

Pathak S, Mishra BK, Misra P, Misra P, Joshi VK, Shukla S, Trivedi PK (2012) High frequency somatic embryogenesis, regeneration and correlation of alkaloid biosynthesis with gene expression in Papaver somniferum. Plant Growth Regul 68:17–25. https://doi.org/10.1007/s10725-012-9689-z

Patra NK, Kumar B (2005) Improved varieties and genetic research in medicinal and aromatic plants (MAPs). In: Proceedings of 2nd national interactive meet on medicinal and aromatic plants. CIMAP, Lucknow, India 29–30 October, 2004, pp 53–61

Pei L, Wang B, Ye J, Hu X, Fu L, Li K, Ni Z, Wang Z, Wei Y, Shi L, Zhang Y, Bai X, Jiang M, Wang S, Ma CH, Li S, Liu K, Li W, Cong B (2021) Genome and transcriptome of *Papaver somniferum* Chinese landrace CHM indicates that massive genome expansion contributes to high benzylisoquinoline alkaloid biosynthesis. Hortic Res 8:5. https://doi.org/10.1038/s41438-020-00435-5

Pienkny S, Brandt W, Schmidt J, Kramell R, Ziegler J (2009) Functional characterization of a novel benzylisoquinoline O-methyltransferase suggests its involvement in papaverine biosynthesis in opium poppy (*Papaver somniferum* L). Plant J 60(1):56–67

Rahimi A, Arslan N, Pour Rezaeieh KA, Gurbuz (2015) Variation in fatty acid composition of four turkish registered poppy (*Papaver somniferum* L.) seeds in two locations (Ankara and Boldavin) of Turkey. Eur Online J Nat Soc Sci 4(1):183–190

Rod J (ed) (1982) Šlechtení rostlin [Plant Breeding]. Státní zemnědělské nakladatelství Praha [State Agricultural Publishing House Prague], 352. [In the Czech and Slovak language]

Salavert A, Zazzol A, Martin L, Antolín F, Gauthier C, Thil F, Tombret O, Bouby L, Manen C, Mineo M, Mueller-Bieniek A, Piqué R, Rottoli M, Rovira N, Toulemonde F, Vostrovská I (2020) Direct dating reveals the early history of opium poppy in western Europe. Sci Rep 10:20263. https://doi.org/10.1038/s41598-020-76924-3

Selale H, Celik I, Gültekin V, Allmer J, Doganlar S, Frary A (2013) Development of EST-SSR markers for diversity and breeding studies in opium poppy. Plant Breeding 132:344–351. https://doi.org/10.1111/pbr.12059

Sharma JR, Lal RK, Gupta AP, Misra HO, Pant V, Singh NK, Pandey V (1999) Development of non-narcotic (opiumless and alkaloid-free) opium poppy, *Papaver somniferum*. Plant Breeding 118:449–452

Sharma JR, Lal RK, Gupta MM, Verma RK, Misra HO (2002) A superior gum variety Sampada of opium poppy. J. Med. Aromat. Plant Sci. 24:478–480

Schreier J, Zájeda J (1994) Technologie výroby máku [Poppy production technology]. Metodiky pro zavádění výsledků výzkumu do zemnědelské praxe [Methodologies for introducing research results into agricultural practice] ÚZPI Praha [ÚZPI Prague] 32. [In the Czech language]

Singh SP, Khanna KR, Dixit BS, Srivastava SN (1990) Fatty acid composition of opium poppy (*Papaver somniferum*) seed oil. Indian J. Agri. Sci. 60:358

Singh R, Pandey RM (2011) Combining ability and heterosis in opium poppy (*Papaver somniferum* L.). Curr Adv Agric Sci 3(2):130–134

Sinskaja EN (1973) Historická geografie kulturních rostlin. [Historical geography of cultivated plants] Academia Praha [Academia, Prague], pp 367–368. [In the Czech language]

Straka P, Nothnagel T (2002) A genetic map of *Papaver somniferum* L. Based on molecular and morphological markers. J Herbs, Spices & Med Plants 9:2–3, 235–241. https://doi.org/10.1300/J044v09n02_a

Shukla S, Maurya KN, Mishra BK, Rastogi A. (2016) Identification of heterotic crosses based on the combining ability of novel genotypes in opium poppy (*Papaver somniferum* L.). Russ Agric Sci 42(2):137–144. https://doi.org/10.3103/S1068367416020142

Svoboda P, Vašek J, Vejl P, Ovesná J (2020) Genetic features of Czech blue poppy (*Papaver somniferum* L.) revealed by DNA polymorphism. Czech J Food Sci 38(3):198–202. https://doi.org/10.17221/23/2020-CJFS

Tasmanian Alkaloids pty. Ltd. (2004) Top ten growers. In: Poppy growers, Bulletin 45:3
Thangavel T, Wilson CR, Jones S, Scott JB, Voglmayr H (2017) First report of systemic downy mildew of opium poppy caused by *Peronospora somniferi* in Australia. Plant Dis 101(2):392. https://doi.org/10.1094/PDIS-06-16-0796-PDN
Tisserat B, Berhow M (2009) Production of pharmaceuticals from papaver cultivars in vitro. Eng Life Sci 9(3):190–196. https://doi.org/10.1002/elsc.200800100
Vašák J (ed) (2010) Mák [Poppy]. Powerprint, Praha, [Powerprint, Prague], 352. [In the Czech language]
Vašek J, Čílová D, Melounová M, Svoboda P, Zdeňková K, Čermáková E, Ovesná J (2021) OpiumPlex is a novel microsatellite system for profiling opium poppy (*Papaver somniferum* L.). Sci Rep 11:12799. https://doi.org/10.1038/s41598-021-91962-1
Verma N, Jena SN, Shukla S, Yadav K (2016) Genetic diversity, population structure and marker trait associations for alkaloid content and licit opium yield in India-wide collection of poppy (*Papaver somniferum* L.). Plant Gene 7:26–41. https://doi.org/10.1016/j.plgene.2016.08.001
Vlk R, Kosek Z, Šimek P (2011) Výnosy máku, makoviny a obsah morfínu [Yield of the poppy seeds and poppy straw, morphine content]. Sborník odborných seminářů "Mák v roce 2011" 10: [Proceedings of scientific seminars "Poppy in the year 2011" 10:], 20–22. [In the Czech language]
Voškeruša J (ed) (1965) Pěstování olejnin v ČSSR [Growing oilseeds in the Czechoslovak Socialist Republic]. Státni zemnědělské nakladatelství Praha [State Agricultural Publishing House Prague], 178–227. [In the Czech language]
Wang W, Xu J, Fang H, Li Z, Li M (2020) Advances and challenges in medicinal plant breeding. Plant Sci 298:110573. https://doi.org/10.1016/j.plantsci.2020.110573
Xu Y (2010) Molecular plant breeding. CAB International 2010, 734. ISBN: 978 1 84593 392 0
Yadav HK, Singh SP (2011) Inheritance of quantitative traits in opium poppy (*Papaver somniferum* L). Genetika 43(1):113–128
Yazici L, Yilmaz G (2020) Hybrid vigor for alkaloid contents in opium poppy (*Papaver somniferum* L.). Agric Sci Dig 40(3):290–294. https://doi.org/10.18805/ag.A-496
Yazici L, Yilmaz G (2021) Investigation of alkaloids in opium poppy (*Papaver somniferum* L.) varieties and hybrids. J Agric Sci 27(1):62–68. https://doi.org/10.15832/ankutbd.612506
Yoshimatsu K, Shimoura K (1996) Genetic transformation in *Papaver somniferum* L. (Opium Poppy) for enhanced production of morphinan. Biotechnol Agric For 38; In: Bajaj YPS (ed) Plant protoplasts and genetic engineering, vol 7, pp 243–252

Internet sources

https://www.itis.gov/servlet/SingleRpt/SingleRpt?search_topic=TSN&search_value=18894#null
https://www.upov.int/test_guidelines/en/list.jsp
https://www.uksup.sk/storage/app/uploads/public/605/855/8aa/6058558aaa030423659630.pdf
https://eagri.cz/public/web/file/668246/Cesky_mak_ENG.pdf
https://en.wikipedia.org/wiki/Golden_Crescent

Chapter 7
Advance in Breeding Strategies of Patchouli (*Pogostemon cablin* Benth)

Versha Dixit, Chole Pranjali Bajrang, Wudali Narasimha Sudheer, Praveen Nagella, V. L. Vasantha, and B. T. Manjunath

Abstract *Pogostemon cablin* Benth, also known as Patchouli, belongs to the family Lamiaceae indigenous to south and southeast Asia. Patchouli is now widely cultivated in Caribbean countries, India, China, Mauritius, Philippines, Malaysia, Indonesia, Vietnam, and West Africa. This aromatic and essential oil-yielding plant has many therapeutic and industrial importance. The essential oil of this plant is used in the perfumery, aromatherapy, and fragrance and has medicinal attributes. Patchouli plant is mainly cultivated for its essential oil; its consumption is around 1500 tons worldwide. Patchouli's oil production in India is only 10 tons, and India imports more than 150 tons of its essential oil. This less amount is due to unfavorable environmental conditions. *P. cablin* is mainly grown through vegetative propagation using stem cuttings, but this plant is prone to many pest diseases like viruses, fungi, and nematodes. Infection of pests harms the crop and decreases the essential oil content drastically. New and improved varieties are needed for disease resistance and higher oil yield. To overcome these challenges and conserve its germplasm and biodiversity, extensive study is required for different breeding techniques. Conventional breeding methods like; selection of pure lines, mutation breeding, hybridization, backcross, and cell fusion methods are now being used with advanced techniques

V. Dixit · C. P. Bajrang · W. N. Sudheer · P. Nagella · V. L. Vasantha · B. T. Manjunath (✉)
Department of Life Sciences, School of Sciences, CHRIST (Deemed to be University), Bengaluru, Karnataka, India
e-mail: manjunath.bt@christuniversity.in

V. Dixit
e-mail: versha.dixit@res.christuniversity.in

C. P. Bajrang
e-mail: cholepranjali.bajrang@res.christuniversity.in

W. N. Sudheer
e-mail: wudali.sudheer@res.christuniversity.in

P. Nagella
e-mail: praveen.n@christuniversity.in

V. L. Vasantha
e-mail: vasantha.vl@christuniversity.in

J. M. Al-Khayri et al. (eds.), *Biodiversity and Genetic Improvement of Medicinal and Aromatic Plants II*, Advances in Plant Breeding Strategies 10,
https://doi.org/10.1007/978-3-031-81857-8_7

like molecular approaches and other modern genomic tools for developing new varieties. This chapter describes recent advances in genetic improvement, some essential conservation, and breeding methods for *P. cablin*.

Keywords *Pogostemon cablin* · Patchouli · Germplasm conservation · Essential oil · Plant breeding

7.1 Introduction

Plants have been used in medication for thousands of years. Humans are dependent on plants for several requirements, such as foodstuffs, medicines, fertilizers, and other industrial applications (Singh 2015; Zarrelli 2022). Despite the boom in synthetic drug development, plants and plant products are considered as the primary source of medicines and are extensively used in the pharmaceutical industry. These industries traditionally use plant and plant products with advanced technologies. Medicinal plants are widely used in Ayurveda, Unani, Siddha, and other ancient medicinal practices (Swamy et al. 2015). Asia is famed for its medicinal and aromatic plants. It is also known as 'the land of spices,' 'the land of aromatic plants'and 'the land of traditional perfumes' as it has favorable conditions for developing such plants. These plants have been used as spices or as a raw material for several essential oils or pharmaceutical industries at the commercial level. Aromatic plants have fragranced volatile compounds in the form of essential oil; these compounds are particular in their structure and function (Ramya et al. 2013). Lamiaceae (mint family) is an essential medicinal plant family with therapeutic and prophylactic properties used in flavoring, food preservation, and medicinal purposes. All the species are well known for their bioactive compounds, with specific therapeutic importance. Lamiaceae family contains around 236 genera and 6000 species (Carovic et al. 2016). Plants of Lamiaceae are usually herbs or shrubs having aromatic smell due to the good amount of essential oil. The mint family is commonly found in Maltese islands and some other Mediterranean countries (Raja 2012). Few small groups from this family are locally grown in some parts of Australia, South America, and southwest Asia. This family has many dietary and seasoning herbs commonly used as tea, spices, and mainly indigenous medicine. The main therapeutic component is their essential oil, usually obtained from water or steam distillation of their fresh or dried plant parts (Kokkini et al. 2003).

Pogostemon cablin Benth, also known as Patchouli, belongs to the Lamiaceae family. It is a perennial herb indigenous to South and Southeast Asia, and it is now widely grown in Caribbean countries, India, China, Mauritius, Philippines, Malaysia, Indonesia, Vietnam, and West Africa (Lu et al. 2011; Swamy et al. 2016). Twenty-five species of *Pogostemon* are cultivated in India. Patchouli is known differently in different languages like; Tamala patra in Sanskrit, Patchouli in Hindi, patche tene in Kannada, pachilai in Tamil, patchilla in Malayalam, patchapan or patcha in Marathi and Guang how Xiang in Chinese, phimsen in Thailand and nilam in Malaysia and Indonesia. *P. cablin* is tolerant to hot and humid climates. Patchouli plant have upright stem and broad leaves which can attain a height of 1–1.2 m. leaf margins are lobed, and the dorsal surface is covered with hairs. It is mainly grown for its leaves, containing essential oil in the glandular trichomes. It has small pale pink and white flowers (Swamy et al. 2015).

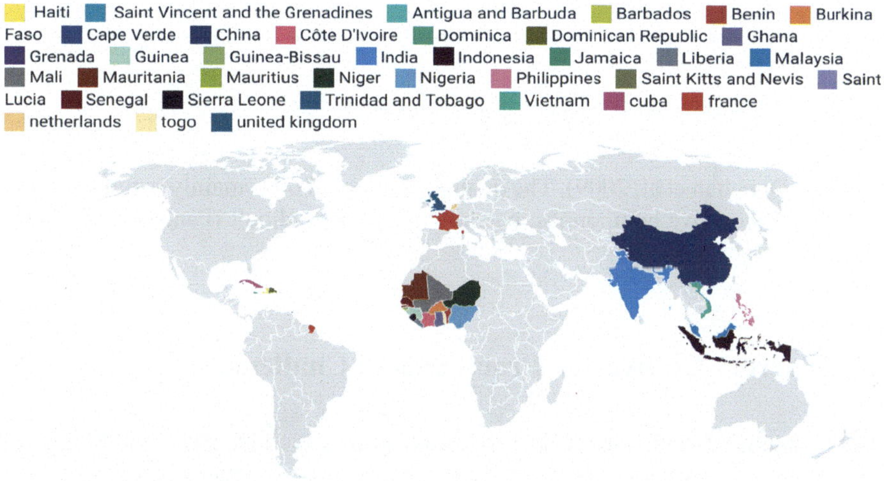

Taxonomic position (Swamy et al. 2016).
Kingdom-Planta.
Division-Magnoliophyta.
Class-Magnoliopsida.
Order-Lamiales.
Family-Lamiaceae (Labiatae).
Genus-*Pogostemon.*
Species-*cablin.*

Patchouli is well known for its essential oil, which is obtained from its leaves, and this oil is the main component of the aromatherapeutic and perfume industries. Indonesia is the chief producer of patchouli oil, but its supply is quite asymmetrical because of unfavorable environmental conditions. Around 1500 tons of patchouli oil is consumed worldwide and India imports more than 150 tons of patchouli oil. The current production of this oil in India is approx 10 tons (Patil et al. 2012).

Patchouli oil helps in decreasing depression, anxiety and soothe nerves. This plant has several therapeutic properties as; antimicrobial, analgesic, anti-inflammatory, antioxidant, antiplatelet, aphrodisiac, antithrombotic, antidepressant, antimutagenic, fibrinolytic, antiemetic, and cytotoxic activities (Swamy et al. 2016). Patchouli oil is quite distinctive and complex in characteristics as it consists of more than 24 different sesquiterpenes instead of the mixture of mono-, di-, and sesquiterpenes. The significant component of patchouli is sesquiterpene patchoulol which is mainly responsible for its aroma. α-/β-/γ- patchoulenes, α-guaiene, seychellene, α-himachalene and pogostone are some sesquiterpene hydrocarbons that are also present in good

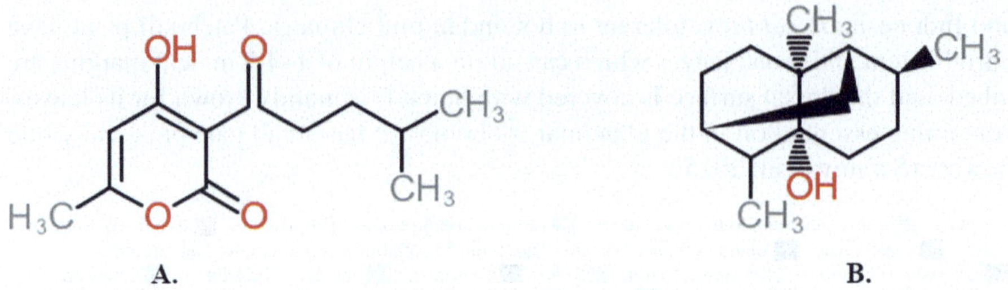

Fig. 7.1 Chemical structure of **a** Pogostone and **b** Patchouli alcohol

numbers (Donelian et al. 2009). The quality of essential oil is mainly regulated by its two critical bioactive compounds, patchoulol and α-patchouli (Pandey et al. 2021) (Fig. 7.1).

7.2 Current Cultivation Practices and Challenges

Plant is commonly cultivated using vegetative propagation by 10–12 cm long stem cuttings having 3–4 nodes. Seeds of the patchouli plant are unavailable, so its stem cutting with leaves is used as it has higher rooting and shooting capacity. Previous studies suggested using apical and medium meristematic regions of stem cutting with two leaves for better rooting and shooting. It is a heliophobic plant that requires fertile and well-drained soil with slightly acidic deep loamy soil (Pandey et al. 2021). This plant grows well from the lowest plains to relatively high land areas up to 2000 msl. Plants from plains give a higher yield of essential oil when compared to highlands. Patchouli plants grow well in the temperature of 22–28 °C or 22–28 water vapor capacity (g/m^3) with more than 75% humidity. For ideal growth, patchouli requires water from the starting point of the planting till the plant reaches optimal growth. When plants reach the age of more than three months, it requires proper sunlight till harvesting (Ritawati et al. 2019). March–May and September to October are preferred for its planting, with an average precipitation of 300–400 cm. Proper plant growth requires suitable nutrient doses from organic and inorganic sources. The essential nutrient dose of Patchouli is 50 kg phosphorus pentoxide, 50 kg potassium oxide, and 25 kg nitrogen in addition to 25 kg nitrogen at the two-month interval (Pandey et al. 2021).

The main challenge faced during patchouli cultivation is the infection of various diseases causing pests like viruses, fungi, and nematodes. Yellow mosaic virus, Patchouli mosaic virus, and Peanut strip virus are some major groups that infect Patchouli. Nematode infection is also a severe problem for patchouli cultivation. *Meloidogyne incognita*, commonly known as Root-knot nematode, is the most dangerous nematode for patchouli plants, decreasing plant growth and essential oil yield. This nematode can be controlled by treating ethion and furadan granule (Pandey et al. 2021).

Continuous cropping is also a problem for mass cultivation industries as it changes soil properties and the soil microbiome. Mixed cropping with turmeric and ginger with Patchouli improves soil properties and enzyme activities (Zeng et al. 2020). Tissue culture methods are also standardized for mass cultivation of the patchouli plant. This method was specific for mass cultivation and virus-free plant production (Kukreja et al. 1990).

7.3 Germplasm Biodiversity and Conservation

Detailed studies on genetic diversity are essential for efficient germplasm management and utilization, genotype selection, and genetic fingerprinting. Many methods are used for germplasm biodiversity, and among them, DNA fingerprinting with genomic DNA nucleotide variations based on molecular markers gives accurate and fast population identification. Molecular-level studies are essential for understanding ecological and genetic aspects of species and the design and utilization of management strategies (Wu et al. 2011). A group of researchers investigated genome-wide diversity through SLAF-seq technology that reveals the interconnection between patchouli accession from its indigenous place and center of origin like; Vietnam, China, and Indonesia (Huang et al. 2016) (Table 7.1).

Information on genetic diversity and phylogeny can be obtained from sequence information which also plays a significant role in conserving plant genetic resources (Huang et al. 2016). As discussed above, Patchouli is propagated vegetatively, so conservation of Patchouli can be done by maintaining improved lines in vivo or in vitroconditions. In vivo conservation of select germplasm makes it susceptible to diseases and other unfavorable environmental conditions. On the other hand, in vitro conservation can be done through somaclonal variations and synthetic seed technology (Swamy et al. 2009a, b, c). Conservation of genetic resources involves maintaining collections out of their occurrence place (ex-situ) or in their occurrence (in situ). In an ex-situ conservation, seeds, embryos, pollen, or other plant parts can be preserved at a particular temperature, whereas in greenhouses, fields are used for in situ conservation. Nowadays, synthetic seed technology has also gained importance for conserving scattered hybrids, genetically engineered, and selected genotypes of patchouli plants. An ex-situ conservation, this technique is known as a viable, potential, and valuable system for commercially essential plants.

In vitro conservation provides genetic stability with minimum subculture frequency and maximum survival rate. In vitro technique shows two basic systems; slow growth and cryopreservation. In a slow-growth system, plant material can be stored for a short or medium duration and is performed by reducing temperature stress and light intensity (Santos et al. 2015). It will not affect the cellular viability of the plant preserved. In long-term conservation, the cryopreservation method is used commonly. In cryopreservation, storage is done at very low temperature (-196 °C

Table 7.1 Accession, Sampling locations of *Pogostemon cablin* and amount of Patchouli alcohol, pogostone and its ratio (Huang et al. 2016)

Accession	Location	Patchouli alcohol* (mg/g)	Pogostone (mg/g)	Ratio
YN1	Bengkulu, Indonesia	2510	21.6	120
YN2	Muaradua, South Sumatra, Indonesia	2280	52.8	43
YN3	Sarik, West Sumatra, Indonesia	3080	34.2	90
YN4	Simpang Timbu Abu, West Sumatra, Indonesia	1660	18.6	89
YN5	Subulussalam, Aceh, Indonesia	2600	86.1	30
YN6	Tapaktuan, Aceh, Indonesia	1940	21.3	91
YN7	Tapaktuan, Aceh, Indonesia	2010	2.83	710
YN8	Sehdes Isez, Aceh, Indonesia	3460	10.8	320
YN9	Aceli, Aceh, Indonesia	2570	ND	–
YN10	Aceli, Aceh, Indonesia	1610	18.7	86
VT1	Van Lam County, Hung Yen, Vietnam	704	4.03	170
VT2	Van Lam County, Hung Yen, Vietnam	769	286	2.7
KK1 (Zhanxiang)	Leizhou, Zhanjiang, China	1940	15.6	120
KK2 (Zhanxiang)	Leizhou, Zhanjiang, China	1410	3.73	380
KK3 (Zhanxiang)	Leizhou, Zhanjiang, China	1920	ND	–
YC1 (Zhanxiang)	Yangchun, Yangjiang, China	3050	9.47	320
YC2 (Zhanxiang)	Yangchun, Yangjiang, China	3480	ND	–
YC3 (Zhanxiang)	Yangchun, Yangjiang, China	2310	6.90	330
GY1 (Zhaoxiang)	Gaoyao, Zhaoqing, China	1670	ND	–
GY2 (Zhaoxiang)	Gaoyao, Zhaoqing, China	148	1390	0.11

(continued)

Table 7.1 (continued)

Accession	Location	Patchouli alcohol* (mg/g)	Pogostone (mg/g)	Ratio
SP1 (Paixiang)	Shipai, Guangzhou, China	43.9	756	0.06
SP2 (Paixiang)	Shipai, Guangzhou, China	46.9	3750	0.01

in liquid nitrogen). For long-term preservation, buds are encapsulated and stored at a freezing temperature which provides maximum genotypic as well as phenotypic stability and causes minimum maintenance and storage space. In patchouli plants, nodal segments are encapsulated, and germplasm conservation is done (Swamy et al. 2009a, b, c). Santos et al. (2015) concluded that the viability of all genotypes of Patchouli at 25 °C was maintained in micro plants. In an MS medium with 20 g/L of sucrose genotype, POG002 can be conserved for upto three months. Genotypes POG021 and POG014 in MS medium with 5 g/L of mannitol + 10 g/L sucrose for upto six months of cultivation.

7.4 Traditional Breeding

With respect to the EU (European Union) Directive 2001/18/EC, those breeding methods that do not include genetic modification are traditional plant breeding methods. Here the term traditional may be ambiguous because it does not mean the absence of advanced techniques; it includes both basic and advanced methods. More appropriately, it includes all the methods that do not involve recombinant DNA technology or any genetic modifications. However, there are a few exceptions where genetic material changed somehow, but these methods does not include recombinant DNA technology, for example; mutagenesis and cell fusion (including protoplast fusion), but it counts as traditional breeding methods as it is traditionally used before the introduction of recombinant DNA technology (Van de Wiel et al. 2010). In traditional breeding, commonly desirable traits are selected from different but closely related plants using crossing over. It usually targets a particular trait of a species without introducing a new gene. Methods used in traditional breeding are; (Acquaah 2015).

- Selection of advance desirable plants
- Artificial pollination
- Hybridization
- Wide crosses (from different species or genera)
- Embryo culture
- Chromosomal doubling
- Doubled haploids

- Bridge crossing
- Protoplast fusion
- Seedlessness
- Genetic markers
- Mutagenesis

Polyploidy induction can help improve the medicinal value of *P. cablin*. In vitro colchicine treatment for the tetraploid induction was performed, and the chromosome number of the tetraploid plantlet was 2n = 4x = 56. These plants have a higher yield of essential oil (Wu and Li 2013). In vitro induction of octoploid was done using colchicine treatment in *Pogostemon cablin*. Chromosome number of octoploid was 2n = 8x = 128 (Yan et al. 2016). Some high-yielding mutagenic type was also identified using gamma radiation (Rekha et al. 2009; Lal et al. 2018).

7.5 Molecular Breeding

In the molecular breeding technique, DNA markers are used that are tightly linked to the particular phenotypic trait in a selection strategy. The important aspect of molecular breeding technique is the selection, identification, and characterization of suitable genetic markers like; random amplified polymorphic DNA (RAPD), inter simple sequence repeats (ISSRs), and amplified fragment length polymorphism (AFLP) (Jaradat 2016). Exciting genetic variations within a species can be characterized with the help of molecular markers and other genomic applications. Plant biotechnology creates new genetic diversity that can expand the species' genetic boundaries. Scientists are making efforts to develop lots of molecular markers, high-density genetic maps, and suitably structured mapping populations now become so regular and usual for many crop plants having the ability to interpret gene action and breeding values varying from hundreds to thousands of loci that are distributed uniformly over the entire genome (Moose and Mumm 2008).

Simple sequence repeats (SSRs) or microsatellites obtained from expressed sequence tags (ESTs) are commonly used for molecular assisted-selection breeding, comparative genomic analysis, and genetic diversity studies. In a previous study, 54,546 ESTs were investigated for identifying and developing SSR markers in *Pogostemon cablin*. 1219 stars were recognized from 1144 SSR- containing ESTs (Ouyang et al. 2018). First de Novo assembly of the 1.15-Gb draft genome sequence was done using next-generation sequence technology. Whole-genome sequence analysis identified 3,147,33 heterozygous single-nucleotide polymorphism and 490,407 insertions and deletions, giving an estimated heterozygosity rate of 0.274% (He et al. 2016a, b, c).

7.5.1 Tissue Culture Applications

Patchouli is a notable aromatic medicinal plant with various economic and pharmacological applications. Conventional methods of production are having less feasibility as there are limitations in flowering, frequent virus and pest attacks. Vegetative stem cutting alone will not be sufficient to meet the demand of this plant. So, tissue culture applications and advanced biotechnological applications can help in mass production and multiplication of *Pogostemon cablin* (Swamy and Sinniah, 2016).

Biotechnological studies helped in production of virus free plantlets from meristem tip culture of patchouli. MS medium with 0.2 ppm BAP helped in induction of multiple shoots. According to ELISA-based studies, plantlets were free of the patchouli mild virus, that is the major cause of improper cultivation, and the essential oil yield increased (Sugimura et al. 1995). Nodal explants when cultured on MS medium supplemented with 0.5 mg/L BAP resulted in formation of multiple shoots which on further subculturing on BAP and 0.5 mg/L KN enriched medium increased the mean shoot number. Roots were induced on half strength MS medium supplemented with 100 mg/L activated charcoal (Kumaraswamy and Anuradha, 2010). Leaf explants from second node of in-vivo plants showed best regeneration efficiency when cultured on MS medium supplemented with 2.5 μM BAP along with 0.5 μM NAA, and significant elongation of the shoots was seen when 1.0 μM GA_3 was supplemented onto the MS medium (Paul et al. 2010). Multiple shoots were induced from leaf explants when cultured on MS medium supplemented with 3.33 μM BAP (Fitriana, 2011). Multiple shoots were induced from leaf and nodal explants when cultured on MS medium supplemented with 2.5 mg/L KN along with 0.1 mg/L IBA (Sharma and Sarma, 2015). MS medium supplemented with 1.0 mg/L BAP along with 0.1 mg/L NAA showed best regeneration and multiplication efficiency (Rajkarnikar and Rajbahak 2016). Similarly, 0.1 mg/L of NAA along with 0.5 mg/L of BAP when supplemented to the basal medium showed increased shooting efficiency (Mazlan and Karim 2018). MS medium supplemented with 0.6 mg/L BAP along with 0.12 mg/L IAA induced shoots when nodal explants were inoculated and the multiplication efficiency increased when sub cultured onto the MS medium enriched with 1.5 ml/L BAP along with 0.3 ml/L IAA. Rooting was best seen in MS medium supplemented with 0.5 m//L IAA along with 1 m/L IBA (Darunde and Pandey 2019). Optimal growth and development of the shoot buds from nodal explants was reported in MS medium supplemented with 0.5 mg/L KN and 1.0 mg/L KN along with 0.5 mg/L BAP and 1.0 mg/L BAP (Mayerni et al. 2020). Highest number of shoots with best height was reported from leaf and nodal explants when cultured on MS medium supplemented with 0.3 ppm BAP (Yusniwati et al. 2021). Recent studies also support the previous literature for the in vitro regeneration from shoot explants when cultured on MS medium supplemented with combination of BAP and NAA. It is seen that best shoot growth, height and highest number of leaves were reported from MS medium supplemented with 1.0 mg/L BAP along with 2.0 mg/L NAA. Best rooting was observed in MS medium supplemented with 2.0 mg/L BAP and better length was reported from BAP in combination with 1.0 mg/

Fig. 7.2 Different growth of shoot from; A. controlled plant B. Colchicine treated plant (*Source* Widoretno 2016: DOI: 10.1007/s11240-016-0946-0, accessed on 27 July 2022; reproduced with permission from the publisher; License No: 5357030738782)

L NAA (Maulia et al. 2021). Usage of natural supplements in the growth medium will help in in-vitro propagation and regeneration of the patchouli. It is seen that MS medium supplemented with 0.5 mg/L BAP along with 0.5 mg/L KN showed best results with respect to biomass, shoot length, multiple shoot induction. Incorporation of coconut water at 10% concentration increased biochemical parameters like protein content and carbohydrate content (Swamy et al. 2014). Incorporation of sugarcane juice at 20% to the MS medium replacing carbon source resulted in formation of the highest number of multiple shoots with optimal chlorophyll content (Swamy et al. 2010) (Fig. 7.2).

MS medium supplemented with 2 mg/L NAA along with 0.5 mg/L BAP helped in induction of callus and subsequent culturing on MS medium supplemented with 1 mg/L BAP showed increased shoot induction and multiplication. Rooting was seen best in MS medium supplemented with 0.5 mg/L NAA (Misra 1996). Leaf and nodal explants when cultured on MS medium supplemented with 0.1 mg/L NAA along with 0.1 mg/L KN and 100 mg/L of casein hydrolysate induced callus and further subculturing helped in multiple shoots formation (Sharma and Sarma 2015).

7.5.2 Synthetic Seeds

Encapsulation of various explants will aid in in vitro regeneration of plantlets when synthesized and stored at appropriate optimal conditions. Nodal explants collected from in-vitro shoots were encapsulated using 4% sodium alginate which helped in formation of synthetic seeds. MS medium supplemented with 2.22 μM BAP showed the highest percentage of germination efficiency of the synthetic seeds (Swamy et al. 2009a, b, c). Apart from direct explant encapsulation, protoplast encapsulation also helped in plant regeneration. Liquid MS medium supplemented with 10 μM NAA along with 10 μM BAP showed best results with regeneration of plants upon transferring to semi solid medium (Kageyama et al. 1995).

7.6 Genetic Engineering and Gene Editing

Genetic engineering, also known as Recombinant DNA technology, includes classes of techniques that include cutting and joining DNA fragments. Genetic engineering includes modification in genetic material for improved traits. That foreign DNA is generally obtained from different species and introduced into the host DNA resulting in the formation of hybrid DNA that forms a new combination of heritable genetic material and changes one or more traits of the organism (Rosenberg 2017). Genome editing techniques are developed based on engineered or bacterial nuclease. With the development of genome editing techniques, the possibilities of directly targeting and modifying the genome sequence of eukaryotic cells are increased. Gene functions are assessed, and cellular behavior and function can be manipulated accurately using targeted gene modification through chimeric genome editing tools (e.g., ZFNs, TALENs, and CRISPR/Cas9) (Li et al. 2020).

Disease-resistant plants can be obtained by genetic engineering technology. The organism with a modified genome is known as a Genetically modified organism (GMO) or transgenic organism. Transgenic plant for Patchouli was also produced with Patchouli mild mosaic virus (PaMMV) coat protein precursor (CP-P) gene by Agrobacterium-mediated gene transformation. These transgenic plants were resistant to Patchouli mild mosaic virus (Kadotani and Ikegami 2002). A WRKY transcription factor is well known for regulating the expression of pathogen-causing infectious diseases. OsWRKY76 gene from rice has been identified for its broad spectrum of plant resistance. This gene from rice can be introduced into Patchouli for resistance to some major diseases. One trial used *Agrobacterium tumefaciens* mediated gene transfer (Santoso et al. 2017).

Whole transcriptome-wide analyses of Patchouli were done using next-generation sequencing technology. A group of scientists generated 108,996 unigenes using RNA sequencing data; among them, 65,536 were annotated in the NR database. 40,670 intact open reading frames (ORFs) were obtained, about 69% of the 58,820 predicted ORFs (He et al. 2016a, b, c). Entire chloroplast genome sequencing of *P. cablin* was also performed. The chloroplast genome with 38.24% GC content is 152,460 base pairs in length. The chloroplast genome encodes 127 genes, out of which 107 are single copies, including 79 protein-coding genes, 4 rRNA genes, and 24 tRNA genes (He et al. 2016a, b, c).

7.7 Mutation Breeding

Mutation breeding helps to develop new varieties in plant breeding. Induced mutation breeding is an alternative technique that allows a broader patchouli diversity. Mutation breeding can be done by using various physical and chemical mutagens. They help to induce genotypic as well as phenotypic variability. Physical mutagens like X-rays, gamma rays, or neutrons are commonly used, whereas colchicine is the most frequently used chemical mutagen. Nowadays, chemical mutagens are commonly used to enhance plant genetic diversity. Colchicine will not allow spindle formation during metaphase hence the number of doubling chromosomes known as polyploidy induction (Suhesti et al. 2020). It increases the level of polyploidy in a plant. Polyploidy induction is known as a potential way to enhance productivity in plants. Patchouli mutants developed with colchicine concentrations of 0.02% and 0.06% showed high structural diversity (Suhesti et al. 2020). Mutants showed four morphological characters which are related to stem (base color, shoot color, growth type, and middle color). Mutated leaf shows six morphological characters such as; leaf tip, leaf base, leaf tip, leaf shape, leaf surface, leaf surface color, and leaf edge (Suhesti et al. 2020). Several scientists have studied the genetic variability in Patchouli. In vitro fusion of colchicine and protoplast induction showed promising results (Tahir et al. 2019).

Various mutation studies have been conducted in patchouli clones with gamma-ray irradiation, resulting in several morphological changes in leaf properties (Tahir et al. 2019). The genetic base has putatively broadened. Patchouli stem with a single leaf and an axillary bud cutting treated with gamma irradiation resulted in 'PM-12', an erect plant mutant with dark, shiny, thick leaves (Rekha et al. 2009). It is also proven that effectiveness can be provided within different mutant clones for breeders. These mutant plants slightly decreased leaf number, branches, plant height, length of leaf, plant spread, and dry and fresh foliage yield. The percentage of oil present in parental plants and mutant plants varied from 3.3% and 3.0-to 3.2%, respectively (Rekha et al. 2009). Using gamma-ray irradiation (Cobalt60 as radiation source), an applied dose of 30 Gamma irradiation treatment (Gy) developed a mutant variety of Patchouli, rich in alcohol and more commercial viability. In Patchouli, the mutant replicate showed a high alcohol percentage (60.3%) than the parental replicate (47.8%) (Rekha et al. 2009). Mutational variations can occur in morphological or anatomical traits (Swamy and Sinniah 2016). In Patchouli, anatomical traits like stem color, leaf shape, and trichome were observed to be modified along with nutrient and climatic changes (Li et al. 2020).

The main application of selective breeding, tissue culture, hybridization, and somatic hybridization is the production of polyploidy cell lines. In vitro, polyploidy induction can be done by artificial colchicine treatment (Widoretno 2016). There are very few reports on hybridization experiments patchouli plant as it is vegetatively propagated so not much hybridization studies have been done on this plant.

7.8 Conclusion and Future Prospects

Patchouli plant is an essential oil-yielding plant and has immense commercial importance. To increase patchouli production on a wide scale, it is recommended to apply a variety of propagation tactics, such as crop management and cultivation procedures. Additionally, it is recommended to use biotechnological methods such as genetic modification, cell culture, micropropagation, and molecular markers for the validation of these approaches in the future in order to meet the demand for patchouli oil on the worldwide market through large-scale production. The data presented here may open up new avenues for the large-scale micropropagation of elite patchouli germplasm using inexpensive tissue culture techniques. Additionally, there is potential to enhance patchouli plant characteristics including quality and yield by utilizing recent advancements in genome manipulation research and in vitro mutations. However, in order to increase yields and improve profits, growers need to choose a high-yielding variety that is suited for the area and adhere to a set of scientific guidelines. So, there is a need to study further on different breeding techniques and other approaches for the development of new cultivars with desired traits. Molecular and genomic level studies will be needed to explore more about its secondary metabolites and further biochemical studies.

References

Acquaah G (2015) Conventional plant breeding principles and techniques. In: Advances in plant breeding strategies: breeding, biotechnology and molecular tools. Springer, Cham, pp 115–158

Carović-Stanko K, Petek M, Grdiša M, Pintar J, Bedeković D, Ćustić MH, Satovic Z (2016) Medicinal plants of the family Lamiaceae as functional foods—a review. Czech J Food Sci 34:377–390

Darunde D, Pandey S (2019) Standardization and Optimization of Initiation, Shooting and Rooting Medium for Patchouli Micro-Propagation. Int J Trend Sci Res Dev 3:895–902. https://doi.org/10.31142/ijtsrd23033

Donelian A, Carlson LH, Lopes TJ, Machado RA (2009) Comparison of extraction of patchouli (*Pogostemon cablin*) essential oil with supercritical CO_2 and by steam distillation. J Supercritical Fluids 48(1):15–20

Fitriana S (2011) Efficient direct regeneration of true-to type *Pogostemon cablin* Benth from leaf explant and profile of essential oils. Universiti Malaysia Pahang

He Y, Deng C, Xiong L, Qin S, Peng C (2016a) Transcriptome sequencing provides insights into the metabolic pathways of patchouli alcohol and pogostone in *Pogostemon cablin* (Blanco) Benth. Genes & Genomics 38(11):1031–1039

He Y, Xiao H, Deng C, Xiong L, Nie H, Peng C (2016) Survey of the genome of *Pogostemon cablin* provides insights into its evolutionary history and sesquiterpenoid biosynthesis. Sci Rep 20;6(1):1–0

He Y, Xiao H, Deng C, Xiong L, Yang J, Peng C (2016) The complete chloroplast genome sequences of the medicinal plant *Pogostemon cablin*. Int J Molecular Sci 6;17(6):820

Huang HR, Wu W, Zhang JX, Wang LJ, Yuan YM, Ge XJ (2016) A genetic delineation of Patchouli (*Pogostemon cablin*) revealed by specific-locus amplified fragment sequencing. J Syst Evol 54(5):491–501

Jaradat AA (2016) Breeding oilseed crops for climate change. In: Breeding oilseed crops for sustainable production 1. Academic Press, pp 421–472

Kadotani N, Ikegami M (2002) Production of patchouli mild mosaic virus resistant patchouli plants by genetic engineering of coat protein precursor gene. Pest Manage Sci: Formerly Pesticide Sci 58(11):1137–1142

Kageyama Y, Honda Y, Sugimura Y (1995) Plant regeneration from patchouli protoplasts encapsulated in alginate beads. Plant Cell Tissue Organ Culture 41:65–70. https://doi.org/10.1007/BF00124088

Kokkini S, Karousou R, Hanlidou E (2003) HERBS herbs of the labiatae

Kukreja AK, Mathur AK, Zaim M (1990) Mass production of virus-free patchouli plants (*Pogostemon cablin* (Blanco) Benth.) by in vitro culture. Tropical Agriculture, Trinidad and Tobago 67(2):101–104

Kumaraswamy M, Anuradha M (2010) Micropropagation of *Pogostemon cablin* Benth. through direct regeneration for production of true to type plants. Plant Tissue Culture and Biotech 20:81–89. https://doi.org/10.3329/ptcb.v20i1.5971

Lal M, Pandey SK, Dutta S, Munda S, Baruah J, Paw M (2018) Identification of high herbage and oil yielding variety (Jor Lab P-1) of *Pogostemon cablin* (Blanco) Benth through mutation breeding. J Essent Oil Bearing Plants. 21(1):131–138

Li CG, Wu YG, Guo QS (2011) Floral and pollen morphology of *Pogostemon cablin* (Lamiaceae) from different habitats and its taxonomic significance. Procedia Engineering. 18:295–300

Li H, Yang Y, Hong W, Huang M, Wu M, Zhao X (2020) Applications of genome editing technology in the targeted therapy of human diseases: mechanisms, advances and prospects. Signal Trans Targeted Therapy 5(1):1–23

Lu TC, Liao JC, Huang TH, Lin YC, Liu CY, Chiu YJ, Peng WH (2011) Analgesic and anti-inflammatory activities of the methanol extract from *Pogostemon cablin*. Evidence-Based Complementary and Alternative Med 2011:671741. https://doi.org/10.1093/ecam/nep183

Maulia E, Zuyasna BB (2021) Growth of patchouli shoots (*Pogostemon cablin* Benth) with several concentrations of growth regulator substances in vitro. IOSR J Agric Vet Sci 14:38–46. https://doi.org/10.9790/2380-1401013846

Mayerni R, Warnita, Armansyah, Chan SROS (2020) The direct organogenesis. In: Local clones of patchouli plant (*Pogostemon cablin* Benth) In Vitro. JERAMI Indones J Crop Sci 3:16–19. https://doi.org/10.25077/jijcs.3.1.16-19.2020

Mazlan FI, Karim KA (2018) Response of plant growth regulators on the growth of different types of in vitro cultures of *Pogostemon cablin* (Blanco) benth for patchouli alcohol production. J Eng Sci Technol 13:2820–2828

Misra M (1996) Regeneration of patchouli (*Pogostemon cablin* Benth.) plants from leaf and node callus, and evaluation after growth in the field. Plant Cell Rep; 15:991–4. https://doi.org/10.1007/BF00231602.

Moose SP, Mumm RH (2008) Molecular plant breeding as the foundation for 21st century crop improvement. Plant Physiol 147(3):969–977

National Research Council (US) Committee on Identifying and Assessing Unintended Effects of Genetically Engineered Foods on Human Health. Safety of Genetically Engineered Foods: Approaches to Assessing Unintended Health Effects. Washington (DC): National Academies Press (US); 2004. 2, Methods and Mechanisms for Genetic Manipulation of Plants, Animals, and Microorganisms. Available from: https://www.ncbi.nlm.nih.gov/books/NBK215771/

Ouyang P, Kang D, Mo X, Tian E, Hu Y, Huang R (2018) Development and characterization of high-throughput EST-based SSR markers for *Pogostemon cablin* using transcriptome sequencing. Molecules 23(8):2014

Pandey SK, Bhandari S, Sarma N, Begum T, Munda S, Baruah J, Gogoi R, Haldar S, Lal M (2021) Essential oil compositions, pharmacological importance and agro technological practices of Patchouli (*Pogostemon cablin* Benth.): a review. J Essential Oil Bearing Plants. 24(6):1212–1226

Patil S, Nayak G, Barve S, Tembe R, Khan R, Trivedi M (2012) Impact of biofield treatment on growth and anatomical characteristics of *Pogostemon cablin* (Benth.). Biotechnology; 3(11):154–162

Paul A, Thapa G, Basu A, Mazumdar P, Kalita MC, Sahoo L (2010) Rapid plant regeneration, analysis of genetic fidelity and essential aromatic oil content of micropropagated plants of Patchouli, *Pogostemon cablin* (Blanco) Benth. - An industrially important aromatic plant. Ind Crops Prod; 32:366–374. https://doi.org/10.1016/j.indcrop.2010.05.020

Raja RR (2012) Medicinally potential plants of Labiatae (Lamiaceae) family: an overview. Res J Med Plant 6(3):203–213

Rajkarnikar K, Rajbahak S (2016) Micropropagation of *Pogostemon cablin* (Benth.) (Patchouli). Countrys Mediev Engl 38:97–101

Ramya HG, Palanimuthu V, Rachna S (2013) An introduction to patchouli (*Pogostemon cablin* Benth.)–A medicinal and aromatic plant: It's importance to mankind. Agricultural Engineering International: CIGR J 15(2):243–250

Rekha K, Bhan MK, Dhar AK (2009) Development of erect plant mutant with improved patchouli alcohol in patchouli [*Pogostemon cablin* (Blanco) Benth.]. J Essential Oil Res 21(2):135–137

Ritawati S, Utama P, Pancaningsih E (2019) Evaluation of land suitability for cultivation of java patchouli (*Pogostemon heyneanus* Benth) in the dry land of northern Serang, Banten province. In: IOP conference series: earth and environmental science 1, vol 383, No 1. IOP Publishing, pp 012045

Robinson RK (2014) Encyclopedia of food microbiology. Academic press, vol 2

Rosenberg E (2017) Genetic engineering. It's in Your DNA, pp 81–93

Santos AV, Tavares FF, Blank AF (2015) In vitro conservation of patchouli genotypes. Acta Hort, pp 499–506

Santoso TJ, Siharmini A, Apriana A, Sirait N (2017) Gene transformation on patchouli for resistance to major diseases mediated by *Agrobacterium tumefaciens*. Buletin Penelitian Tanaman Rempah Dan Obat/bulletin of Res Spice and Med Crops 28(1):37–46

Sharma P, Sarma J. (2015) *Pogostemon cablin* (Blanco) Benth. (Lamiaceae): It's Ethnobotany & in vitro regeneration. Pharmacogn J 7:152–156. https://doi.org/10.5530/pj.2015.3.2

Singh R (2015) Medicinal plants: a review. J Plant Sci. Special Issue: Medicinal Plants 3(1–1):50–55. https://doi.org/10.11648/j.jps.s.2015030101.18

Sugimura Y, Padayhag BF, Ceniza MS, Kamata N, Eguchi S, Natsuaki T et al (1995) Essential oil production increased by using virus free patchouli plants derived from meristem tip culture. Plant Pathol 44:510–515. https://doi.org/10.1111/j.1365-3059.1995.tb01673.x

Suhesti S, Susilowati M, Sirait N, Amalia, Hadipoentyanti E. (2020) Morphological variations of patchouli putative mutants. IOP Conf Ser Earth Environ Sci 418(1)

Swamy MK, Balasubramanya S, Anuradha M. (2009) Germplasm conservation of patchouli (*Pogostemon cablin* Benth.) by encapsulation of in vitro derived nodal segments. Int J Biodiversity Conservat 1(8):224–230.

Swamy MK, Balasubramanya S, Anuradha M (2009b) Patchouli-Germplasm Conservation-Patchouli 1:224–230

Swamy MK, Balasubramanya S, Anuradha M (2009) Germplasm conservation of patchouli (*Pogostemon cablin* Benth .) by encapsulation of in vitro derived nodal segments 1(8):224–30.

Swamy MK, Mohanty SK, Anuradha M (2014) The effect of plant growth regulators and natural supplements on in vitro propagation of *Pogostemon cablin* Benth. J Crop Sci Biotechnol 17:71–78. https://doi.org/10.1007/s12892-013-0038-1

Swamy MK, Sinniah UR. (2015) A comprehensive review on the phytochemical constituents and pharmacological activities of *Pogostemon cablin* Benth.: an aromatic medicinal plant of industrial importance. Molecules 20(5):8521–8547

Swamy MK, Sinniah UR. (2016) Patchouli (*Pogostemon cablin* Benth.): Botany, agrotechnology and biotechnological aspects. Ind Crops Prod 87:161–176. https://doi.org/10.1016/j.indcrop.2016.04.032

Swamy MK, Sudipta KM, Balasubramanya S, Anuradha M (2010) Effect of dfferent carbon sources on in vitro morphogenetic response of patchouli (*Pogostemon cablin* Benth). J Phytol 2:11–17

Tahir M, Riniarti D, Ersan E, Kusuma J (2019) Genetic and leaf characteristic diversity on 10 mutant progenies of patchouli (*Pogostemon cablin*) provide insights to selection strategies. AGRIVITA, J Agric Sci 41(1):139–148

Van de Wiel CC, Schaart JG, Niks RE, Visser RG. (2010) Traditional plant breeding methods. Wageningen UR Plant Breeding

Widoretno W. (2016) In vitro induction and characterization of tetraploid Patchouli (*Pogostemon cablin* Benth.) plant. Plant Cell, Tissue and Organ Culture (PCTOC). 125(2):261–267

Wu L, Wu Y, Guo Q, Li S, Zhou K, Zhang J (2011) Comparison of genetic diversity in *Pogostemon cablin* from China revealed by RAPD, morphological and chemical analyses. J Med Plants Res 5(18):4549–4559

Wu Y, Li M (2013) Induction of tetraploid plants of *Pogostemon cablin* (Blanco) and its quality evaluation. Pharmacognosy J 5(6):281–285

Yan HJ, Xiong Y, Zhang HY, He ML (2016) In vitro induction and morphological characteristics of octoploid plants in *Pogostemon cablin*. Breed Sci 66(2):169–174

Yusniwati Y, Setiawan RB, Syarif Z, Yusmarni Y, Fitriawati F (2021) Direct organogenesis of patchouli [*Pogostemon cablin* Benth]. IOP Conf Ser Earth Environ Sci 741. https://doi.org/10.1088/1755-1315/741/1/012048

Zarrelli A (2022) Plants as biofactories to produce food, medicines, and materials for a true green revolution. Int J Mol Sci 23(10):5827. https://doi.org/10.3390/ijms23105827

Zeng J, Liu J, Lu C, Ou X, Luo K, Li C, He M, Zhang H, Yan H (2020) Intercropping with turmeric or ginger reduce the continuous cropping obstacles that affect *Pogostemon cablin* (patchouli). Front Microbiol 2526

Chapter 8
Advancements in Improvement Strategies for Plumbagin Enhancement in *Plumbago zeylanica* L.

Yogesh Chandrakant Suryawanshi, Saurabha Bhimrao Zimare, and Prachi Saurabha Zimare

Abstract *Plumbago zeylanica* L. (Plumbaginaceae) is a widely accepted medicinal plant. It is an important constituent in various formulations used in Ayurveda, Unani, and Siddha. These earliest documented evidences have now become the basis of modern medicine and an important source for future medicine and therapeutics. Roots of *P. zeylanica* are a rich source of a naphthoquinone 'plumbagin' known to have numerous traditional and ethnobotanical uses, and therapeutic and pharmacological properties. The demand for the roots of *P. zeylanica* is increasing for plant-based medicines, health products, pharmaceuticals, cosmetics, etc. in the national and international markets. This growing demand is creating a heavy pressure on the natural plant populations in the wild due to over-harvesting for roots. There is a need to focus on the conservation and cultivation of such medicinal plants used by the pharma-companies and industries considered to be endangered or threatened. There is a need to develop an alternative to the intact plant for production of plumbagin in order to make available plumbagin for pharmaceutical industries without disturbing and depleting natural plant resources. For the enhancement of plumbagin from *P. zeylanica* several strategies are employed such as media optimization, elicitation, biotransformation, *Agrobacterium*-mediated transformation and scale-up. Various plant tissue culture (PTC) strategies have proved useful for the enhancement of secondary metabolites from *P. zeylanica.* This chapter brings to light an overview of the origin, distribution, taxonomical classification, economic importance, cultivation

Y. C. Suryawanshi
Department of Yoga and Naturopathy, Vishwakarma University, Pune, India
e-mail: yogesh.suryawanshi@vupune.ac.in

S. B. Zimare
Department of Botany, Sir Parashurambhau College (Autonomous), Pune, India

P. S. Zimare (✉)
Department of Botany, Comrade Godavari Shamrao Parulekar College of Arts, Commerce and Science, Talasari, Palghar 401606, Maharashtra, India
e-mail: prachik16@gmail.com

J. M. Al-Khayri et al. (eds.), *Biodiversity and Genetic Improvement of Medicinal and Aromatic Plants II*, Advances in Plant Breeding Strategies 10,
https://doi.org/10.1007/978-3-031-81857-8_8

practices, genetic resources conservation approaches, and molecular breeding and their applications for the conservation and plumbagin enhancement in *P. zeylanica.*

Keywords Adventitious roots · Chitrak · Micropropagation · Plant tissue culture · Plumbagin

8.1 Introduction

The genus *Plumbago* (family Plumbaginaceae) includes 10 genera and 280 species (Shukla et al. 2021). *Plumbago indica* L., *Plumbago auriculata* L. and *Plumbago zeylanica* L. are the important species. *P. zeylanica* L. is one of the most promising species due to its medicinal properties. It is found growing as a weed in the wild and has a wide distribution throughout the tropical and sub-tropical regions of the world (POWO 2022) in the deciduous woodland, savannas' and scrublands from sea-level up to 2000 m altitude (Pant et al. 2012). It is commonly referred as Ceylon leadwort, doctor bush and wild leadwort, while in India it is known as '*Chitraka*'. *P. zeylanica* is a widely accepted and important medicinal plant used in Ayurveda, Unani, Siddha and many traditional systems of medicines. It is enlisted in the Ayurvedic Pharmacopoeia of India (2001) and Quality Standards of Indian Medicinal Plants (2008) which states that *P. zeylanica* (Śveta Citraka) roots are commonly used. *P. zeylanica* has been mentioned and extensively used in Ayurvedic classics (*Saṁhitās*) and Ayurvedic drug lexicons (*Nighaṇṭus*) (Akhilraj and Rukmini 2021). In traditional system of medicine, root and root bark have been used in several preparations to cure various diseases and disorders like loss of appetite, indigestion, helminthiasis, peptic ulcers, dysentery, inflammations, piles, worms, colitis, ascites, anaemia, epilepsy, hysteria, rheumatic affections, bronchitis, elephantiasis, obesity, and used in the treatment of liver diseases, rheumatism, laryngitis, scabies, disease of spleen, itchy skin, leukoderma, leprosy, psoriasis, prurigo and other skin diseases, and for amenorrhoea and contraception (Akhilraj et al. 2019; Shukla et al. 2021). These earliest documented evidence of the medicinal uses of *P. zeylanica* became the basis of modern medicine and an important source for the future medicine and research in pharmacotherapeutics. Numerous investigations revealed the presence of several phytoconstituents of interest in *P. zeylanica* plant parts. Among the profuse therapeutic phytoconstituents, 'plumbagin' (5-hydroxy-2-methyl-1,4-naphthoquinone), a naphthoquinone is the active pharmaceutical ingredient found in the root possessing diverse pharmacological properties (Singh et al. 2020). Thus, the demand for the *P. zeylanica* roots is increasing for plant-based medicines, health products, pharmaceuticals, cosmetics etc. in the national and international markets. This growing demand is creating a heavy pressure on the natural plant populations in the wild due to over-harvesting the plant for roots. Also, uprooting of the plant has made the population vulnerable and in near future will soon be threatened. Besides this, the chemical synthesis of plumbagin is not economically feasible because of its compound structure, specific stereo-chemical requirement of the compound and high cost of chemicals (Ahmad et al. 2015). Thus, there is a need to focus on the conservation and cultivation of such medicinal plants used by the pharma-companies and industries. Different strategies should be developed not only for the enhancement of such valuable phytochemicals of interest in order to suffice the demand, but also for its conservation to avoid over-exploitation and depletion of natural plant resources.

8.1.1 Taxonomy, Distribution and Description

As Plumbaginaceae show affinities with both Centrospermae and Primulales, its taxonomic position is ambiguous. According to the Bentham and Hooker's classification, taxonomic profile of *Plumbago zeylanica* L. has been given as follows (Naik 1984; POWO 2022) – (Table 8.1).

P. zeylanica is found in the tropical and sub-tropical regions of the world. It is native to South Asian countries and is widely distributed in India, Andaman Islands, Sri Lanka, Bangladesh, Nepal, South-Central and Southeast China, Pakistan, Oman, Saudi Arabia, Africa, South America, and Australia (POWO 2022). *P. zeylanica* has its occurrence throughout India (Kakade 2017) (Fig. 8.1).

P. zeylanica is an evergreen, perennial undershrub, 2 to 4 feet in height. Leaves 3–8.5 × 2–4.5 cm, ovate, or ovate-lanceolate, glabrous, rounded at the base, acute, petiole with two auricled bases, and reticulated vein in shapes. Inflorescence of terminal spikes. Calyx long, persistent, densely covered with stalked glandular hairs. This tube-shaped calyx having glandular trichomes secret sticky mucilage, which attracts insects, ants, etc. Corolla white, lobes oblong, apiculate, style glabrous, 5-fid. Capsules oblong, pointed, enclosed in stalked glandular persistent calyx. The pericarp is thin below and hardened. Seeds solitary, reddish black. Flowering and fruiting between March to December (Singh et al. 2001; Bhagat et al. 2008) (Fig. 8.2). Vernacular names of *P. zeylanica* are presented in Table 8.2.

Table 8.1 Taxonomical classification of *P. zeylanica*

Kingdom	Plantae
Division	Angiospermae
Class	Dicotyledonae
Sub-class	Gamopetalae
Order	Caryophyllales
Family	Plumbaginaceae
Genus	*Plumbago*
Species	*zeylanica* L

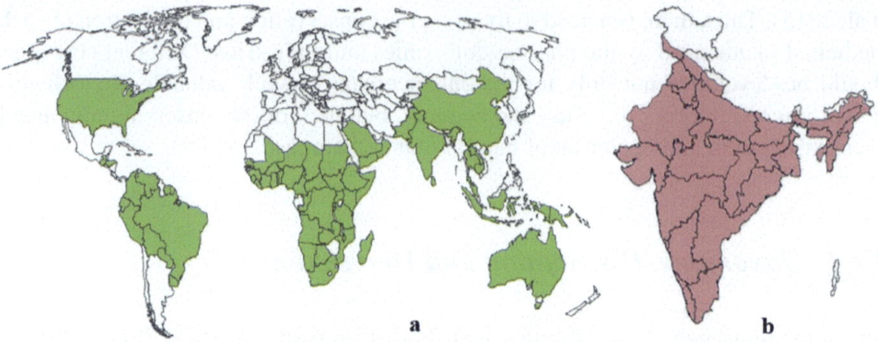

Fig. 8.1 **a** Geographical distribution of *Plumbago zeylanica* in **a:** the world (POWO, 2022), and **b** India (https://www.nmpb.nic.in)

Fig. 8.2 *Plumbago zeylanica* **a:** habitat, **b:** inflorescence axis, **c:** flowers

8.1.2 Economic Importance and Health Benefits

Roots of *P. zeylanica* are extensively employed in the preparations of numerous commercial medicinal herbal products such as Dashmularishtha, Chitrakadivati, Chitrak Capsules, Chitrak Haritaki, Chitrak powder, and many more (Chitrak Products Sources 2022). The projected annual trade for the roots or whole plant of *P. zeylanica* is approximately 500–1000 metric tons, thus, the National Medicinal Plants Board (NMPB), India has listed *P. zeylanica* as high demanded medicinal plant (https://www.nmpb.nic.in/medicinal_list).

Table 8.2 Vernacular names of *Plumbago zeylanica*

Language	*Vernacular names of P. zeylanica*
English	Leadwort, White plumbago, Wild plumbago
Hindi	Cheeta, Chitra, Chiti
Marathi	Chitramul, Chitraka
Tamil	Chittiri
Telugu	Chitramulam
Malayalam	Vellakeduveli
Punjabi	Chitra
Bengali	Chita, Chitu
Arabic	Sheetaraj
Spanish	Bella Emilia, Pegosa, Centella
German	Ceylo-nische Bleiwurz, Bleiwurz, Zahnkraut
Mandarin	Snowflakes, Báixuě huā

Source Ganesan and Gani (2013), Peng and Zhang (2000), POWO (2022), Choudhary et al. (2021)

The noteworthy phytochemicals in *P. zeylanica* makes it a highly priced pharmaceutically important medicinal plant. It is known to exhibit a broad spectrum of pharmacological activities like anti-bacterial, anti-fungal, anti-inflammatory, anti-atherogenic, anti-diabetic, anti-hyperglycemic, anti-cancer, anti-oxidant, anti-carcinogenic, anti-fertility, anti-atherosclerotic, hepatoprotective, neuroprotective, central nervous system stimulatory, cytotoxic, cardiotonic, wound healing, and radio-modifying property (Singh et al. 2020; Sundari et al. 2020; Shukla et al. 2021).

Numerous phytochemicals have been reported from different plant parts of *P. zeylanica* which have been reviewed in detail (Kishore et al. 2012; Roy and Bharadvaja 2017; Shukla et al. 2021). The roots of *P. zeylanica* contains the most significant compound 'plumbagin'. Apart from plumbagin, literature reports major naphthoquinones (chitranone, 3-biplumbagin, chloroplumbagin, maritinone, elliptinone), flavonoids, alkaloids, glycosides, saponins, steroids, tannins, triterpenoids, coumarins (seselin, 5-methoxy seselin, xanthyletin, suberosin), carbohydrates, phenolic compounds, fixed oils, fats and proteins. Other compounds include plumbagic acid, β-sitosterol, 2-dimethyl-5-hydroxy-6-acetylchromene, saponaretin, isoaffinetin, zeylinone, isozeylinone, plumbazeylanone, glucose, fructose vanilic acid, isoshinanolone, and trans-cinnamic acid (Roy and Bharadvaja 2017; Shukla et al. 2021). Essential oil (0.11%) is obtained from the aerial parts of *P. zeylanica.* The GC–MS analysis identified presence of 48 bioactive compounds from the essential oil which include carotol, dauca-4(11),8-diene, germazone, viridiflorol, and longifolene (Liu et al. 2022).

Plumbagin (5-hydroxy-2-methyl-1, 4-napthoquinone; formula: $C_{11}H_8O_3$) (Fig. 8.3) is an organic compound, a yellow dye derived from a simple hydroxy-naphthoquinone. It is named after the plant genus *Plumbago*, from which it was originally isolated from the roots and aerial parts of *Plumbago pearsonii* L. Bolus,

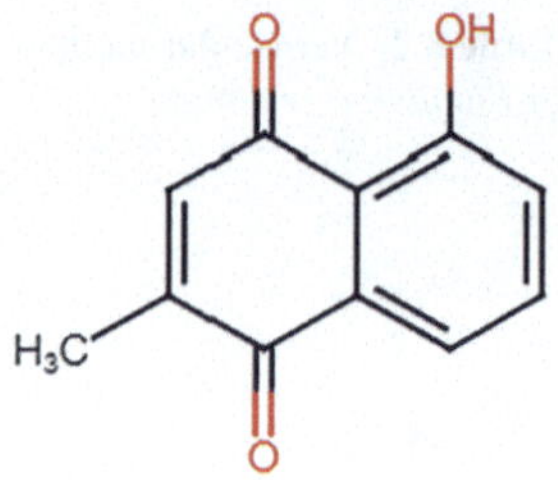

Fig. 8.3 Structure of Plumbagin

Dyerophytum africanum (Lam.) Kuntze., and in roots of *P. auriculata* Lam. (van der Vijver 1972). Plumbagin has been reported from the families Plumbaginaeace (van der Vijver 1972), Drosophyllaceae (Crouch et al. 1990), Droseraceae (Jayaram and Prasad 2005), Nepenthaceae (Raj et al. 2011) Ebenaceae (Padhye et al. 2012) and Juglandaceae, Iridaceae, Dioncophyllaceae, and Anicistrocladaceae (Checker et al. 2009). Plumbagin has numerous therapeutic, traditional, ethnobotanical and pharmacological uses such as anti-microbial, anti-fungal, anti-bacterial, anti-leishmanial, anti-parasitic, anti-malarial, anti-inflammatory, anti-oxidant, anti-diabetic, anti-tumor, anti-fibrotic, anti-depressant, anti-carcinogenic, cyto-protective, neuroprotective, anti-fertility, anticoagulant, hypolipidemic, and hyperglycemic activities, and has revealed anti-cancer activity against numerous types such as lung, liver, pancreatic, prostrate, ovarian, breast, cervical, colon, esophageal and gastric cancers; and acute promyelocytic leukemia, lymphoma, myeloma, osteosarcoma, melanoma, tongue and oral squamous cell carcinomas (Badwaik et al. 2019; Kakade et al. 2022).

8.1.3 Domestication, Selection and Early Improvements

Although *P. zeylanica* is extensively distributed throughout Asia, its evolutionary record is unclear. Nevertheless, the wide distribution signifies its comprehensive genetic base (Panda et al. 2015). Three *Plumbago* species are now brought into cultivation and propagation namely, *P. zeylanica, P. indica, P. auriculata,* of which *P. zeylanica* is widely cultivated. *P. zeylanica* is native to South East Asia and it is mostly cultivated in India. A list of major research institutes engaged in *P. zeylanica* research is listed in Annexure—I.

8.2 Current Cultivation Practices and Challenges

8.2.1 Current Cultivation Practices

Karnick et al. (1982) for the first-time reported cultivation technique of *P. zeylanica*. Plants collected from the wild were grown in polybags and later transferred to soil beds made up of three layers viz. lower thick layer of black cotton soil, covered with a middle layer of loamy humus and top layer of river sand. Fertilizers N:P:K (10:10:10) was applied thrice a year and watered daily.

The 10–15 cm stem-cuttings having 2–3 nodes are used for propagation in March–April to raise the stock. For best result, cuttings are treated with 500 ppm Naphthalene Acetic Acid (NAA) for fast rooting and maintained in the mist chamber with proper humidity, with a survival rate of 80–100%. In open conditions, low humidity can result in increased mortality rate of 70–90%. Root initiation is observed within one month of planting. To cultivate *P. zeylanica* on one hectare of land, approximately 80,000 rooted stem-cuttings are required. *P. zeylanica* can grow in a wide range of soils, varying from red soil, with very little topsoil, to deep black soil, and since *P. zeylanica* cannot withstand in waterlogged conditions, it prefers well drained sandy-loam to clay-loam soil having high organic content (Pant et al. 2012). Rooted cuttings or plantlets developed in nursery are planted in the beginning of monsoon, and transplantation is done after 60–90 days of growth with a recommended spacing of 50 $\times$ 25 cm in order to produce maximum root biomass.

In nursery, seeds are sown in polybags containing sand: soil: FYM (1:1:1) in March. Seeds are scarified or cut at the micropylar end prior sowing, since it shows poor germination. Seed sprouting is observed in 21–30 days followed by transplantation of seedlings in field at a density of 58 $\times$ 58 cm. Intercropping of *P. zeylanica* is also practiced along with many fruit and medicinal trees (Tyagi and Menghani 2014).

The conventional method of propagation using seeds is not a reliable approach, because of poor seed setting, germination and early and high mortality rate of seedlings under normal conditions after transplantation. While these agro-techniques are problematic and insufficient to meet the growing the global demand, advanced propagation approaches for mass production should employed.

8.2.2 Current Agricultural Challenges

After transplanting the stem cuttings in the field, the plant attains its maturity within 10–12 months. Harvesting generally takes place after 12 months. The period of cultivation is comparatively higher than other food crops. Seed collection and its use for cultivation are comparatively rare than the stem cutting method. Proper water management is necessary. The plant is widely present in Asia and Africa where water problems in agriculture are severe. Because of urbanization and industrialization, the soil quality has worsened. Although *P. zeylanica* grows in a wide variety of soil, it is important to check the soil health before cultivation.

P. zeylanica is mostly cultivated for the extraction of plumbagin from its roots and it is difficult to know the exact timing of the plant where it produces the highest amount of plumbagin in its life cycle. The concentration of plumbagin changes according to the environmental conditions. The biotic and abiotic factors also affect the secondary metabolite production in the plant. The pharmaceutical industries depend on the roots obtained from *P. zeylanica* which leads to rapid depletion, raising concerns about its probable vulnerability (Roy and Bharadvaja 2019). To overcome these challenges genetic engineering, plant tissue culture, molecular breeding, gene editing, micropropagation, hybridization etc. tools find wide application.

8.2.3 Genetic Improvement Objectives

Regardless of the advantages, applications and significance, medicinal plants have been positioned in the last ring of domestication. Plant domestication is an evolutional process which develops new and productive varieties of plants from wild species through genetic modification (Niazian 2019). Little is known about the domestication of medicinal plants, and a small number of medicinal plants are cultivated whereas majority of them are still collected from wild. The phytochemical constituents are directly affected due to the genetic, environmental, and ecological variables (Naik and Al-Khayri 2016; Mankar et al. 2021) and not only influence their organoleptic characteristics, but also contributes to the phenotypic variation within and between populations. The common set of traits in the domesticated plants that distinguished them from the wild progenitors is referred to as domestication syndrome. Medicinal plants are deprived of this domestication syndrome. These deficiencies need to be compensated using alternative biotechnological approaches as shown in Fig. 8.4 (Niazian 2019). Researchers can overcome these ambiguities using biotechnology and genetics. Biotechnological approaches like micropropagation, cell culture and bioreactor production, gene transfer in plant and microbes, modification of biosynthetic pathways, etc. The entire genomic and transcriptomic data can be obtained using next-generation sequencing (NGS) techniques, which can further be incorporated with proteomics and metabolomics information to characterize the unidentified compounds in these medicinally important plants (Zhao et al. 2018). NGS along with DNA barcoding can facilitate in the revealing the genetic diversity and classification at the genus and species level (Techen et al. 2014). An understanding of the patterns of genetic variation within and among populations of medicinal plants is essential for formulating optimal genetic resource management approaches for their conservation, sustainable utilization and genetic improvements.

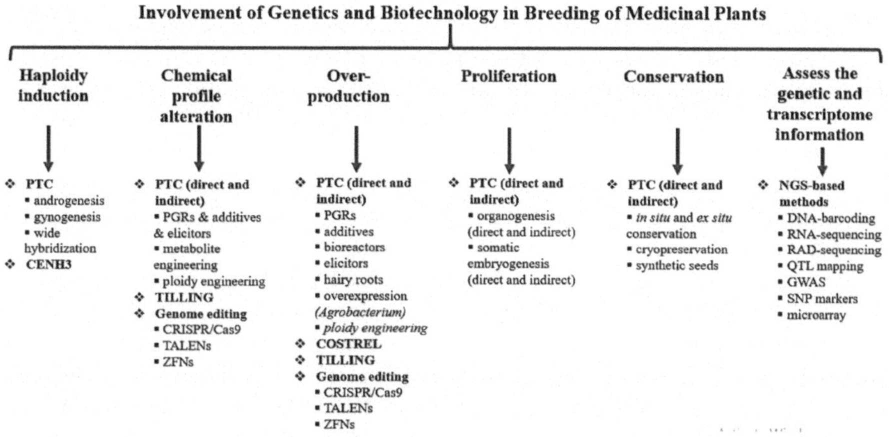

Fig. 8.4 Different applications of Next-Generation Sequencing (NGS) and plant tissue culture (PTC) techniques in improvement of medicinal plants. *CENH3*: centromere histone H3; *COSTREL*: combinatorial supertransformation of transplastomic recipient lines; *CRISPR*: clustered regularly interspaced short palindrome repeats; *Cas*: CRISPR-associated; *GWAS*: genome-wide association; *PGRs*: plant growth regulators; *TALENs*: transcription activator-like effector nucleases; *TILLING*: targeting-induced local lesions in genomes; *ZFNs*: zinc-finger nucleases. Source: Niazian (2019)

8.3 Germplasm Biodiversity and Conservation

Seeds, plants, plant parts, tissues that are beneficial for breeding, research and conservation, when maintained or stored, is known as germplasm. The germplasm is the collection of current and obsolete varieties, inbred lines obtained from hybrids or varieties, wild species, elite lines consisting special combinations of traits, breeding lines, superior varieties, natural hybrids, primitive cultivars, and inter-mated populations of elite and selected plant material. Germplasm delivers knowledge about the genetic information of plants, rescues rare, threatened category and important plant species, and thus, plays a key role in plant diversity conservation. Advanced biotechnological approaches, a few to mention, such as tissue culture, embryo culture, cryopreservation, DNA banks have opened new avenues for germplasm conservation (Priyanka et al. 2021; National Research Council 1991).

8.3.1 Germplasm Diversity

8.3.1.1 Morphological and Genetic Diversity

Germplasm diversity study is important for breeders and germplasm curators to understand the utilization of genetic resources and to formulize conservation strategies. Concerns over sustainable usage, conservation, and authenticity of *P. zeylanica* have arisen as a result of the continued exploitation and increased demand. The lack of centralized and trustworthy information on the plants' genetic variety in commercial trade heightened the worry. As a result, accurate information of *P. zeylanica* genetic diversity is required. Genetic diversity is present in *P. zeylanica* as a part of the evolutionary process. Environmental factors also contribute to genetic diversity. *P. zeylanica* is widely present in India which is divided into 12 agro-climatic zones. Plants form different agro-climatic zones are prone to possess different genetic make-up (Mittal and Hariharan 2016). Gadge and Nathar (2014) in their genetic diversity study using random amplified polymorphic DNA (RAPD) markers revealed that *P. rosea* and *P. zeylanica* were closely related to each other compared to *P. auriculata*. For conservation and management purposes it is essential to understand the diversity among the populations of *P. zeylanica*. Nalina et al. (2017) evaluated 42 accessions of *P. zeylanica* collected from different parts of India like Tamil Nadu, Kerala and Maharashtra to identify the elite genotype for yield and plumbagin content based on the investigation of variability in plant growth, root yield and quality. RAPD markers were used to evaluate the genetic integrity of micro-propagated *P. zeylanica* plants, where no polymorphism was detected within the micro-propagated plants (Rout and Das 2002). Haji et al. (2014) employed SSR, SRAP, and ITS markers to determine the inter- and intra-population genetic relationships among the accessions of *P. zeylanica* from different geographical regions of India. Low genetic diversity was observed among populations, while high genetic diversity was observed at accession level, which could probably be due to the geographical isolation and genetic drift. High level of genetic diversity using inter-simple sequence repeat (ISSR) markers was seen in the *P. zeylanica* populations of Southern Western Ghats of India (Dharmar and De Britto 2012). Genotypic and seasonal variation in plumbagin content in roots of *P. zeylanica* from various populations in India and its correlation with expression analysis of polyketide synthase (*PKS*) gene was studied. Plumbagin content varied from different localities and was genotype-dependent. The *PKS* gene correlated with the plumbagin content indicating its role in plumbagin biosynthesis (Panda and Kamble 2014). Panda et al. (2015) assessed genetic structured variation within and among *P. zeylanica* populations by ISSR and RAPD markers. Higher genetic diversity and genetic differentiation was observed at the species and population level. Furthermore, their findings indicated higher genetic diversity in northern and eastern populations as compared to the southern and western populations implying the probable loss of genetic diversity in the course of the dissemination of *P. zeylanica* to Southern India. A recommendation was made for large-scale domestication for commercial purpose along with an urgent need for conservation of prevailing natural populations. Roy et al. (2022) demonstrated the use of fifteen SCoT (Start codon targeted) and twenty CBDP (CAAT Box Derived Polymorphism) markers in the genetic diversity analysis of thirteen accessions of *P. zeylanica*. *P. zeylanica* was collected from the natural habitat of untouched geographical zones of Gangetic plains of India to identify the elite chemotypes on the basis of plumbagin content. Significant difference in plumbagin content among the populations validated chemotaxonomic variation among the populations which could be as a consequence of the shift in the environmental as well as phytogeographical aspects (Shukla et al. 2020).

8.3.2 *Genetic Resources Conservation Approaches*

Conservation of *P. zeylanica* species is becoming more important because of the increasing demand and limited supply, over-exploitation, indiscriminate collection, habitat destruction, and uncontrolled deforestation. Although, sustainable utilization of wild resources can be an effective conservation alternative, several sets of recommendations concerning to its conservation such as in situ and ex situ conservation are being developed. The conservation strategies vary in different plants, depending on their specificity of habitat, distribution range, population size, species diversity, growth rate, and reproductive system (Chen et al. 2016). Figure 8.5 demonstrates the strategies beneficial in the conservation of *P. zeylanica.*

8.3.2.1 In Situ Conservation

In situconservation maintains natural communities accompanied by their complex network of associations, makes stronger the connection between resource conservation and sustainable usage, and enhances the amount of diversity to be conserved. Natural reserves like sacred groves, national parks, and gene sanctuaries are protected areas created to preserve and restore biodiversity of important wild resources. In India, *P. zeylanica* is conserved in its natural habitat in number of sacred groves of Chhindwara (Madhya Pradesh), Pune, Dapoli, and Kolhapur (Maharashtra), Belgavi (Karnataka), Madhya Pradesh, Ariyalur and Thanjavur (Tamil Nadu); and national parks like Chandoli National Park (Maharashtra), Dudhwa National Park (Uttar Pradesh), Eravikulam National Park (Kerala), Guindy National Park (Tamil Nadu).

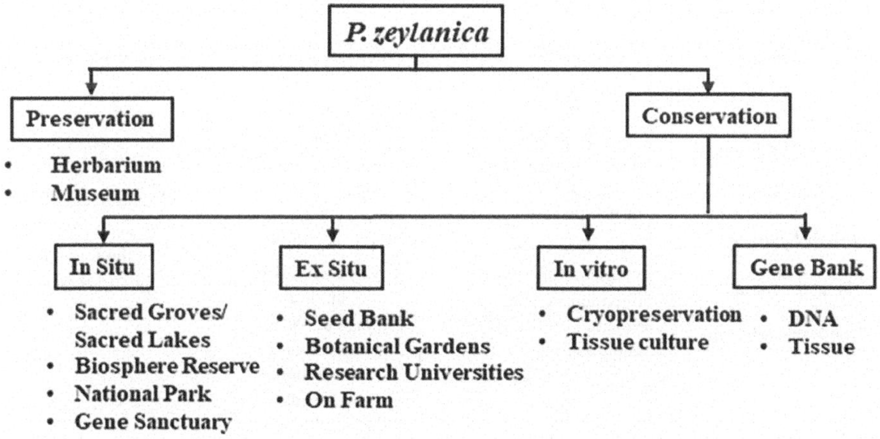

Fig. 8.5 Strategies for the conservation of *Plumbago zeylanica*

8.3.2.2 Ex Situ Conservation

Ex situconservation means the conservation of genetic resources or germplasms outside the natural habitat in facilities which have provision for storage and/or preservation having maintained conditions best suited for the viability and genetic constitution. Ex situ conservation should be looked upon as a vital complementary approach for conservation and use of the biological diversity, aiming at those populations that are challenging to conserve in situ.

In India *P. zeylanica* is well preserved in various centres like Indian Council of Agricultural Research (ICAR), Council of Scientific and Industrial Research (CSIR), National Medicinal Plants Board (NMPB), sub-centres like Regional-Cum-Facilitation Centre (RCFC), etc., along with botanic gardens, nurseries, seed banks, research institutes, and experiment stations. It effectively safeguards and caters the required supply of germplasm for research and breeding. Figure 8.6 shows various possible ex situ approaches for the conservation of *P. zeylanica.*

Botanic Gardens

Botanic gardens have a unique feature of growing huge variety of plant species belonging to taxonomically and ecological diverse environments together in common conditions. Numerous institutions having large areas of botanic gardens have conserved *P. zeylanica* for research and breeding programmes.

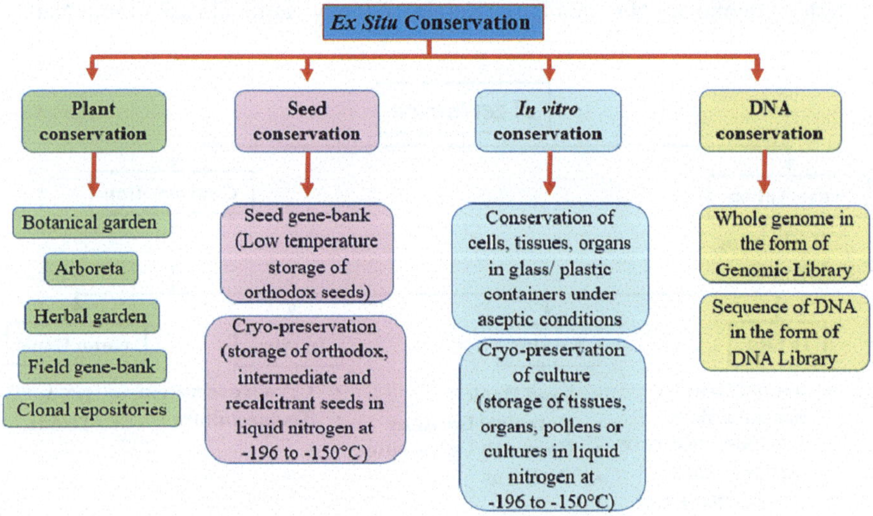

Fig. 8.6 Ex situ approaches for *Plumbago zeylanica* conservation

Cryopreservation

In the cryopreservation method, the plant material is stored at extremely low temperature (-196 to -150 °C) in liquid nitrogen to arrest the metabolic activities and cell division, and plant material can be stored for long period without genetic changes. By using cryopreservation method ICAR-National Bureau of Plant Genetic Resources (NBPGR) has conserved the 19 accessions of *P. zeylanica* (Agrawal et al. 2019).

DNA Bank

DNA bank is an emerging technique in genetic resources conservation. Routine extraction of DNA or RNA sequences, frozen, viable cell cultures and derivatives such as RNA and cDNA are counted as genetic resources and preserved in DNA banks. DNA storage is the most notable alternative for conventional methods for germplasm storage. Significant advancement in genome analysis of various species has been possible due to the latest advances in DNA sequencing technology. The nucleotide sequence data produced are submitted and registered in National Centre for Biotechnology Information (NCBI/GenBank), DNA Data Bank of Japan (DDBJ) and European Bioinformatics Institute (EMBL/EBI). *P. zeylanica* DNA extraction has been has been demonstrated by researchers (Panda et al. 2015; Dharmar and De Britto 2012; Roy et al. 2022).

8.3.2.3 Cytogenetics

Cytogenetical data is essential for developing the chromosome database. Previous studies showed that *P. zeylanica* contains 16 chromosomes (Dahlgren 1964; Taxon 1979), however recent studies confirmed that *P. zeylanica* has somatic chromosome number of $2n = 28$ (Paiva and Leitao 1987; Khatoon and Ali 1993). The total chromosome length of *P. zeylanica* is observed as 92.81 ± 2.98 μm. The difference between the smallest and largest chromosome in *P. zeylanica* was approximately 3.5 μm, indicating heterogenous nature of chromosome length. The fluorescent banding analysis using CMA and DAPI revealed eleven CMA bands and six DAPI bands (Khatun et al. 2019). Majumder et al. (2022) presented a comparative cytological analysis of in vitro and in vivo grown *P. zeylanica* plants, which confirmed the somatic chromosome number as $2n = 28$. Of the fourteen basic chromosomes, three chromosomes were submetacentric while eleven were metacentric chromosomes.

8.4 Molecular Breeding

8.4.1 Molecular Marker-Assisted Breeding

Metabolic engineering aims in the enhancement of production of a specific metabolite through overexpressing transcription factors or inhibition of metabolic enzymes that control the metabolic pathways involved in the accumulation of target metabolites. Molecular breeding was performed using the plumbagin single-chain variable fragment (PL-scFv) that was constructed for the enhancement of plumbagin production in *P. zeylanica* hairy roots. Significantly positive correlation between the PL-scFv expressed in hairy roots might influence the plumbagin biosynthesis pathway and increase accumulation of plumbagin in recombinant plants (Sakamoto et al. 2012).

8.4.2 Functional Genomics

Decoding of novel gene functions in medicinally important plant constituents has now become effortless due to the functional genomics approach in consort with targeted metabolite and transcriptome profiling. In the recent years, genome sequencing and transcriptomics of medicinal plants has escalated due to the cost efficient high-performance next generation sequencing technologies. (Sundari BKR et al. 2017) performed RNA sequencing and transcriptome analysis in the root tissues of *P. zeylanica* to contemplate possible pathways and phytoconstituents having pharmacological functions. The candidate gene transcripts of terpenoids and quinines characterized in transcriptome analysis of root was involved in the biosynthesis of naphthoquinone, one of the most noteworthy constituent of *P. zeylanica.*

8.5 Tissue Culture Applications

Plumbagin is has been exploited from the roots of *P. zeylanica, P. indica*, and *P. auriculata.* However, the growth of these plants is slow and it requires 2–3 years to obtain a good amount of plumbagin from the roots. The traditional methods are time-consuming and less profitable. It is important to find alternative methods for enhanced plumbagin production. Although there are other plants from families like Droseraceae, Ebenaceae, Nepenthaceae, and Drosophyllaceae from which plumbagin has been reported, but the concentration is very low and these are more complicated in cultivation than *P. zeylanica* (Badwaik et al. 2019). New methods like aquaponics, hydroponics and plant tissue culture are now gaining attention for maximizing the secondary metabolite production (Suryawanshi 2021). Plant tissue culture is the essence of biotechnology in medicinal plants and helps in the conservation of medicinal plants, enhances production of active phytoconstituents/ secondary metabolites,

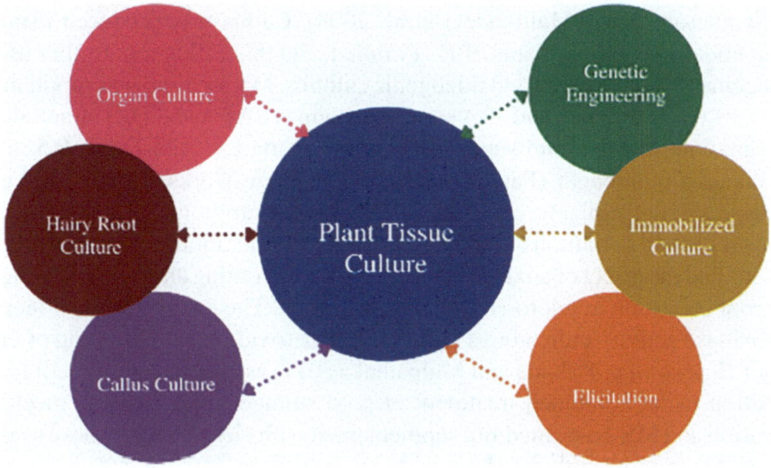

Fig. 8.7 Different techniques used in plant tissue culture for enhancement of secondary metabolites in *Plumbago zeylanica*

and helps to develop novel forms of secondary metabolites that are important for food, pharmaceutical, and other industries. It is an alternative and effective method for plant metabolite synthesis without impacting the natural flora and independent of environmental or geographical variations. For the enhancement of these valuable phytoconstituents several strategies are employed such as media optimization, elicitation, biotransformation, *Agrobacterium*-mediated transformation and scale-up. Various PTC strategies useful for the enhancement of secondary metabolites from *P. zeylanica* have been shown in Fig. 8.7.

8.5.1 Micropropagation Approaches

In vitro micropropagation can be achieved through organogenesis and somatic embryogenesis, both via direct and indirect pathways. It is an alterable process and numerous factors like plant genotype, explant type and size, types of plant growth regulators (PGRs) and their concentrations, age of explant, environmental factors of culture medium such as light intensity, temperature, photoperiod, pH of medium, carbohydrate source, additives, and gelling agent can influence the outcome of *in vitro* propagation in different medicinal plants (Niazian 2019).

Initial work on direct regeneration of *P. zeylanica* has been reported by Rout et al. (1999) and Das and Rout (2002). Further, many researchers established direct regeneration protocols in *P. zeylanica* (Chaplot et al. 2006; Gbadamosi and Egunyomi 2010; Sivanesan and Jeong 2010; Chandravanshi et al. 2014; Ceasar et al. 2013; Vijay et al. 2016; Majumder and Rahman 2016). Effective methods of indirect plant regeneration from leaf callus cultures of *P. zeylanica* have been described (Rout et al.

1999; Sivanesan, 2007; Mallikadevi et al. 2008). Callus was produced using leaf disc and nodal segment explants of *P. zeylanica*, and this callus was further used for embryogenic cell suspension and rhizogenic cultures. MS medium was supplemented with auxins (2, 4-D, NAA and IBA) and cytokinins (BAP and TDZ) either alone or in combination. MS medium supplemented with 3 mg L^{-1} 2, 4-D and 0.5 mg L^{-1} IBA enhanced plumbagin (Patidar et al. 2015). These works included the use of different sources of explants, effects of different concentrations and combinations of PGRs, and growth conditions to attain optimized regeneration system. Saxena et al. (2000) studied the effect of auxins and photoperiod on rooting and peroxidase activity during root induction in micro-shoots of *P. zeylanica*. The auxin treatment increased the peroxidase activity indicating a major role of peroxidase in the rooting of in vitro shoots of *P. zeylanica*. Kakade and Malpathak (2017) established a protocol for seed germination using combined treatment of seed storage, seed soaking, mechanical scarification and MS-basal medium supplemented with GA_3. Further, these seedlings were used for in vitro multiple shoot and root induction.

8.5.2 Embryo Rescue

Gbadamosi and Egunyomi (2010) studied the effect of culture media and growth regulators on the in vitro response of embryo and nodal cuttings of *P. zeylanica*. 20–80% embryos showed viability in MS medium supplemented with NAA and BAP.

8.5.3 Synthetic Seeds

A synthetic seed is defined as encapsulated plant tissues, such as somatic embryos, shoot buds, axillary buds, shoot tips, cell aggregates, and any other micro propagules that have the potential to grow like a plant, under in vivo or in vitroconditions when sown as seeds. Simply meaning, encapsulated somatic embryos that functionally resemble seeds and which can grow into seedlings under sterile environment. Synthetic seed technology has significant utilization in large-scale generation of commercially important plants that are difficult to propagate via conventional vegetative techniques.

Synthetic seeds of *P. zeylanica* were developed with an objective of conservation. Leaf explants of *P. zeylanica* were inoculated of MS medium supplemented with different PGRs. Significant callusing response was obtained on MS medium supplemented with NAA (2 ppm). Embryoid callus was produced which was used for the production of synthetic seeds. The embryoid callus was minced and transferred to the liquid MS media-sodium alginate mixture, and then shifted to 50 mM $CaCl_2$ solution for encapsulation. Uniform and firm beads were formed on a short duration exposure to $CaCl_2$, and a high germination rate endorsed practical application of *P. zeylanica* synthetic seeds (Jain et al. 2018).

8.5.4 Enhancement of Plumbagin

"Elicitor can be defined as a substance for stress factors which, when applied in small quantity to a living system, it induces or improves the biosynthesis of specific compound which do have an important role in the adaptations of plants to stressful conditions" (Radman et al. 2003). Elicitation is the most practically viable approach for enhancing the production of the desired secondary metabolites from cell, tissue and organ cultures (Naik and Al-Khayri 2016). Based on the nature of elicitors, these can be divided into abiotic (substances of non-biological origin, and grouped as physical, chemical, and hormonal factors) and biotic (substances of biological origin e.g. chitin, pectin, and cellulose which are polysaccharides from plant cell walls and micro–organisms).

Roots of *P. zeylanica* are the major source of plumbagin, hence number of the plant tissue culture experiments were carried out in root cultures for enhanced plumbagin production. Since plumbagin accumulates in the roots of *P. zeylanica*, root culture happens to be an essential method for mass scale production. Sharma and Agrawal (2018) selected high plumbagin yielding *P. zeylanica* chemotypes using HPLC and established an in vitro regeneration protocol using in vitro roots. They further enhanced maximum plumbagin in root callus of *P. zeylanica* by employing exogenous elicitors namely, 100 mg L^{-1} yeast extract (YE) and salicylic acid (SA) 25 μM. Roy and Bharadvaja (2018) examined the effects of various carbon sources (3% concentration of glucose, sucrose and fructose), nitrogen sources (ammonium nitrate, sodium nitrate and potassium nitrate), and plant tissue culture media (MS medium, Gamborg's B5 medium, White medium and Nitsch medium) on shoot multiplication of five accessions of *P. zeylanica.* The results showed that all accessions developed shoots, but highest shoot multiplication was recorded in accession IC-524441 grown on MS medium supplemented with 3% sucrose and ammonium nitrate. In another investigation by Roy and Bharadvaja (2019) adventitious roots were induced from nodal explants of *P. zeylanica.* Further, culture conditions were optimized for root suspension culture. Various concentrations of abiotic (methyl jasmonate and salicylic acid) and biotic (yeast extract and malt extract) elicitors were used for the enhancement of plumbagin content. Their investigation validated that adventitious root culture can be a substitute for the enhancement of pharmaceutically important phytoconstituents, without hampering the natural population. Singh et al. (2020) induced roots from nodal explants of *P. zeylanica*, and these roots were used for callus induction. The induced root callus was treated with different concentrations of elicitors (biotic- chitosan, yeast extract; abiotic- proline, lysine, salicylic acid; and heavy metals- lead nitrate, and cadmium chloride) and precursors of plumbagin (L-tyrosine, and sodium acetate). HPLC analysis of the elicitors and precursor treated root callus was performed for plumbagin quantification. Their findings showed that root callus had the potential for higher plumbagin production if augmented with appropriate elicitors. Highest plumbagin content was recorded in root callus cultured on MS medium + 5 μM TDZ supplemented with 100 mg L^{-1} yeast extract along in combination with 50 mg L^{-1} chitosan which was about 12 folds higher as compared to control.

Biostimulants are defined as 'materials that, in minute quantities, promote plant growth' (Du Jardin 2015). Seaweed extracts have now gained much wider acceptance as plant biostimulants, because of the presence of compounds that are responsible for plant growth promoting activity (Battacharyya et al. 2015). Kakade et al. (2022) for the first time reported the potential of seaweed liquid extract (SLE) of *Sargassum ilicifolium* for the enhanced growth and plumbagin accumulation in *P. zeylanica*. This study not only established protocols for in vitroseed germination, multiple shoot and root induction of *P. zeylanica*, but has endorsed the use of SLEs for large-scale micropropagation along with enhanced plumbagin production, and use of SLEs as an alternative to the commercially available PGRs.

8.6 *Agrobacterium*-Mediated Genetic Transformation

Verma et al. (2002) established in vitro organogenesis in *P. zeylanica* from nodal explants as a basic prerequisite to conduct *Agrobacterium rhizogenes* (strain A4) mediated hairy-root transformation having potential for in vitro plumbagin production. Wei et al. (2006) intended to produce genetically stable transformants in *P. zeylanica* to assess gene function, for which they optimized an efficient in vitro regeneration system in *P. zeylanica* which was further used for the *A. tumefaciens*-mediated transformation. Hairy-roots were induced in *P. zeylanica* using recombinant *A. rhizogenes* (ATCC15834/SPK). A PL-scFv was constructed as a substitute to elicitor for enhanced plumbagin production in the hairy-roots of *P. zeylanica* (Sakamoto et al. 2012).

8.7 Conclusions and Prospects

8.7.1 An Overview of Current Status

P. zeylanica is a commercially important medicinal plant due to the presence of plumbagin. Numerous reports are available on the extraction techniques of plumbagin from different plant parts of *P. zeylanica*. Immense exploration has been carried out on the pharmacological aspects of plumbagin, that has led to the increasing demand of this phytoconstituent. Natural resources are the prime source of plumbagin, of which *P. zeylanica* is one of the most important sources. The distribution of *P. zeylanica* is throughout the major part of the world, showing its survival in various edapho-climatic conditions. Different plant populations growing in different environmental conditions exhibit variation in their chemical constituents. *P. zeylanica* is hap-hazardly collected from the wild without any prior knowledge of its exact harvesting time when the plumbagin yield is optimum, and lack of information on elite populations having higher plumbagin content. This is leading into rapid depletion of the natural *P. zeylanica* populations raising concern about its existence status. An alternative method of synthesizing plumbagin was through synthetic approach to suffice the increasing demand. This was achieved by several researchers, but the major disadvantages were high expensive procedures, and extensively longer time-consuming reactions resulting in very low yield of the compound.

Thus, it became necessary to develop a more effective and alternative source. The intervention of the advanced biotechnological techniques was the most pragmatic way to amplify plumbagin production in order to fulfil the increasing demand. Plant tissue culture strategies such as in vitro culture techniques, cell culture-based methods, suspension cultures, hairy root cultures, elicitation and genetic engineering were employed successfully which not only enhanced the yields easily, but were at the same time, cost-effective.

8.7.2 Recommendations of Future Research and Utilization

Exploration of novel potentially important plants are still underway that can be used as a source for plumbagin and other important phytoconstituents, and as a substitute for *P. zeylanica.* Attention needs to be given for an extensive bioprospection approach in genomics and transcriptomics of medicinally important phytoconstituents. Although biotechnological techniques have contributed greatly in the facilitation of plumbagin production, advancements would be needed to reap the full benefits of involvement of genetics and biotechnology. The whole genomic and transcriptomic data of *P. zeylanica* can be explored using NGS. Further, this information can be integrated with metabolomics and proteomics to uncover novel phytocompounds. A conjecture of genetic diversity can be provided using NGS data for DNA barcoding that can provide reliable classification and can be able to differentiate morphologically similar life forms at any specific taxon level. A comprehensive data of the entire metabolite profile and transcriptome analysis can bring to light the active gene transcripts involved and responsible for the medicinal properties of *P. zeylanica.* Artificial polyploidy induction plays a vital role in breeding and improvement programmes of medicinal plants. It not only improves quantitative performance, but also affects the qualitative (chemical) performance. It has been substantiated that in medicinal plants polyploid genotypes contain more secondary metabolites than diploid genotypes.

Appendix

List of institutes engaged in Research on *Plumbago zeylanica*

Institute name	Specialization research activities	Contact information and website
An-Najah National University	Extraction, Phytochemistry	Omar Ibn Al-Khattab St., PO Box 7, Nablus, Nablus, West Bank, Palestine https://www.najah.edu/en/#
Arab American University	Extraction, Phytochemistry, Plumbagin enhancement	P.O Box 240 Jenin, 13 Zababdeh, Palestine https://www.aaup.edu/#
Arulmigu Palaniandavar Arts College for Women	Antimicrobial activity, Agar well method, Molecular docking	Department of Botany, Arulmigu Palaniandavar Arts College for Women, Chinnakalayamputhur, Palani, Tamil Nadu, India http://apacwomen.ac.in/page/botany-department
Centre for Advanced Research in Indian system of Medicine (CARISM)	Quality Control, Phytoremediation	CARISM, SASTRA University, Thanjavur, Tamil Nadu, India https://www.sastra.edu/research.html
Centre for Medicinal Plants Research (CMPR)	Phytochemistry, Pharmacognosy	CMPR, AVS Square, Changuvetty, Kottakkal Malappuram, Kerala, India https://cmpr-avs.in/
Indian Council of Agricultural Research -National Bureau of Plant Genetic Resources (ICAR-NBPGR)	Genetic Diversity, Molecular Study	ICAR-NBPGR, Pusa Campus, New Delhi, Delhi, India http://www.nbpgr.ernet.in/
Indian Council of Agricultural Research-Directorate of Medicinal & Aromatic Plants Research (ICAR-DMAPR)	Germplasm enhancement, National Repository of the genetic resources	The Director ICAR-Directorate of Medicinal and Aromatic Plants Research Boriavi - 387 310, Anand, Gujarat, India https://dmapr.icar.gov.in/index.htm
Khon Kaen University	Agrobacterium Transformation, Gene Expression	Khon Kaen University, Khon Kaen, Thailand https://kku.ac.th/research-and-innovation
Kyushu University	Hairy Root Culture, Transformation, Gene Expression	Kyushu University, 3–1-1 Maidashi, Higashi-ku, Fukuoka, Japan https://www.kyushu-u.ac.jp/en/research/information/

(continued)

(continued)

Institute name	Specialization research activities	Contact information and website
Medicinal Plant Biotechnology and Applied Research Laboratory	Tissue Culture, Biochemistry	Medicinal Plant Biotechnology and Applied Research Laboratory, Department of Botany, University of Delhi, India http://www.du.ac.in/index.php?page=department-of-botany
Muhammadiyah University of Palembang	Phytochemistry	MUP, Kec. Seberang Ulu II, Kota Palembang, Sumatera Selatan, Indonesia https://www.um-palembang.ac.id/
Nagasaki International University	Hairy Root Culture, Transformation, Gene Expression	Nagasaki International University, 2825-7 Huis Ten Bosch, Sasebo, Nagasaki, Japan https://www.nagasaki-u.ac.jp/en/research/
Ranchi University	Cytogenetics, Phytochemistry, Plant Breeding	Ranchi University, Morabadi, Ranchi, Jharkhand, India https://www.ranchiuniversity.ac.in/index.php?option=com_content&view=article&id=96&Itemid=280
Regional Ayurveda Institute for Fundamental Research (RAIFR)	Agro-techniques Development, Pharmacognostical evaluation	Regional Ayurveda Research Institute Nehru Garden, Kothrud, Pune, Maharashtra, India http://www.ccras.nic.in/content/about-ccras
Regional Cum Facilitation Centre - Western Region (RCFC-WR)	Germplasm conservation	RCFC-WR, Savitribai Phule Pune University Campus, Pune, Maharashtra, India https://rcfcwestern.org/index.html
Savitribai Phule Pune University	Taxanomy, Mycology, Ecology, Pharmacognosy	Savitribai Phule Pune University, Ganeshkhind, Pune, Maharashtra, India http://www.unipune.ac.in/dept/science/Botany/default.htm
Shandong University	Essential Oil, Cytotoxic, Antibacterial Studies,	Marine College, Shandong University, Weihai, P. R. China https://www.en.sdu.edu.cn/

(continued)

(continued)

Institute name	Specialization research activities	Contact information and website
Ubon Ratchathani University	Hairy Root Culture, Transformation, Gene Expression	Ubon Ratchathani University, 85 Warinchamrap, Ubon Ratchathani, Thailand http://www.english.ubu.ac.th/
University of Chittagong	Karyotyping, Cell Molecular studies	University of Chittagong, Bangladesh https://cu.ac.bd/
University of Oklahoma	Genetic Diversity, Transcript Profiling, Gene Expression, Fertilization	University of Oklahoma, Norman, Oklahoma, USA https://www.ou.edu/cas/mpbio/research

Index Keywords

Plumbago zeylanica	Cell cultures
Chitrak	Suspension culture
Plumbagin	Seed germination
Naphthoquinone	Shoot multiplication
Biotechnology	Root induction
Plant tissue culture	Synthetic seeds
Agrobacterium-mediated transformation	Chemotypes
Conservation	Micropropagation
DNA barcoding	Genetic diversity
Genomics	Seasonal variation
Transcriptomics	Transcriptome analysis
Elicitation	
Biostimulants	
Plant growth regulators	
Adventitious roots	

References

Agrawal A, Singh S, Malhotra EV, Meena DPS, Tyagi RK (2019) In vitro conservation and cryopreservation of clonally propagated horticultural species. In Conservation and utilization of horticultural genetic resources. Springer, Singapore, pp 529–578. https://doi.org/10.1007/978-981-13-3669-0_18

Ahmad N, Fatima N, Ahmad I, Anis M (2015) Efect of PGRs in adventitious root culture in vitro: present scenario and future prospects. Rend Lincei Sci Fis Nat 26:307–321. https://doi.org/10.1007/s12210-015-0445-y

Akhilraj AR, Prabhu S N, Priyalatha B, Vimala KS (2019) Effect of Śodhana (An āyurvedic purification technique) on Citraka (*Plumbago zeylanica* Linn. and *Plumbago rosea* Linn.) with special reference to plumbagin content. J Pharmacogn Phytochem 8(4):630–638

Akhilraj AR, Rukmini S (2021) A Systematic review on synonyms and properties of Citraka (*Plumbago zeylanica* Linn. and *Plumbago rosea* Linn.) from Saṁhitās (Ayurvedic classics) and Nighaṇṭus (Ayurvedic drug lexicons). J Med Plants Stud 9(3):33–40. https://doi.org/10.22271/plants.2021.v9.i3a.1291

Ayurvedic Pharmacopoeia of India (2001) Ministry of health and family welfare. Government of India 1(1):29

Badwaik HR, Kumari L, Nakhate K, Verma VS, Sakure K (2019) Phytoconstituent plumbagin: Chemical, biotechnological and pharmaceutical aspects. In: ur-Rahman A (ed) Studies in natural products chemistry. Elsevier, Amsterdam, pp 415–460. https://doi.org/10.1016/B978-0-12-817901-7.00013-7

Battacharyya D, Babgohari MZ, Rathor P, Prithiviraj B (2015) Seaweed extracts as biostimulants in horticulture. Sci Hortic 196:39–48. https://doi.org/10.1016/j.scienta.2015.09.012

Bhagat RB, Shimpale VB, Deshmukh RB (2008) Flora of Baramati. Prakash offset, Pune, India, p 187

Ceasar SA, Ayyanar M, Ignacimuthu S (2013) An improved micropropagation protocol for *Plumbago zeylanica* L. An important medicinal plant. Asian J Biol Sci 6:214–220

Chandravanshi M, Sahu Y, Agrawal A, Raja W (2014) In vitro micropropagation of important commorcial medicinal plant: *Plumbago zeylanica*. Adv Biol Res 8(3):139–142

Chaplot BB, Dave AM, Jasrai YT (2006) A valued medicinal plant - Chitrak (*Plumbago zeylanica* Linn.): successful plant regeneration through various explants and field performance. Plant Tissue Cult Biotechnol 16(2):77–84. https://doi.org/10.3329/ptcb.v16i2.1108

Checker R, Sharma D, Sandur SK, Poduval KS, TB, (2009) Anti-inflammatory effects of plumbagin are mediated by inhibition of NF-kappaB activation in lymphocytes. Int Immunopharmacol 9(7):949–958. https://doi.org/10.1016/j.intimp.2009.03.022

Chen SL, Yu H, Luo HM, Wu Q, Li CF, Steinmetz A (2016) Conservation and sustainable use of medicinal plants: Problems, progress, and prospects. Chin Med 11:31–10. https://doi.org/10.1186/s13020-016-0108-7

Chitrak products sources (2022) https://www.google.co.in/search?q=chitrak+products&source=lnms&tbm=isch&sa=X&ved=0ahUKEwjhxNGnbnZAhVLG5QKHcO-D1YQ_AUICigB&biw=1440&bih=794

Choudhary S, Kaurav H, Chaudhary G (2021) Citraka (*Plumbago zeylanica*): A potential rejuvenator. Int J Res Appl Sci Biotechnol 8(2):202–212. https://doi.org/10.31033/ijrasb.8.2.26

Crouch IJ, Finnie JF, Staden JV (1990) Studies on the isolation of plumbagin from in vitro and in vivo grown *Drosera* species. Plant Cell Tissue Organ Cult 21(1):79–82. https://doi.org/10.1007/BF00034496

Dahlgren KVO (1964) Chromosome numbers in the genus *Plumbago*. A Review. Revision and Reinvestigation 58:172–177

Das G, Rout GR (2002) Plant regeneration through somatic embryogenesis in leaf derived callus of *Plumbago rosea*. Biol Plant 45(2):299–302. https://doi.org/10.1023/A:1015169311876

Dharmar K, De Britto AJ (2012) DNA fingerprinting of *Plumbago zeylanica* L. using ISSR Markers. Int J Appl Bioresearch 6:23–27

Du Jardin P (2015) Plant biostimulants: Defnition, concept, main categories and regulation. Sci Hortic 196:3–14. https://doi.org/10.1016/j.scienta.2015.09.021

Gadge PJ, Nathar VN (2014) Detection and analysis of genetic variations in species of *Plumbago* L. using RAPD markers. Indian J Biotechnol 13:140–143

Ganesan K, Gani S (2013) Ethnomedical and Pharmacological potentials of *Plumbago zeylanica* LA. Am J Phytomed Clin Therapeut 1(3)313–337. https://www.academia.edu/download/32535442/PA-400163-_6_.pdf

Gbadamosi IT, Egunyomi A (2010) Micropropagation of *Plumbago zeylanica* L. (Plumbaginaceae) in Ibadan, Southwestern, Nigeria. J Med Plant Res 4(4):293–297

Haji RFA, Bhargava M, Akhoon BA, Kumar A, Brindavanam NB, Verma V (2014) Correlation and functional differentiation between different markers to study the genetic diversity analysis in medicinally important plant *Plumbago zeylanica.* Ind Crops Prod 55:75–82. https://doi.org/10.1016/J.INDCROP.2014.01.045

Jain P, Danwr K, Sharma H, Mahato D (2018) *In vitro* tissue culture studies and synthetic seed formation from *Plumbago zeylanica* L. Indian J Exp Biol 56:769–773

Jayaram K, Prasad MNV (2005) Rapidly *in vitro* multiplied *Drosera* as reliable source for plumbagin bioprospection. Curr Sci 89(3):447–448

Kakade PS (2017) Micropropagation, plumbagin production and diversity analysis in *Plumbago zeylanica* L. populations from Maharashtra. Ph.D. Thesis, Savitribai Phule Pune University

Kakade PS, Malpathak N (2017) Micropropagation of *Plumbago zeylanica* L. using *in vitro* germinated seedling explants. Int J Life Sci 5:211–218

Kakade PS, Zimare SB, Malpathak NP (2022) Effects of *Sargassum ilicifolium* seaweed extract on enhanced *in vitro* seed germination, mass propagation, and accumulation of plumbagin in *Plumbago zeylanica* L. Plant Cell Tissue Organ Cult 149(1):399–410. https://doi.org/10.1007/s11240-022-02242-3

Karnick CR, Tiwari KC, Majumber R (1982) Cultivation trials, pharmacognosy and ethnobotanical investigations of *Plumbago zeylanica* L. (Chitraka) of the Indian System of Medicine. Int J Crude Drug Res 20(4):193–199. https://doi.org/10.3109/13880208209083305

Khatoon S, Ali SI (1993) Chromosome atlas of the angiosperms of Pakistan. University of Karachi, Department of Botany

Khatun R, Dash C, Alam S, Sultana S (2019) Chromosomal characterization in *Plumbago zeylanica* and *P. indica* from Bangladesh. Cytologia 84(2):153–155. https://doi.org/10.1508/cytologia.84.153

Kishore N, Mishra BB, Tiwari VK, Tripathi V (2012) An account of phytochemicals from *Plumbago zeylanica* (Family: Plumbaginaceae): a natural gift to human being[#]. Chron Young Sci 3(3):178–198. https://doi.org/10.4103/2229-5186.99564

Liu LC, Li XY, Qin Y, Liu MH, Xing X (2022) Chemical composition and antibacterial and cytotoxic activities of the essential oil of *Plumbago zeylanica.* Chem Nat Compd 58(1):158–160. https://doi.org/10.1007/s10600-022-03622-z

Majumder S, Biswas A, Rahman MM (2022) Comparative karyomorphological analysis of in vivo and in vitro grown plants of *Plumbago zeylanica*: an important medicinal plant. Eur J Biol Biotechnol 3(2):25–29. https://doi.org/10.24018/ejbio.2022.3.2.333

Mallikadevi T, Senthilkumar P, Paulsamy S (2008) In vitro regeneration of the medicinal plant, *Plumbago zeylanica* L. with reference to a unique population in Maruthamalai, the western ghats, India. Plant Tissue Cult Biotechnol 18(2):173–179. https://doi.org/10.3329/ptcb.v18i2.3648

Mankar GD, Gulave AB, Datkhile K, SB Zimare (2021) Altitudinal gradients influence the accumulation of pharmaceutically important phenolic compounds in the leaves of *Lobelia nicotianifolia* Roth. and regulates its antioxidant and anticancer property. Ind J Biochem Biophys 58(3):253–60

Mittal S, Hariharan VK (2016) Crop diversification by agro-climatic zones of India-Trends and drivers. Indian J Econ Dev 12(1):132. https://doi.org/10.5958/2322-0430.2016.00014.7

Naik PM, Al-Khayri JM (2016) Abiotic and biotic elicitors-role in secondary metabolites production through in vitro culture of medicinal plants. In: Abiotic and biotic stress in plants—recent advances and future perspectives. Rijeka: InTech, pp 247–277. https://doi.org/10.5772/61442

Naik VN (1984) Taxonomy of angiosperms. Tata McGraw-Hill Education

Nalina L, Shanthanu K, Rajamani K, Joel AJ, Dhanani T, Singh R, Kumar S (2017) Morphological, yield and quality variability among the accessions of *Plumbago zeylanica.* Electron J Plant Breed 8(2):680–687. https://doi.org/10.5958/0975-928X.2017.00103.X

National Medicinal Plants Board (NMPB), Ministry of AYUSH, Government of India. Indian medicinal plants database. Distribution map of *Plumbago zeylanica* L. in India. Available at https://www.nmpb.nic.in

National Medicinal Plants Board, Ministry of AYUSH, Government of India. High Demanded Medicinal Plants > Medicinal List. Available at https://www.nmpb.nic.in/medicinal_list

National Research Council (US) (1991) Committee on managing global genetic resources: agricultural imperatives. Managing global genetic resources: The U.S. National Plant Germplasm System. Washington (DC): National Academies Press (US). Available from: https://www.ncbi.nlm.nih.gov/books/NBK235632/

Niazian M (2019) Application of genetics and biotechnology for improving medicinal plants. Planta 249(4):953–973. https://doi.org/10.1007/s00425-019-03099-1

Padhye S, Dandawate P, Yusufi M, Ahmad A, Sarkar FH (2012) Perspectives on medicinal properties of plumbagin and its analogs. Med Res Rev 32(6):1131–1158. https://doi.org/10.1002/med.20235

Paiva J, Leitao MT (1987) Números cromosomáticos de plantas de Africa tropical. Fontqueria 14:37–44

Panda S, Kamble A (2014) Genotypic and seasonal variation of plumbagin content from different populations of *Plumbago zeylanica* L. in India. Proc Natl Acad Sci India Sect B Biol Sci 86:165–169. https://doi.org/10.1007/S40011-014-0432-6

Panda S, Naik D, Kamble A (2015) Population structure and genetic diversity of the perennial medicinal shrub *Plumbago*. AoB Plants 7:1–12. https://doi.org/10.1093/aobpla/plv048

Pant M, Lal A, Rana S, Rani A (2012) *Plumbago zeylanica* L.: A mini review. Int J Pharm Res Appl 3(3):399–405

Patidar SL, Tiwari G, Tripathi MK, Patel RP, Mishra SN (2015) *In vitro* biosynthesis and quantification of plumbagin in cell suspension culture of *Plumbago zeylanica*. Med Plants - Int J Phytomed Relat Ind 7(1):60–67. https://doi.org/10.5958/0975-6892.2015.00009.X

Peng Z, Zhang Y (2000) 54. Plumbaginaceae. In Higher Plants of China Vol. 4: Angiospermae, vol 538, pp 538–551

POWO (2022) Plants of the World Online, Facilitated by the Royal Botanic Gardens, Kew. Published on the Internet; http://www.plantsoftheworldonline.org/Retrieved 04 July 2022. https://powo.science.kew.org/taxon/urn:lsid:ipni.org:names:687109-1

Priyanka V, Kumar R, Dhaliwal I, Kaushik P (2021) Germplasm conservation: Instrumental in agricultural biodiversity - A Review. Sustainability 13(12):6743. https://doi.org/10.3390/su13126743

Quality Standards of Indian Medicinal Plants (2008) Medicinal plants unit. Indian Council of Medical Research, New Delhi 7:208–210

Radman R, Saez T, Bucke C, Keshavarz T (2003) Elicitation of plant and microbial systems. Biotechnol Appl Biochem 37:91–102. https://doi.org/10.1042/BA20020118

Raj G, Kurup R, Hussain AA, Baby S (2011) Distribution of naphthoquinones, plumbagin, droserone, and 5-O-methyl droserone in chitin-induced and uninduced *Nepenthes khasiana*: Molecular events in prey capture. J Exp Bot 62(15):5429–5436. https://doi.org/10.1093/jxb/err219

Rout G, Das G (2002) An assessment of genetic integrity of micropropagated plants of *Plumbago zeylanica* by RAPD markers. Biol Plant 45(1):27–32. https://doi.org/10.1023/A:1015186709691

Rout GR, Saxena C, Samantaray S, Das P (1999) Rapid plant regeneration from callus cultures of *Plumbago zeylanica*. Plant Cell Tissue Organ Cult 56(1):47–51. https://doi.org/10.1023/A:1006224404734

Roy A, Bharadvaja N (2017) A review on pharmaceutically important medical plant: *Plumbago zeylanica*. J Ayurvedic Herb Med 3(4):225–228

Roy A, Bharadvaja N (2018) Effect of various culture conditions on shoot multiplication and GC–MS analysis of *Plumbago zeylanica* accessions for plumbagin production. Acta Physiol Plant 40(11):1–11. https://doi.org/10.1007/s11738-018-2766-9

Roy A, Bharadvaja N (2019) Establishment of root suspension culture of *Plumbago zeylanica* and enhanced production of plumbagin. Ind Crops Prod 137:419–427. https://doi.org/10.1016/J.INDCROP.2019.05.007

Roy A, Sharma N, Bharadvaja N (2022) Assessment of phytochemical and genetic diversity analysis of *Plumbago zeylanica* L. accessions. Genet Resour Crop Evol 69(1):209–219. https://doi.org/10.1007/S10722-021-01220-6

Sakamoto S, Putalun W, Pongkitwitoon B, Juengwatanatrakul T, Shoyama Y, Tanaka H, Morimoto S (2012) Modulation of plumbagin production in *Plumbago zeylanica* using a single-chain variable fragment antibody against plumbagin. Plant Cell Rep 31(1):103–110. https://doi.org/10.1007/s00299-011-1143-6

Saxena C, Samantaray S, Rout GR, Das P (2000) Effect of auxins on *in vitro* rooting of *Plumbago zeylanica*: Peroxidase activity as a marker for root induction. Biol Plant 43(1):121–124. https://doi.org/10.1023/A:1026519417080

Sharma U, Agrawal V (2018) In vitro shoot regeneration and enhanced synthesis of plumbagin in root callus of *Plumbago zeylanica* L.- An important medicinal herb. In Vitro Cell Dev Biol Plant 54(4):423–435. https://doi.org/10.1007/s11627-018-9889-y

Shukla B, Saxena S, Usmani S, Kushwaha P (2021) Phytochemistry and pharmacological studies of *Plumbago zeylanica* L.: a medicinal plant review. Clin Phytoscience 7(1):1–11. https://doi.org/10.1186/s40816-021-00271-7

Shukla PK, Misra A, Patra KK, Srivastava S (2020) Study of metabolite variability in *Plumbago zeylanica* Linn. collected from different localities of the Gangetic plains of India. JPC-J Planar Chromat 33(2):179–189. https://doi.org/10.1007/s00764-020-00018-z

Singh NP, Lakshminarasimhan P, Karthikeyan S, Prasanna PV (2001) Flora of Maharashtra State, vol 1, Series 2. Botanical Survey of India, Calcutta

Singh T, Sharma U, Agrawal V (2020) Isolation and optimization of plumbagin production in root callus of *Plumbago zeylanica* L. augmented with chitosan and yeast extract. Ind Crops Prod 151:112446. https://doi.org/10.1016/J.INDCROP.2020.112446

Sivanesan I (2007) Shoot regeneration and somaclonal variation from leaf callus cultures of *Plumbago zeylanica* Linn. Asian J Plant Sci 6:83–86

Sivanesan I, Jeong BR (2010) Micropropagation of *Plumbago zeylanica* L. Afr J Biotechnol 8(16):3761–3768. https://doi.org/10.4314/ajb.v8i16.62057

Sundari BKR, Telapolu S, Dwarakanath BS, Thyagarajan SP (2017) Cytotoxic and antioxidant effects in various tissue extracts of *Plumbago zeylanica*: implications for anticancer potential. Pharmacogn J 9(5):706–712. https://doi.org/10.5530/pj.2017.5.111

Sundari BRK, Budhwar R, S, Dwarakanath BS, Thyagarajan SP (2020) *De novo* transcriptome analysis unravels tissue-specifc expression of candidate genes involved in major secondary metabolite biosynthetic pathways of *Plumbago zeylanica*: implication for pharmacological potential. 3 Biotech 10:271 https://doi.org/10.1007/s13205-020-02263-9

Suryawanshi YC (2021) Hydroponic cultivation approaches to enhance the contents of the secondary metabolites in plants. *In*: Biotechnological approaches to enhance plant secondary metabolites, CRC Press, pp 71–80. https://doi.org/10.1201/9781003034957

Taxon ÁL (1979) IOPB chromosome number reports LXIV. Taxon 391–408. https://www.jstor.org/stable/1219755

Techen N, Parveen I, Pan Z, Khan IA (2014) DNA barcoding of medicinal plant material for identification. Curr Opin Biotechnol 25:103–110. https://doi.org/10.1016/j.copbio.2013.09.010

Tyagi R, Menghani E (2014) A review on *Plumabgo zeylanica*: a compelling herb. Int J Pharma Sci Res 5(4):119–126

van der Vijver LM (1972) Distribution of plumbagin in the plumbaginaceae. Phytochemistry 11(11):3247–3248. https://doi.org/10.1016/S0031-9422(00)86380-3

Verma P, Singh D, Rahman L, Gupta M, Banerjee S (2002) In vitro-studies in *Plumbago zeylanica*: Rapid micropropagation and establishment of higher plumbagin yielding hairy root cultures. J Plant Physiol 159(5):547–552. https://doi.org/10.1078/0176-1617-00518

Vijay R, Shukla J, Saxena R (2016) Micropropagation of *Plumbago zeylanica* L.: An important medicinal plant of India. Int J Pure App Biosci 4(6):159–167

Wei X, Gou X, Yuan T, Russell SD (2006) A highly efficient *in vitro* plant regeneration system and *Agrobacterium*-mediated transformation in *Plumbago zeylanica*. Plant Cell Rep 25(6):513–521. https://doi.org/10.1007/s00299-006-0114-9

Zhao X, Park SY, Yang D, Lee SY (2018) Synthetic biology for natural compounds. Biochemistry 58(11):1454–1456. https://doi.org/10.1021/acs.biochem.8b00569

Chapter 9
Breeding Approaches for the Improvement of *Hypericum perforatum* L.

Wudali Narasimha Sudheer, Chole Pranjali Bajrang, Akshatha Banadka, Rashmi Ramakrishnan, V. L. Vasantha, Jameel M. Al-Khayri, and Praveen Nagella

Abstract St. John's wort, scientifically labelled as *Hypericum perforatum* L. belongs to the Hypericaceae family. The plant body contains numerous secondary metabolite compounds with diverse pharmacological effects such as anti-inflammatory, antioxidant, anticancer and anti-microbial properties. The metabolites hypericin and hyperforin are those that have achieved economic relevance due to their capacity to treat cancer and alleviate patient depression. Numerous studies attest to their effectiveness. In this regard, this chapter aims to consolidate information about medicinal properties, breeding techniques, cultivation strategies, and conservation techniques. The chapter also discusses various contemporary biotechnological methods like tissue culture techniques for regeneration, genetic transformation for

W. N. Sudheer · C. P. Bajrang · R. Rashmi · V. L. Vasantha · P. Nagella (✉)
Department of Life Sciences, School of Sciences, Christ University, Bengaluru, India
e-mail: praveen.n@christuniversity.in

W. N. Sudheer
e-mail: wudali.sudheer@res.christuniversity.in

C. P. Bajrang
e-mail: cholepranjali.bajrang@res.christuniversity.in

R. Rashmi
e-mail: rashmir@res.christuniversity.in

V. L. Vasantha
e-mail: vasanth.vl@christuniversity.in

A. Banadka
Department of Biotechnology and Genetics, School of Sciences, JAIN (Deemed to be University), Bengaluru 560027, India
e-mail: b.akshatha@jainuniversity.ac.in; akshatha.rlt95@gmail.com

J. M. Al-Khayri
Department of Agricultural Biotechnology, College of Agriculture and Food Sciences,, King Faisal University, Al-Ahsa 31982, Saudi Arabia
e-mail: jkhayri@kfu.edu.sa

J. M. Al-Khayri et al. (eds.), *Biodiversity and Genetic Improvement of Medicinal and Aromatic Plants II*, Advances in Plant Breeding Strategies 10,
https://doi.org/10.1007/978-3-031-81857-8_9

the production of metabolites, their limitations, and hybridization for the conception of elite germplasm for increased yield of biomass and corresponding phytochemicals in *Hypericum perforatum* L.

Keywords *Hypericum perforatum* · Conservation · Breeding · Tissue culture · Hybridization

9.1 Introduction

In the traditional ayurvedic medical system, various medicinal and aromatic plants were used to treat chronic diseases and various health issues. *H. perforatum* is one such aromatic plant used for its medicinal property. It is commonly called St John's wort as the flowers of this plant bloom on St John's Day, i.e., the birthday (24th June) of Saint John (Notable Baptist) (Barnes et al. 2001). It is a perennial herb having a tap root system with well-developed shoots bearing oblong shaped leaves with smooth margins. This medicinal herb usually has flowers with yellow-colored petals that have black spots towards the petal margins. Flowers modify into reddish brown capsules bearing several brown seeds (Shrivastava and Dwivedi 2015). Taxonomically, *H. perforatum* belongs to the family Hypericaceae of the order Malpighiales (Budantsev et al. 2021). There are around 400 species that come under the genus Hypericum. The detailed taxonomic classification of hypericum is given below:

H. perforatum (HP) is considered as a problematic weed in countries like Australia (Buckley et al. 2003), and is abundantly grown in Europe, Asia, and a few North African and American countries (Erdelmeier, 1998) (Fig. 9.1). The usage of this herb as medicine goes way back to the 1st AD century by the Greek physician Pedanios Dioskourides and his associates. Also, Paracelsus, a Swiss physician, used this herb for medicinal purposes, and slowly it gained a reputation all over the world for its medicinal use (Istikoglou et al. 2010).

St. John's wort has been used for various medicinal purposes for a significant period of time. This plant was utilised in the Greek medicinal system by eminent physicians, including Hippocrates, Pliny, and Dioscorides. Dioscorides is used to treat menstrual-related disorders, wound healing, and snake bites. The genus is called Hypericum because it is said that HP has magical properties and the capacity to drive away bad spirits. Hyper in Greek means over, while eikon is apparition (Klemow et al. 2011). Because HP may alleviate sadness, agitation, and sleeplessness, it was widely employed by ancient intellectuals and healers. It also turned out to be quite effective at reducing inflammation, discomfort, and skin problems. The plant's pharmacological and therapeutic characteristics were disseminated around the world in the eighteenth and nineteenth centuries due to its remarkable medical capabilities (Clement et al. 2006).

HP is rich in diversified groups of phytochemicals like phenylpropanes, flavonol glycosides, biflavones, xanthones, phloroglucinols, naphthodianthrones, tannins and proanthocyanidins (Nahrstedt and Butterweck 1997). Hypericin, a notable secondary

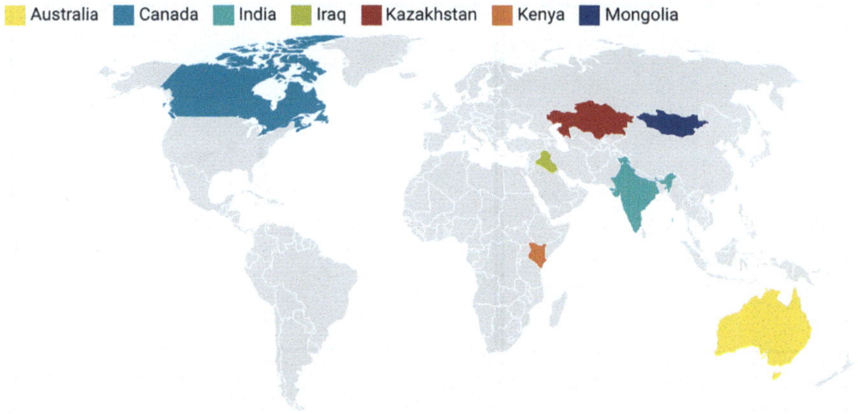

Fig. 9.1 Distribution of *H. perforatum* L.

metabolite from Hypericum, belongs to Naphthodianthrones, and is found in the black dots along the petals of the flower (Klemow et al. 2011). The plant body contains a variety of flavonoids with a range of pharmacological properties. A few significant flavonoids have been identified, including kaempferol, quercetin, rutin, myricetin, hyperin, luteolin, and miquelianin (Barnes et al. 2001). HP dry biomass that has undergone distillation is known for producing essential oils. There have been reports of limonene, myrcene, and caryophyllene traces as well as major constituents including pinene, terpineol, geraniol, 2-methyl-octane, and n-nonane (Reuter 2002). There are a few other lipophilic molecules with medicinal properties in HP, including hyperforin, adhyperforin, and furohyperforin (Klemow et al. 2011).

Hypericin (Fig. 2a), an important plant secondary metabolite from HP, is known for its ability to work as an antidepressant. Studies reveal that hypericin has the capacity to compete with monoamine oxidases and inhibit their action. It is also seen that hypericin plays a key role in regulating dopamine levels (Suzuki et al. 1984). By competing with inhibitors of serotonin, noradrenaline, and GABA from the synaptic cleft, hyperforin from HP is also said to have antidepressant capabilities (Cervo et al. 2002; Zanoli 2004). Studies suggest that HP has a rich phytochemical profile, which is attributed to anti-cancer activity. Hyperforin (Fig. 2B) showed potential anti-cancer activity on cancer cell lines of K562 and U937, which are leukemia cancer cell lines, and LN229 (Glioblastoma) cell lines (Hostanska et al. 2003). Similarly, the synergistic effect between 5-aminolevulinic acid, hypericin with optimum illumination showed anti-cancer activity on cancer cells found in endometrial cancer (Schneider-Yin et al. 2009). In pheochromocytoma cell lines, a flavonoid component from HP demonstrated considerable anti-oxidant action against hydrogen peroxide-induced apoptosis, which may eventually lead to the treatment of neurodegenerative disorders (Zou et al. 2010). Antioxidants from the HP extracts reduce superoxide dismutase activity and shield the neurons from oxidative injury, aiding in the treatment of neurological conditions (Sánchez-Reus et al. 2007). The anti-inflammatory efficiency of

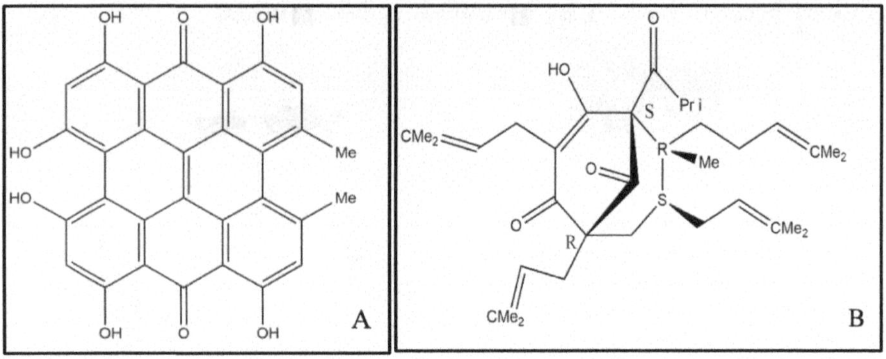

Fig. 9.2 Structure of pharmacologically important molecules from *Hypericum perforatum.* A. Hypericin, B. Hyperforin

HP extracts was monitored on colitis induced male wistar rats. In colonic mucosa, inflammatory markers were shown to be significantly decreased, and HP's antioxidant activity was also noted (Dost et al. 2009). The fibroblast collagen production was increased when chicken embryonic fibroblasts were cultured with HP extracts and the efficiency of wound healing was compared with dexpanthenol (Öztürk et al. 2007). The phytochemical rich components from HP proved to have anti-bacterial and anti-fungal activities (Saddiqe et al. 2010). Antiviral activity against retroviral organisms was observed in a light-dependent approach. It was shown that 3-hydroxy lauric acid from HP extracts exhibited anti-HIV activity and might be advised for therapy in the future (Maury et al. 2009).

HP wild varieties were domesticated for their therapeutic benefits. Later, with the development of contemporary biotechnology and breeding techniques, the elite clones and germplasm underwent improvement. Parameters like plant breeding, phytochemical diversity, and ideal agronomic characteristics helped in the improvement of HP varieties (Poutaraud and Girardin, 2005). The market for trading HP is quite promising because of its extensive pharmacological efficacy. For instance, the global market for HP's raw materials was roughly 570 million US dollars in 1998. As a result of increased demand, the market has increased and thousands of hectares of HP are currently being grown (Bruni and Sacchetti 2009; Pluhár et al. 2002).

9.2 Current Cultivation Practices and Challenges

HP is considered a weed in many regions, and is mostly grown in Australia, California, Europe, North Africa, North America, Oregon, South Africa, South America, and Western Asia. It is found in countries like Hungary, Italy, France, Sweden, Turkey, the USA, and Washington. It grows in various environmental conditions, which include barren savannas, dry sand prairie communities, forest, railways, rangeland, river banks, roadsides, pastures, waste areas, and woodlands (Weeds of Australia 2016) (Zouhar 2004). The optimum temperature for the growth of HP ranges from 19 to 25 °C in summer season and can tolerate up to −3–1 °C in the winter season. HP requires minimal watering once established and grows well in regions with an annual rainfall of greater than 30 inches (Zouhar 2004). St. John's wort prefers coarse textured well drained soil. It can also be adjusted to sandy, loamy and clayey soil too. The plant grows in mild acidic, neutral, and mild basic soil and prefers a pH range of 5.5 to 7. The plant can grow across different seasons. The seeds germinate during autumn, winter, or spring, and flowering is observed during the summer season (Zouhar 2004).

HP can be propagated from seeds, divisions, or stem cuttings. Propagation by seeds is the most widely used method. It is a hermaphrodite, and the seeds are naturally pollinated by insects. The best time to sow the seeds of HP is either early spring or in the beginning of autumn. The sown seeds have to be lightly covered and have to be slightly exposed to sunny and partly shady areas. The seeds can also be germinated indoors in small pots. It germinates within 1–3 months at 10 °C. Once germinated, it can be transferred outdoors for further growth. It has been noted that the small seedlings of St. John's wort are slow to grow in comparison to seedlings of associated species. The competition for nutrients, light, moisture, and space can be attributed to its slow growth. However, inoculation with arbuscular mycorrhizal fungi has shown a positive growth response in the seedlings (Zouhar 2004).

It can be propagated from stem cuttings (wood cuttings), preferably winter cuttings. About 10–20 cm stem cuttings have to be planted in 15 cm pots with loose soil and rooting media. It has to be watered well. The root formation starts within a few weeks of watering. Once the roots are well developed and stable, the cutting with developed roots is transferred outdoors. It can also be propagated by root division, where a portion of the root with a small portion of top growth has to be separated from the existing root ball. It has to be further subdivided into pieces and transferred into small pots for the growth of small plants. Once well-developed it can be transferred outdoors. Pruning of HP to 1/3rd of its height is necessary to manage the cultivation, which would otherwise grow uncontrollably. The damaged and dead branches have to be removed. Trimming should be performed in early spring before the onset of the flowering season. The HP is preferably harvested in early summer, immediately after summer for better taste (Furbeck 2019).

HP, an important source of active metabolites such as hypericins, rutin, and mangiferin, is prone to biotic and abiotic stresses during its cultivation, resulting in both quantitative and qualitative variations of these bioactive compounds. Plant diseases, frequent changes in the climate, soil differences, and genotypic changes are some of the challenges faced during the cultivation of HP (Gaid et al. 2017). The primary challenge faced in cultivating HP is its uncontrolled growth and spreading. St. John's Wort possesses extensive rhizomes, resulting in its spreading to unwanted underground areas. Because of its highly invasive nature, it is capable of replacing native plants in natural ecosystems. During winter, the hypericum foliage is prone to damage due to severe windburn. The damage can be minimized by surrounding plants such as burlap or agricultural fleece. The plant is also attacked by scale insects which appear as white, cottony masses or brownish bumps on hypericum stems. Hypericum gets infested with whitish wormlike root knot nematodes, resulting in wilting, yellowing of leaves, and stunted growth. The nematodes can be controlled by adding an excess of compost. Rust disease is commonly found to affect the plant. It is a fungal disease caused by *Melampsora hypericorum* causing brownish powdery spots and sometimes showing white fungus caused by *Microsphaera* species. The fungal attack can be controlled by spraying sulfur (Yardener 2003).

9.3 Conservation of Genetic Resources

Gene preservation can be done either in-situ or ex-situ techniques. In both techniques, optimal storage strategies need some information about genetic patterns and their variations. The plant's genetic resources will be protected outside their natural habitats. This technique involves the preservation of pollen, seeds, tissue culture in specific facilities, and scion. Different techniques like cryopreservation, seed bank, and in-vitro culture can be used; based on the type of plants, merits and demerits. Some of the most common methods which are used for the conservation of HP are discussed below;

(a) **Seed Bank**:

The most economical and simplest method for conserving genetic resources is seed storage. Seed banks help to maintain the seed viability with 5% of seed moisture (Pérez-García et al. 2006). Several species have been maintained successfully by using ultra-dry seed methods (Ellis et al. 1995). In this method, seed moisture will be maintained below 5% and conserved. Some studies showed that HP seeds can be conserved with low moisture and showed good germination.

(b) **In-Vitro Culture**:

Nowadays, in vitro conservation techniques have gained more importance for conserving genetic resources. Considering the many medicinal advantages of St. John's wort, conservation of this plant is considered as important. In vitro conservation shows three types of techniques;

1. **Short-term conservation:** This technique is also known as optimal growth storage. It involves growing HP plant material at an ideal rate and storing it under optimal to moderate physical conditions. (Syahid 2021).
2. **Medium-term conservation:** In the medium-term conservation method, the storage of plant material is done under medium and outside physical circumstances. In this, explants show minimal growth, so this technique is also known as the minimal growth method. The minimal growth technique is widely practiced in countries like Indonesia because it risks contamination, is efficient and shows low mutations (Syahid 2021).
3. **Long-term conservation:** In long-term conservation, a non-growth method or freezing is used. The HP plant material is stored in liquid nitrogen freezing

conditions maintained below -196 °C temperature (Engelmann and Engels 2000; Dube et al. 2011).

(c) **Cryopreservation Approaches**:

The encapsulation dehydration process was initially used by Kimakova et al. (1996) for cryopreservation of HP. Vitrification and controlled cooling techniques were used for the *Hypericum spp.* cryoconservation. Controlled cooling is also known as slow cooling. It is mainly based on the induced crystallization principle, which minimizes the ice formation probabilities (Karlson and Toner 1996). Generally, pre-cultured plant parts under specific conditions like above freezing point, treated with osmotic compounds or growth regulators, and treated with cryoprotectant chemicals (Bruňáková and Čellárová 2016). Vitrification is a rapid cooling, direct immersion procedure. Both vitrification and controlled cooling were applied successfully in undifferentiated cells of HP.

9.4 Breeding Techniques

In breeding techniques, the expression of traits is generally moved in a particular direction by using the genetic variability of plant populations. With this, properties like covalence as well as homogeneity will be improved. In breeding technology, conventional breeding is commonly preferred as it is available with a natural population with a high success rate, and cost-effective and simple techniques are involved. Some general breeding objectives for *Hypericum spp* are;

1. Stable and high yield
2. Development of disease and pest resistance varieties
3. High safety and functional value

HP is a pseudogamous, facultative, apomictic plant (Mayo and Langridge 2003). Tools like restriction fragment length polymorphism (RFLP), fingerprinting, and randomly amplified polymorphic DNA (RAPD) are commonly utilized to study the reproduction mode in HP. Polymorphism, hybridization, and mutation breeding showed good results for the new varieties in *H. perforatum* development (Poutaraud and Girardin 2005). In HP, double haploid plants regenerated using an anther culture technique, and a breeding line was developed. Different factors like genotype, media composition (carbohydrates, growth factors), different species (*H. tetrapterum*, *H. hirsutum*), and donor plant conditions were evaluated (Schulte et al. 1996). Callus formation was noted in *H. perforatum*, and *H. tetrapterum* but was not noted in *H. hirsutum* whereas shoot formation was absent in both *H. tetrapterum* and *H. hirsutum* (Ferrie 2007). For the improvement of genetic problems, both induced and spontaneous mutational breeding methods are used in medicinal plants. For example, the application of gamma-irradiation in *Solanum laciniatum* results in dwarf mutants. Chemical mutagens also worked successfully (Bhardwaj et al. 2020). By using mutagenic chemicals or radiation, spontaneous mutation can be induced in plants. Use of a mutation-inducing agent results in a change of the genome, which forms a reaction between the non-mutated original plant and the mutation-induced plant. Mutation breeding results in the segregation of mutant progeny with existing plant genotypes or the development of new variability (Pank 2007). The main disadvantage of mutation breeding is that trait expression-related changes occur by chance.

9.5 Tissue Culture Applications

The phytochemistry and pharmacological applications of HP have increased the demand for the raw materials for pharmaceutical formulations. The employment of modern biotechnological approaches like tissue culture and gene manipulation strategies helps in deriving the optimum yield of biomass and desired metabolites (Al-Khayri et al. 2024).

Hypocotyl explants of HP when cultured on Murashige and Skoog medium (MS medium) along with 5 to 10 μmol/L Thidiazuron (TDZ) assisted in regeneration to full plantlet (Murch et al. 2000). Similarly, the effectiveness of shoot induction in various genotypes was analysed. It is seen that explants of leaf when cultured on MS medium supplemented with 1.0 mg/L BAP has shown more number of shoots (Akhtar and Aharizad 2013). Research indicated that using MS medium with increased levels of BAP and reduced levels of IAA resulted in the most effective shoot induction (Sharifi et al. 2014). MS medium with 0.1 mg/L BAP showed the highest number of shoots and respective shoot lengths when shoot tips explants were cultured (Abdollahpoor et al. 2017). Shoot proliferation was observed from nodal explants cultured on MS medium supplemented with 0.1 mg/L BAP, and rooting was then induced by subsequent culturing on a medium rich in IBA (Parsamanesh et al. 2018). Similarly, nodal explants showed optimal shoot induction and multiplication when cultured on MS medium supplemented with 7.5 μM of BAP along with 2.0 μM of IAA, and optimal rooting was observed when in-vitro shoots were cultured on 4.0 μM of IBA (Mir et al. 2019). Petal explants induced the maximum number of shoots when cultured on MS medium supplemented with 1.0 mg/L IAA along with 0.1 mg/L KN followed by 1.0 mg/L IBA along with 0.1 mg/L KN (Palmer and Keller 2011).

Indirect organogenesis was observed in HP when explants of leaves were inoculated on MS medium supplemented with 2,4-D and BAP. Optimal cell proliferation is observed on MS medium supplemented with 4.4 μM BAP and 4.5 μM 2,4-D when maintained under dark conditions. Subculturing of callus on MS medium supplemented with 4.6 μM KN and 0.45 μM 2,4-D induced the maximum number of shoots and rooting was induced when cultured on half or full-strength MS medium supplemented with or without 4.9 μM IBA (Raquel Pretto and Romanato Santarém 2000). Callus induction from leaf explants was observed when cultured on an MS medium supplemented with various auxins and cytokinins. It is seen that when callus was subcultured on MS medium supplemented with 2.85 μM Indole- 3-acetic acid (IAA) and 4.44 μM BAP induced shoots (Fig. 9.3). Rooting of the in-vitro induced shoots was observed when cultured on MS medium supplemented with or without IAA

(Wójcik and Podstolski 2007). Optimal callus induction and growth were observed when explants of leaf were cultured on MS medium supplemented with 0.25 mg/L of 2,4-D along with 1.0 mg/L KN, and subsequent shoot induction from callus was recorded when cultured on MS medium supplemented with 0.5 mg/L BAP. Rooting of the in-vitro shoots was seen when cultured on MS medium supplemented with 0.5 mg/L of IBA (Abdollahpoor et al. 2017).

Fig. 9.3 Formation and growth of *H. perforatum* shoots from leaf explants on MS medium supplemented with IAA and BAP. Formation of the shoots (2.85 μM IAA and 4.44 μM BAP) from the callus (**a**, **b**). Regenerating shoots with dark glands on the young leaves (**b**). Longitudinal section through young shoot (**c**). Shoots after 49 days from the onset of the regeneration experiment—2.85 μM IAA and 4.44 μM BAP (**d**); 2.85 μM IAA and 2.22 μM BAP (**e**), 2.85 μM IAA and 8.88 μM BAP (**f**), 2.85 μM IAA and 13.32 μM BAP (**g**). Scale bars indicate 1 cm. (*Source* Anna Wojcik and Andrzej Podstolski 2002: 10.1007/s11738-006-0019-9, accessed on 11 July 2022; reproduced with permission from the publisher; License No: 5355180260652)

9.6 Genetic Engineering and Transcriptome Studies

The science of gene transformation has advanced significantly, and many medicinal plants have been genetically engineered to meet the requirements of the plantlets. Franklin et al. (2007) suggested that the particle bombardment mode of gene transfer is successful when compared to Agrobacterium mediated transfer in HP. Callus cultures that are resistant to hygromycin due to genetic modification were subcultured and genetically modified individual transgenic plants were propagated. PCR studies also showed positive results for the expression of β-glucuronidase and hygromycin phosphotransferase genes (Franklin et al. 2007). De novo transcriptome analyses of 260 unigenes were identified as involved in the synthesis of phytochemicals of HP like hyperforin, hypericin, and melatonin. These unigenes were taken from among the 59,184 unigenes examined by the Illumina/Solexa deep sequencing platform (Miao et al. 2012). The differentially expressed genes from leaves with and without dark glands were evaluated in light of research showing that hypericin is generated in the dark glands of HP aerial sections. It can be noted that 263 enzymes were actively contributing to the biosynthesis of hypericin, while 799 genes were upregulated in the process of metabolite production (Soták et al. 2016). *Agrobacterium tumefaciens* and rhizogenes have been used in a series of attempts to investigate gene transformation. When HP interacts with Agrobacterium, it is discovered that phytoalexins are synthesized in the cultures as a plant defence mechanism (Hou et al. 2016). Researchers have attempted to mass produce the desired metabolites in different model systems, including fungi and bacteria, due to rising demand and the medicinal significance of the plant secondary metabolites from HP. Because, the biosynthesis and production processes for metabolites like hypericin and hyperforin are complicated and site-specific (in dark glands). Researchers are working to understand the function of genes like AGL6 and MYB38, which are crucial for the differentiation of black glands in the aerial regions of HP (Rizzo et al. 2020).

9.7 Hybridization

It has been established that HP is a facultative apomictic with apospory and pseudogamy (Galla et al. 2011). It can be found in diploid ($2n = 16$) or tetraploid form ($2n = 32$) (Galla et al. 2011). According to Pank et al. (2003), the hypericum plants which are apomictically produced will be in tetraploid and diploid forms by sexual reproduction. When HP produces viable pollen grains or, less commonly, when unreduced embryo sacs are fertilized, it results in the formation of hybrids between HP and other Hypericum species (Mártonfi et al. 1996).

Since apomixis excludes recombination and segregation, traditional breeding methods are not suitable for Hypericum. Crossing between obligate sexual and apomictic plants is the only way to produce variabilities (Mártonfi and Mártonfiová, 2011). Interspecific hybridization was reported between Hypericum species

Table 9.1 List of cultivars of *H. perforatum* L.:

Cultivars	Institute/ Place	References
Topas	North America	Kirakosyan et al. (2003)
Anthos	North America	Kirakosyan et al. (2003)
New Stem	North America	Kirakosyan et al. (2003)
Elixir	North America	Kirakosyan et al. (2003)
Helos	Canada	Kirakosyan et al. (2003) Kwiecień et al. (2015)

with HP. Crackles (1990) found a Natural hybrid named '*Hypericum x desetangsii nm. Desetangsii*', a hybrid between *H. perforatum* and *H. maculatum Crantz* from Yorkshire. The natural hybridization between *H. perforatum* and autotetraploid *H. maculatum subsp, obtusiusculum* resulted in the tetraploid hybrids (2n = 32) being also reported, but there's a chance that a hexaploid hybrid (2n = 48) could form if a *H. perforatum* embryo sac with aposporous characteristics is fertilized. Hybrids of Hypericum were recognized as intermediates between their parents, spreading partly through restricted vegetative propagation and partly through limited seed fertility (Mártonfi et al. 1996).

Intraspecific hybridization of the 'Topas' variety of HP was also found to be successful (Stefanovits-Bányai et al. 2002). The possibility of obtaining sexual offspring by intraspecific hybridization was confirmed by Steck et al. (2001) and also found that RAPD markers are effective for identifying diploid and tetraploid offspring of intraspecific hybridization of Hypericum. Table 9.1 and 9.2 presents different cultivars and hybrids of *H. perforatum*.

9.8 Conclusions and Prospects

HP is a medicinal plant rich with plant secondary metabolites like hypericin, an important anti-cancer drug, and hyperforin, a notable antidepressant drug with high market value. It is produced in the distinctive black spots on oblong leaves and yellow flowers. The conventional breeding techniques will not be sufficient for the increased demand of the plant secondary metabolites from HP. Selective breeding has refined its traits, optimizing phytochemical diversity and agronomic characteristics to enhance its therapeutic potential. This has led to a burgeoning market, indicative of its widespread pharmacological efficacy, with global trade values soaring. Cultivating HP presents both opportunities and challenges. While its adaptability to various habitats and propagation methods offers flexibility, its invasive nature and susceptibility to biotic and abiotic stresses necessitate careful cultivation practices. Strategies like seed banking and in vitro conservation techniques ensure the preservation of its genetic resources, safeguarding its botanical legacy for future generations. In the realm of breeding, traditional methods blend with modern genetic engineering techniques to steer trait expression towards desired outcomes. From hybridization to

Table 9.2 Hybrids of *H. perforatum* L.

Male parent	Female parent	Hybrid	References
H. tetrapterum L. (n)	*H. maculatum* subsp, *maculatum* (2n = 24)	Triploid (2n = 24). Intermediate between parents	Mártonfi et al. (1996)
H. maculatum subsp, *maculatum* (2n = 24)	*H. perforatum L.*	Pentaploid (2n = 40) Dissimilar to parents	Mártonfi et al. (1996)
H. perforatum L.	*H. maculatum* subsp, *obtusiusculum* (2n = 32)	Tetraploid (2n = 32) Intermediate to parents	Mártonfi et al. (1996)
H. maculatum subsp, *obtusiusculum* (2n = 32)	*H. perforatum* L. (aposporous egg)	Hexaplod (2n = 48) Intermediate to parents	Mártonfi et al. (1996)
H. maculatum Crantz subsp. *obtusiusculum* (Touriet) Hayek	*H. perforatum* L.	*Hypericum x desetangsii Lamotte nm. Desetangsii* (2n = 24) Intermediate to parents	Crackles (1990), Lihova et al. (2000)
H. maculatum subsp. *maculatum* (2n = 32)	*H. perforatum* (2n = 16)	*H. x desetangsii nothosubsp. carinthiacum* (2n = 40)	Lihova et al. (2000)
H. maculatum CRANTZ subsp, *maculatum*	*H. perforatum* L.	*H. x desetangsii nm. perforatiforme* (A. FROHL.) N. ROBSON Pentaploid, more similar to H. *perforatum* L.	Mártonfi et al. (1996)

mutational breeding, each approach leverages the genetic variability of HP populations to enhance traits like yield, disease resistance, and medicinal potency. Moreover, tissue culture applications have revolutionized biomass production and metabolite yield, offering precise control over growth and phytochemical synthesis. Transcriptome studies shed light on the genetic basis of St. John's wort's medicinal properties, guiding efforts to optimize metabolite production through genetic manipulation.

References

Abdollahpoor M, Kalantari S, Azizi M, Saadar YA (2017) In vitro shoot proliferation of *Hypericum perforatum* L. through indirect and direct plant regeneration. J Med Plants By-Product 6:81–9

Akhtar N, Aharizad S (2013) In vitro shoot regeneration and hypericin production in four *Hypericum perforatum* L. genotypes. Int J Agric Res Rev 3:887–93.

Al-Khayri JM, Narasimha SW, Vennapusa AR, Nagella P, Shehata WF, Al-Mssallem MQ (2024) Biotechnological approaches for the production of hypericin and other important metabolites from the genus Hypericum. Plant Cell Tiss Org Cult 156:100

Barnes J, Anderson LA, Phillipson JD (2001) St John's wort (*Hypericum perforatum* L.): a review of its chemistry, pharmacology and clinical properties. J Pharm Pharmacol 53:583–600.

Bhardwaj R, Sharma K, Sharma DK, Prakash P (2020) Conventional genetic improvement methods in medicinal and aromatic plants: A review. Int J Econ Plants 7:156–61. https://doi.org/10.23910/2/2020.0382.

Bruňáková K, Čellárová E (2016) Conservation strategies in the genus hypericum via cryogenic treatment. Front Plant Sci 7:1–12. https://doi.org/10.3389/fpls.2016.00558

Bruni R, Sacchetti G (2009) Factors affecting polyphenol biosynthesis in wild and field grown St. John's Wort (*Hypericum perforatum* L. Hypericaceae/Guttiferae). Molecules 14:682–725. https://doi.org/10.3390/molecules14020682

Buckley YM, Briese DT, Rees M (2003) Demography and management of the invasive plant species *Hypericum perforatum* L. Using multi-level mixed-effects models for characterizing growth, survival and fecundity in a long-term data set. J Appl Ecol 40:481–93. https://doi.org/10.1046/j.1365-2664.2003.00821.x

Budantsev AL, Prikhodko VA, Varganova IV, Okovityi S V (2021) Biological activity of *Hypericum perforatum* L. (hypericaceae): A review. Farmatsiya i Farmakol 9:17–31. https://doi.org/10.19163/2307-9266-2021-9-1-17-31.

Cervo L, Rozio M, Ekalle-Soppo CB, Guiso G, Morazzoni P, Caccia S (2002) Role of hyperforin in the antidepressant-like activity of *Hypericum perforatum* extracts. Psychopharmacology 164:423–428. https://doi.org/10.1007/s00213-002-1229-5

Clement K, Covertson CR, Johnson MJ, Dearing K (2006) St. John's wort and the treatment of mild to moderate depression: A systematic review. Holist Nurs Pract 20:197–203. https://doi.org/10.1097/00004650-200607000-00008

Crackles FE (1990) *Hypericum x desetangsii* Lamotte nm. *desetangsii* in Yorkshire, with special reference to its spread along railways. Watsonia 18:63–7

Dost T, Ozkayran H, Gokalp F, Yenisey C, Birincioglu M (2009) The effect of *Hypericum perforatum* (St. John's Wort) on experimental colitis in rat. Dig Dis Sci 54:1214–21. https://doi.org/10.1007/s10620-008-0477-6

Dube P, Gangopadhyay M, Dewanjee S, Ali MN (2011) Establishment of a rapid multiplication protocol of *Coleus forskohlii* Briq. and in vitro conservation by reduced growth. Indian J Biotechnol 10:228–31

Ellis R, Hong T, Roberts E (1995) Survival and vigour of lettuce (*Lactuca sativa* L.) and sunflower (*Helianthus annus* L.) seeds stored at low and very-low moisture contents 521–134

Engelmann F, Engels JMM (2000) Technologies and strategies for ex situ conservation. Manag Plant Genet Divers Proc an Int Conf Kuala Lumpur, Malaysia 12–16(2009):89–103. https://doi.org/10.1079/9780851995229.0089

Erdelmeier CAJ (1998) Hyperforin, possibly the major non-nitrogenous secondary metabolite of *Hypericum perforatum* L. Pharmacopsychiatry 31:2–6. https://doi.org/10.1055/s-2007-979339

Ferrie AMR (2007) Doubled haploid production in nutraceutical species: a review. Euphytica 158:347–357. https://doi.org/10.1007/s10681-006-9242-0

Franklin G, Oliveira M, Dias ACP (2007) Production of transgenic *Hypericum perforatum* plants via particle bombardment-mediated transformation of novel organogenic cell suspension cultures. Plant Sci 172:1193–1203. https://doi.org/10.1016/j.plantsci.2007.02.017

Furbeck A (2019) How to plant, grow, harvest, and use st. John's wort [Internet]. The Homesteading Hippy. [cited 2022 Apr 15]. Available from: https://thehomesteadinghippy.com/how-to-grow-st-johns-wort/

Gaid M, Wucherpfennig T, Scholl S, Beerhues L, Krull R (2017) Challenges for the cultivation of plant cells on the example of *Hypericum perforatum* and *Taxus chinensis*. In: Pavlov A, Bley T (eds) Bioprocessing of plant in vitro systems. Springer International Publishing, Cham, pp 1–26

Galla G, Barcaccia G, Schallau A, Molins MP, Bäumlein H, Sharbel TF (2011) The cytohistological basis of apospory in *Hypericum perforatum* L. Sex Plant Reprod 24:47–61. https://doi.org/10.1007/s00497-010-0147-7

Hostanska K, Reichling J, Bommer S, Weber M, Saller R (2003) Hyperforin, a constituent of St John's wort (*Hypericum perforatum* L.) extract induces apoptosis by triggering activation of caspases and with hypericin synergistically exerts cytotoxicity towards human malignant cell lines. Eur J Pharm Biopharm 56:121–132. https://doi.org/10.1016/S0939-6411(03)00046-8

Hou W, Shakya P, Franklin G (2016) A perspective on *Hypericum perforatum* genetic transformation. Front Plant Sci 7:1–12. https://doi.org/10.3389/fpls.2016.00879

Istikoglou CI, Mavreas V, Geroulanos G (2010) History and therapeutic properties of *Hypericum perforatum* from antiquity until today. Psychiatriki 21:332–338

Karlson JO, Toner M (1996) Long-term storage of tissues by cryopreservation: critical issues. Biomaterials 17:243–256

Kimáková K, Čellárová E, Hončariv R (1996) A study of regeneration capacity of encapsulated meristems of *Hypericum perforatum* L. after cryopreservation. Biologia 51(83)

Kirakosyan A, Gibson DM, Sirvent T (2003) A Comparative study of *Hypericum perforatum* plants as sources of hypericin and hyperforins. J Herbs, Spices Med Plants 10:73–88. https://doi.org/10.1300/J044v10n04_08

Klemow KM, Bartlow A, Crawford J, Kocher N, Shah J, Ritsick M (2011) Medical attributes of St. John's wort (*Hypericum perforatum*). Herb Med Biomol Clin Asp Second Ed 211–237. https://doi.org/10.1201/b10787-12

Kwiecień I, Szydłowska A, Kawka B, Beerhues L, Ekiert H (2015) Accumulation of biologically active phenolic acids in agitated shoot cultures of three *Hypericum perforatum* cultivars: 'Elixir', 'Helos' and 'Topas.' Plant Cell Tissue Organ Cult 123:273–281. https://doi.org/10.1007/s11240-015-0830-3

Lihova J, Pavolmartonfi, Martonfiova L (2000) Experimental study on reproduction of *Hypericum* X *desetangsii* notho subsp carinthiacum (A. Frohl.) N. Robson (Hypericaceae). Caryologia 53:127–132. https://doi.org/10.1080/00087114.2000.10589187

Mártonfi P, Mártonfiová L (2011) Reproduction mode in hypericum and its consequences. Med Aromat Plant Sci Biotechnol 5:48–52

Mártonfi P, Repcăák M, Mihoková L (1996) *Hypericum maculatum* crantz subsp. maculatum X *H. perforatum* L. (Hypericaceae): Corroboration of natural hybridization by secondary metabolite analysis. Folia Geobot 31:245–250. https://doi.org/10.1007/bf02812067

Maury W, Price JP, Brindley MA, Oh C, Neighbors JD, Wiemer DF, Wills N, Carpenter S, Hauck C, Murphy P, Widrlechner MP (2009) Identification of light-independent inhibition of human immunodeficiency virus-1 infection through bioguided fractionation of *Hypericum perforatum*. Virol J 6:1–2. https://doi.org/10.1186/1743-422X-6-101

Mayo GM, Langridge P (2003) Modes of reproduction in Australian populations of *Hypericum perforatum* L. (St. John's wort) revealed by DNA fingerprinting and cytological methods. Genome 46:573–579. https://doi.org/10.1139/g03-038

Miao H, Ying W, Hua W, Zhang Y, Wang Z (2012) De novo sequencing of *Hypericum perforatum* transcriptome to identify potential genes involved in the biosynthesis of active metabolites. PLoS ONE 7:1–10. https://doi.org/10.1371/journal.pone.0042081

Mir MY, Kamili AN, Hassan QP, Rafi S, Parray JA, Jan S (2019) In vitro regeneration and free radical scavenging assay of *Hypericum perforatum* L. Natl Acad Sci Lett 42:161–167. https://doi.org/10.1007/s40009-018-0699-x

Murch SJ, Choffe KL, Victor JMR, Summon TY, KrishnaRaj S, Saxena PK (2000) Thidiazuron-induced plant regeneration from hypocotyl cultures of St. John's wort (*Hypericum perforatum*. cv 'Anthos'). Plant Cell Rep 19:576–581. https://doi.org/10.1007/s002990050776

Nahrstedt A (1997) V B. Biologically active and other chemical constituents of the herb of *Hypericum perforatum* L. Pharmacopsychiatry 30:129–134. https://doi.org/10.1055/s-2007-979533

Öztürk N, Korkmaz S, Öztürk Y (2007) Wound-healing activity of St. John's wort (*Hypericum perforatum* L.) on chicken embryonic fibroblasts. J Ethnopharmacol 111:33–39. https://doi.org/10.1016/j.jep.2006.10.029

Palmer CD, Keller WA (2011) Plant regeneration from petal explants of *Hypericum perforatum* L. Plant Cell Tissue Organ Cult 105:129–134. https://doi.org/10.1007/s11240-010-9839-9

Pank F (2007) Use of breeding to customise characteristics of medicinal and aromatic plants to postharvest processing requirements. Stewart Postharvest Rev. 4:1

Pank F, Matzk F, Kästner U, Blüthner WD, Foltys De Garcia E, Meister A et al (2003) Reproductive diversity and strategies for breeding in St. John's wort (*Hypericum perforatum* L.). Euphytica 134:77–784. https://doi.org/10.1023/A:1026115508883

Parsamanesh Z, Hedayat M, Bayat F (2018) Effects of different media factors on in vitro proliferation of St. John's wort (*Hypericum perforatum.*) J Crop Breed 10:75–83

Pérez-García F, Huertas M, Mora E, Peña B, Varela F, González-Benito ME (2006) *Hypericum perforatum* L. seed germination: Interpopulation variation and effect of light, temperature, presowing treatments and seed desiccation. Genet Resour Crop Evol 53:1187–198. https://doi.org/10.1007/s10722-005-2012-3

Plants for A future (2010) *Hypericum perforatum* L [Internet]. Plants for A future. 2010 [cited 2022 Apr 14]. Available from: https://pfaf.org/user/Plant.aspx?LatinName=Hypericum+perforatum

Pluhár Z, Bernáth J, Neumayer É (2002) Morphological, production biological and chemical diversity of St. John's Wort (*Hypericum perforatum* L.). Acta Hortic 576:33–40. https://doi.org/10.17660/ActaHortic.2002.576.4

Poutaraud A, Girardin P (2005) Improvement of medicinal plant quality: a *Hypericum perforatum* literature review as an example. Plant Genet Resour 3:178–189. https://doi.org/10.1079/pgr200567

Raquel Pretto F, Romanato SE (2000) Callus formation and plant regeneration from *Hypericum perforatum* leaves. Plant Cell Tissue Organ Cult 62:107–113. https://doi.org/10.1023/A:1026534818574

Reuter HD. Chemistry and biology of *Hypericum perforatum*. Phytomedicines Eur. Lawson, L et al (2002) ACS symposium series. American chemical society. Washington, DC, 1998, 2002, p 303–316

Rizzo P, Altschmied L, Ravindran BM, Rutten T, D'auria JC (2020) The biochemical and genetic basis for the biosynthesis of bioactive compounds in *Hypericum perforatum* L., one of the largest medicinal crops in Europe. Genes (Basel) 11:1–21. https://doi.org/10.3390/genes11101210

Saddiqe Z, Naeem I, Maimoona A (2010) A review of the antibacterial activity of *Hypericum perforatum* L. J Ethnopharmacol 131:511–521. https://doi.org/10.1016/j.jep.2010.07.034

Sánchez-Reus MI, Gómez del Rio MA, Iglesias I, Elorza M, Slowing K, Benedí J (2007) Standardized *Hypericum perforatum* reduces oxidative stress and increases gene expression of antioxidant enzymes on rotenone-exposed rats. Neuropharmacology 52:606–16. https://doi.org/10.1016/j.neuropharm.2006.09.003

Schneider-Yin X, Kurmanaviciene A, Roth M, Roos M, Fedier A, Minder EI et al (2009) Hypericin and 5-aminolevulinic acid-induced protoporphyrin IX induce enhanced phototoxicity in human endometrial cancer cells with non-coherent white light. Photodiagnosis Photodyn Ther 6:12–18. https://doi.org/10.1016/j.pdpdt.2009.02.001

Schulte J, Büter B, Schaffner W, Berger K (1996) Gametic embryogenesis in *Hypericum spp.* International symposium on breeding research on medicinal and aromatic plants, Quedlinburg pp 307–310

Sharifi I, Fotokian MH, Kordenaeij A, Ramazani S, Hasanlu T, Khayyam Nikoei SM et al (2014) Effects of growth regulators and explants on Callus induction and organogenesis in *Hypericum perforatum.* Acta Hortic 1023:241–248. https://doi.org/10.17660/actahortic.2014.1023.35

Shrivastava M, Dwivedi LK (2015) Therapeutic potential of *Hypericum perforatum*: a Review. Int J Pharm Sci Res 6:1000. https://doi.org/10.13040/IJPSR.0975-8232.6(12).1000-07

Soták M, Czeranková O, Klein D, Nigutová K, Altschmied L, Li L, Jose A, Wurtele ES, Čellárová E (2016) Differentially expressed genes in hypericin-containing *Hypericum perforatum* leaf tissues as revealed by de novo assembly of RNA-Seq. Plant Mol Biol Report 34:1027–1041. https://doi.org/10.1007/s11105-016-0982-2

Steck N, Messmer M, Schaffner W, Bueter KB (2001) Molecular markers as a tool to verify sexual and apomictic off-spring of intraspecific crosses in *Hypericum perforatum.* Planta Med 67:384–385

Stefanovits-Bányai É, Nagy A, Sárdi É, Végvári A, Pluhár Z. Morphological, chemical and biochemical characterization of St. John's Wort (*Hypericum perforatum* L.) hybrids. Acta Biol Szeged 2002;46:259–61.

Suzuki O, Katsumata Y, Oya M, Baldt S, Wagner H (1984) Inhibition of monoamine oxidase by Hypericin. Planta Med 272–274. https://doi.org/10.1016/0006-2952(59)90039-5

Syahid SF (2021) In vitro conservation of medicinal plant St. John's wort (*Hypericum perforatum* L.) through dilution of basic medium. Agroscript 3:11–18

Wójcik A, Podstolski A (2007) Leaf explant response in in vitro culture of St. John's wort (*Hypericum perforatum* L.). Acta Physiol Plant 29:151–156. https://doi.org/10.1007/s11738-006-0019-9

Yardener (2003) Problems of Hypericum [Internet]. Yardener. [cited 2022 Jun 12]. Available from: https://gardening.yardener.com/Problems-Of-Hypericum

Zanoli P (2004) Role of hyperforin in the pharmacological activities of St. John's Wort CNS Drug Rev 10:203–218. https://doi.org/10.1111/j.1527-3458.2004.tb00022.x

Zou Y, Lu Y-H, Wei D (2010) Protective Effects of a flavonoid-rich extract of *Hypericum perforatum* L. against hydrogen peroxide-induced apoptosis in PC12 cells. Phyther Res 24:6–10. https://doi.org/10.1002/ptr

Zouhar K (2004) Species: *Hypericum perforatum* [Internet]. Fire effects information system (FEIS), USDA. [cited 2022 Apr 10]. Available from: https://www.fs.fed.us/database/feis/plants/forb/hyper/all.html

Chapter 10
Plant Breeding Advances in Candy Leaf (*Stevia rebaudiana Bertoni*)

Rashmi Ramakrishnan, Wudali Narasimha Sudheer, Akshatha Banadka, Chole Pranjali Bajrang, Jameel M. Al-Khayri, and Praveen Nagella

Abstract *Stevia rebaudiana*, also referred to as sugar leaf or sweet leaf, is becoming more and more well-known around the world as a natural zero-calorie sweetener because of its high steviol glycoside (SG) content, specifically Reb A, Reb D, and Reb M. The chapter discusses the role that stevia plays in meeting the growing need for natural sweeteners due to health issues like obesity and diabetes. Despite its potential, stevia production is hindered by problems with large-scale multiplication, inefficient cultivars, and environmental stress. Improvement strategies aim to boost stevia production by creating high-yielding, stress-tolerant cultivars with superior agronomic traits. Increasing SG content with high rebaudioside- stevioside ratio, raising yield, and strengthening resistance to biotic and abiotic stresses are some of the goals of breeding. Stevia production is improved through the use of a variety of breeding techniques, such as marker-assisted breeding, clonal selection, and traditional selection. The genetic makeup of Stevia is understood, important target genes are identified, and breeding strategies are optimized with the use of molecular tools

R. Ramakrishnan · W. N. Sudheer · C. P. Bajrang · P. Nagella (✉)
Department of Life Sciences, Christ University, Bengaluru, Karnataka 560029, India
e-mail: praveen.n@christuniversity.in

R. Ramakrishnan
e-mail: rashmir@res.christuniversity.in

W. N. Sudheer
e-mail: wudali.sudheer@res.christuniversity.in

C. P. Bajrang
e-mail: cholepranjali.bajrang@res.christuniversity.in

A. Banadka
Department of Biotechnology and Genetics, School of Sciences, JAIN (Deemed to be University), Bengaluru 560027, India
e-mail: akshatha.b@res.christuniversity.in

J. M. Al-Khayri
Department of Agricultural Biotechnology, College of Agriculture and Food Sciences, King Faisal University, Al-Ahsa, Saudi Arabia
e-mail: jkhayri@kfu.edu.sa

J. M. Al-Khayri et al. (eds.), *Biodiversity and Genetic Improvement of Medicinal and Aromatic Plants II*, Advances in Plant Breeding Strategies 10,
https://doi.org/10.1007/978-3-031-81857-8_10

like genomic and bioinformatic analyses. Moreover, genetic engineering methods present viable paths for modifying important metabolic pathways to improve stress tolerance and SG production. The chapter also offers a thorough understanding of the difficulties and possibilities in the production of stevia, emphasizing the role that molecular tools and breeding strategies will play in resolving these issues and advancing the cultivation of stevia. Through the integration of advanced genomic and genetic engineering tools with conventional breeding methods, stevia breeders can expedite the development of superior cultivars, thereby satisfying the increasing demand for natural sweeteners in an environmentally friendly manner.

Keywords *Stevia rebaudiana* · Breeding · Hybrids · Mutation · Polyploidy · Cultivation · Transformation

10.1 Introduction

Stevia also referred to as candy leaf, famous for its sweet components, which is 300 times sweeter than sugar. With the increasing nature of metabolic diseases like diabetes, Stevia is getting attention from all over the world, mainly because of the zero-calorie steviol glycosides. Since people started to prefer organic substances rather than synthetic ones, the demand for steviol glycoside also increased and is predicted to continue to rise in the upcoming decades also. However, the amount of Stevia being grown today is insufficient to satisfy the rising demand. In many countries, Stevia is a recently domesticated crop that necessitates the studies on its morphological, biochemical, and agronomic characteristics to aid in breeding experiments. Detailed information on Stevia cultivation is needed by farmers for effective economic production. In this regard, this chapter aims to evaluate the cultivation practices, conventional breeding methods and their limitations, role of biotechnology and molecular tools in improving Stevia from the available data. It also analyzes how modern genetic engineering techniques can aid Stevia breeding to fulfill the industrial need for steviol glycosides with respect to global climatic changes.

10.1.1 Botanical Classification and Distribution

Kingdom	Angiospermae
Division	Tracheophyta
Class	Dicotyledons
Group	Monochlamydeae
Order	Asterales

(continued)

(continued)

Kingdom	Angiospermae
Family	Asteraceae
Subfamily	Asteroideae
Tribe	Eupatorieae
Genus	*Stevia*
Species	*rebaudiana*

It was first named *Eupatorium rebaudianum and* later on, renamed *Stevia rebaudiana* by Moises Santiago Bertoni, a European botanist (Ramesh et al. 2006). *Stevia rebaudiana* is one among the 150–200 species of the genera from the Asteraceae family rooted in Paraguay and the Brazilian border. Due to its origins in Paraguay and Brazil, it is also known as 'the sweet herb of Paraguay (Singh et al. 2019). Around the world, stevia can be found in Mexico, the South American Andes, Central America, Argentina, the southern parts of the United States, and the highlands of Brazil (Gantait et al. 2018). Stevia is recognized as a major shrub/crop in various countries, including Brazil, Canada, Indonesia, India, Japan, Korea, Mexico, the United States, and Tanzania, due to its tremendous medicinal potential (Gantait et al. 2015) (Fig. 10.1).

10.1.2 Importance of Stevia (Economic and Health Benefits)

Stevia has drawn scientific and commercial interest due to its sweet and medicinal qualities, largely because of the presence of ent-kaurene, a diterpene glycoside

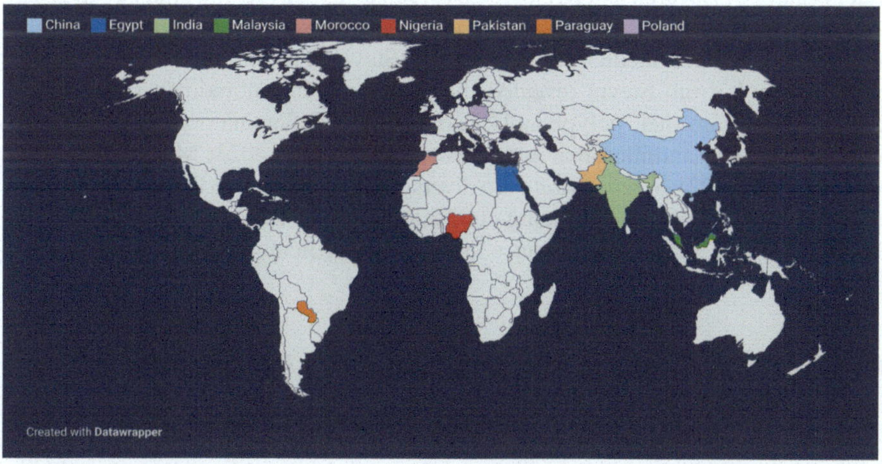

Fig. 10.1 Distribution of *Stevia rebaudiana* in different geographical locations

also known as the steviol glycosides. There are more than 30 distinct glycosides known including stevioside, steviol, steviolbioside, rebaudioside A, B, C, D, M, and dulcoside but stevioside and rebaudioside A are the two abundant and most studied glycosides (Putnik et al. 2020). The major sweet compound in Stevia is the stevioside, but it is also responsible for the after bitterness of Stevia, while rebaudioside A is 300–450 times intense than sugar (Singh et al. 2019; Hernández et al. 2022). The preference for natural compounds over artificial sugars, has led to Stevia becoming a common ingredient in the food industry. The World Health Organization and the Joint Food and Agriculture Organization decided to set a temporary acceptable daily intake (ADI) of 4 mg/kg of steviol glycosides, which may have contributed to the expansion of stevia-based products available on the market (Talevi 2022). The brands like Truvia, pureVia, zevic Stevia, nutri.org, and Steviaworld exploit Stevia as a sweetener in their products (Singh et al. 2019). Stevia finds its application in carbonated beverages, chewing gums, biscuits, candies, pickles, sauces, etc. as a sugar source (Hossain et al. 2017; Góngora Salazar et al. 2018; Arumugam et al. 2020). Because it has technological qualities as an emulsifier, texturizing agent, and coloring agent, as well as antioxidant properties, it is widely used in dairy products, baked goods, jellified products, juices, and other drinks (Schiatti-Sisó et al. 2022).

Apart from the potent sweeteners, they are well known for their medicinal properties too. Some forms of cancers like breast and gastrointestinal cancer, chronic kidney disease, liver fibrosis, cardiac fibrosis, diabetes mellitus, overweight and obesity, fatty liver, inflammatory bowel disease, and hypertension are among the chronic conditions for which Stevia has shown therapeutic benefits as a combined nutraceutical (Ramos-Tovar and Muriel 2017; Zaidan et al. 2019; Singh et al. 2019; Wang et al. 2020). Since it is widely used as a sweetener, its effects on diabetic patients are a matter of concern (Ajami et al. 2020). There are research efforts investigating the antidiabetic features of Stevia by both in vitro and in vivo techniques suggesting its potential lowering blood glucose level or increase insulin production (Ali et al. 2022; Salehi et al. 2019; Kurek and Krejpcio 2019; Jan et al. 2021; Chowdhury et al. 2022).

Other than glycosides, the plant also contains substantial amounts of polyphenols, which will add medicinal properties to Stevia. The studies on the properties of polyphenols from Stevia revealed that it is a potent antioxidant (Myint et al. 2020). Apart from that, polyphenols also have anti-inflammatory, anti-cancerous, anti-diabetic, and antimicrobial activities (Chakma et al. 2023; Myint et al. 2020; Mlambo et al. 2022). *S. rebaudiana* is a diverse and valuable herb that can be securely inferred. And studies are still being conducted in order to completely comprehend the herb. Hence, Stevia's significance in food and medicine has yet to be explored.

10.1.3 Domestication, Selection, and Early Improvements

For centuries, the Guarani Indians and the tribal people of Paraguay have utilized Stevia as a medicinal herb and sweetener. Later multiple nations, including China, Japan, Brazil, Korea, Mexico, the United States, Canada, India, and Indonesia started growing Stevia. When saccharin was banned in Japan, Stevia was first domesticated as a substitute sweetener (Singh et al. 2019). Stevia farming is presently centered in China, while Japan serves as the key consumer (Angelini et al. 2018).

Stevia has the potential to grow under hot climatic conditions as in Japan as well as cold conditions as in Kashmir (Rashid et al. 2021). Even though the steviol glycoside composition and phenotypic characters may vary according to the environmental condition (Ramesh et al. 2006). The large-scale cultivation of Stevia was started in the 1990s only by the Brazilian and Paraguay population. Several reports show that Stevia was extensively used as a sugar substitute during World War 2. Later on, it was introduced into Japan for commercial cultivation and research (Angelini et al. 2018). Later, it was domesticated as a crop in many countries including Brazil, Canada, Indonesia, Korea, Mexico, Tanzania, and the United States (Peteliuk et al. 2021).

Stevia is a plant that has only recently been domesticated (Oliveira et al. 2004). Initially, selection was based on boosting stevioside levels, improved leaf-to-stem ratio, and enhanced plant resilience to stressed environment to increase yields (Badran et al. 2015). Given the industrial preferences for high Reb-A content, the selection was mostly focused on the plants with high Reb-A content (Kumar et al. 2011). Due to a high level of natural variability induced by frequent out-crossing, breeders can increase the level of sweeteners in the leaves and modify the reb A/stv ratio (Kaplan and Turgut 2019). After getting clearance from the world's leading regulatory agencies like FSANZ in 2008; JECFA in 2009; and EFSA in 2010 for the safe use of Stevia-based steviol glycosides, the consumption has been increasing (Mahajan et al. 2021). The global Stevia market was worth USD 650 million in 2020 and is expected to hit USD 1.28 billion by 2028, rising at an annual rate of 8.95% (Judickaitė et al. 2022). To meet the rising demand, more cultivation, as well as improved varieties for satisfying economic needs, would be required.

10.2 Current Cultivation Practices and Challenges

10.2.1 Current Cultivation Practices

Originated from Southern Ontario, Stevia is now domestically cultivated in Pacific Rim countries. It is predominantly found growing in semi-arid habitats such as grasslands, scrub forest and mountain terrain. It is grown in semi-humid, tropical and subtropical climates where temperature ranges from − 6 to 43 °C and the favorable temperature for growing the Stevia ranges from 15 to 30 °C. The relative humidity of 65–80% with annual rainfall of 140 cm is ideal for proper growth and development. It grows well on sandy loam infertile, acidic soils and on more neutral soils (pH 6.5–7.5). It fails to grow on alkaline soil (Sharma et al. 2016b).

The traditional cultivation practices involve seed propagation and vegetative propagation. In Stevia, seed propagation is not much recommended because of its poor viability and germination rates. Since Stevia plants are self-incompatible which means the failure of fertilization by its own pollen, cross-pollination is essential for obtaining viable seeds. There are two morphological types of seeds present in Stevia in which black-colored achenes show more germination than pale or tan-colored achenes (Kumar et al. 2011). Previous reports have demonstrated that, bee-pollinated Stevia plants produced a greater number of seeds with black achene (Özyiğit et al. 2015). Enhanced seed viability can be observed in Stevia plants which are open-pollinated by various pollinators (Martini et al. 2017). As per the study of Martini et al. (2017), the genotypes which flower early and have more inflorescence had more pollinator visits and it boosted seed viability. So, sexual reproduction via viable seeds is possible in Stevia by selecting an appropriate genotype having early flowering and a short reproductive stage. The seeds can be planted in raised nursery beds with predetermined ratios of sand, soil, and organic manure, or they can be planted in pots, wooden boxes, or plastic trays. The germination starts within 7 days of sowing. It is then transplanted after 2 months when the height reaches upto 10–15 cm (Sharma et al. 2016b) (Fig. 10.2).

Stem cuttings are the most commonly used propagation method in Stevia. The stem cuttings of 10–15 cm height with 4–6 nodes from actively growing plants are taken. It is then placed in soil with one node inside the soil. It is grown in partial shade and high humidity which ensures root development within 10–15 days with regular watering (Sharma et al. 2016b). Treating the stem cuttings with *Trichoderma viride* and using soil: sand: vermicompost in a 1:1:1 ratio has proven effective for Stevia propagation (Smitha and Umesha 2012). The treatments of stem cuttings with synthetic auxins like IBA also showed a high percentage of rooting (Abdullateef and Osman 2011; Khalil et al. 2014; Pigatto et al. 2018) and the best season for

Fig. 10.2 **a** Habit of *S. rebaudiana*, **b** inflorescence of *S. rebaudiana*, **c** seeds of *S. rebaudiana*, **d** fertile seeds of *S. rebaudiana*, **e** infertile seeds of *S. rebaudiana*

vegetative propagation by stem cuttings is Feb-July except for June (Castañeda-Saucedo et al. 2020). Various growth stimulators and elicitors like nanoparticles, microbes, oligosacharides has been extensively applied to improve Stevia growth and yield (Abdelsattar et al. 2023; Laha et al. 2023; Andújar et al. 2023; Sharma et al. 2023a; Wang et al. 2023).

10.2.2 Current Agricultural Challenges

Poor seed germination and a preference for cross-fertilization make Stevia difficult to propagate (Hajihashemi, 2021). Despite its drawbacks, such as low rooting rates and high labor costs, vegetative propagation—particularly stem cutting—is recommended for large-scale cultivation (Hajihashemi, 2021). It can be difficult to gather healthy vegetative material for propagation, and pathogenic attacks worsen cultivation by causing diseases like damping-off, leaf spot, and root rot (Basharat et al. 2021).

10.2.3 Genetic Improvement Objectives

The best technique for overcoming agricultural constraints is genetic improvement. In Stevia cultivation, the main objectives for genetic improvement were,

- High leaf yield
- Improved rebaudioside content
- Disease resistance
- Climatic adaptability
- Easy propagation method.

10.3 Germplasm Biodiversity and Conservation

10.3.1 Germplasm Diversity

There is no publicly accessible collection, and only private companies possess collections that are not publicly accessible (Cosson et al. 2019). Since it is believed that Stevia is native to Paraguay, the Perguaen varieties, Criolla and Morita II, are the most experimented germplasms of Stevia. It is known that the morphological and biochemical attributes of Stevia were highly impacted by environmental factors. So, it is necessary to identify the correct germplasm for specific climatic conditions for the effective economic yield. Some of the available germplasms of Stevia are,

- RSIT 94-1306, RSIT 94-751, Sugar High A3, Huny, RSIT 94-1838, RSIT 94-1833, and RSIT 94-1560 from Canada
- Shoutian No. 3, Zhongshan No. 7, Yunri, Yunbing, and Zongping from China
- HimStevia, CIMAP madhu, CIM mithi, Madhuguna, Madhuguni, MDS 13 and MDS 14 from India
- AC BLACKBIRD, Morita SF5 and SF6, from Japan
- Suweon 2 and Suweon 11 from South Korea
- Criolla, Katupyry, Morita II and Eirete from Paraguay
- SW 201, SW 107, SW 129 and AKHL 1, and AKHL 4 from USA.

10.3.2 Cultivars Characterization and Phylogeny

Because of the quality and timeliness of the data obtained, as well as the fact that morphological traits are altered by environmental influences, germplasm fingerprinting and cultivar identification using molecular markers have grown in relevance (Ribeiro et al. 2021). No extensive studies on *S. rebaudiana* have looked into the development of molecular markers to classify germplasm and investigate phylogeny (Cosson et al. 2019). However, some efforts have reported the implementation of molecular markers like microsatellites in the genotype identification and their similarity checking from Stevia accession without revealing the cultivar name (Ribeiro et al. 2021).

10.3.3 Genetic Resource Conservation Approaches

Cryopreservation and sodium alginate-based synthetic seed formation are two in vitro methods used in Stevia germplasm conservation. A synthetic seed production protocol was developed by Lata et al. in 2014 that allows storage at 25 °C for up to 12 months. Techniques like vitrification are used to store germplasm at extremely low temperatures, or cryopreservation (Sakai and Engrlmenn (2007). A successful vitrification protocol was developed by Shatnawi et al. (2011) for Stevia shoot tips, which prevented ice formation and allowed for long-term storage. Recent research by Benelli et al. (2021) has improved Stevia cryopreservation techniques even further, guaranteeing higher growth rates while in storage.

10.3.4 Cytogenetics

Based on cytological studies reported by Frederico et al. (1996), and Oliveira et al. (2004), Stevia is diploid in nature with $2n = 22$ (six metacentric, four submetacentric, and one acrocentric pair) but the occurrence of $2n = 33$ and $2n = 44$, i.e., triploid and tetraploid cytotypes are also possible. They also revealed the presence of a satellite in the short arm of the third chromosome pair. The findings bring up new opportunities for *S. rebaudiana* genetic breeding efforts, as polyploid individuals may contain more stevioside than diploid plants.

The studies on the meiotic behavior of Stevia, give haploid microspores, suggesting that the self-incompatibility assigned to Stevia is not because of the meiotic aberrations (Oliveira et al. 2004). Self-incompatibility is the failure of germination of pollen grains of the same genotype (Yadav et al. 2014). Pollen germination might be hampered by the prevalence of protandry, poor pollen viability, a dearth of nutrients for pollen germination at the stigma surface, and stiff stylar tissue, which restricts pollen germination and pollen tube expansion beyond the stigma surface, eventually inhibiting seed laying (Gantait et al. 2018). The reason for self-incompatibility might be because of the diversion of the GA metabolic pathway into steviol glycoside synthesis, which makes unavailability of sufficient GA for the signaling of pollen tube germination (Tuleja et al. 2021).

10.4 Traditional Breeding

10.4.1 Improvement Strategies

Breeding for high yields and elevated concentrations of desired steviol glycosides, such as Reb A, Reb D, and Reb M, are the main strategies for improving Stevia. In order to meet the rising demand worldwide, efforts are also concentrated on creating cultivars that are efficient at large-scale multiplication and resistant to environmental stress (Xu et al. 2021; Cosola et al. 2021). The previous reports (Al-Taweel et al. 2021) suggests that, the key features of Stevia that can be achieved by traditional and biotechnology breeding processes are as follows:

- Enhance agronomic traits like more leaf: stem ratio and high leaf yield/area
- Achieve quality value in terms of SVgly content such as high Reb A, Reb A—STV ratio, Reb D, Reb M, and other sweet compounds.
- Reduce the bitter taste.
- High regeneration capacity and should be compatible for multiplication via tissue culture, vegetative propagation, and genetic engineering as well.
- Long duration of vegetative growth.
- Develop novel cultivars that are adaptable to changing climate conditions.
- Self-compatible cultivars that can produce viable seeds with genetic uniformity.
- Resilient to biotic and abiotic stresses.
- Develop effective agrotechnologies specific for cultivars.

10.4.2 Traditional Breeding Methodologies

10.4.2.1 Clonal Selection

Since sexual reproduction is the natural reason for genetic variation, there is a high chance of occurrence of hybrid plants with desired qualities in natural open-pollinated Stevia populations. Clonal selection is a technique for selecting good clones from a mixed plant population. In this regard, identification of four plants having desired qualities of glycoside content and other superior agronomic characteristics through the clonal selection method from open-pollinated Stevia cultivation in Turkey was reported. The selected plants were considered as genotypes and suggested for mass production by clonal or vegetative propagation (Kaplan and Turgut 2019).

10.4.2.2 Artificial Selection

It is the selection of plants based on desired agronomic characteristics. Selection of four populations from the cultivated population in the University of Maringa based on various agronomic features like height, glycoside production, branching and early flowering and its propagation by vegetative methods was reported (Luz et al. 2020). In 2018 an artificial selection for self-compatible cultivar 314,018 (US Patent No. US20180070553A1) from a naturally grown population was patented by Markosyan et al. (2018). The identification of self-compatible varieties helps in future breeding programs since it produces viable seeds via self-fertilization (Markosyan and Jing 2018).

Development of Stevia cultivars A01, A03, A05, A06, A07, and A08 through selective breeding (US 2021/0029907 A1): The Stevia plants from Alberta Research Council in the late 2000s are crossed between a succession of half-sibling groups. To yield F1 seeds, the half-sibling families were planted in an isolated field apart from other Stevia plants. The F1 seeds were then planted to produce F1 plants, which were then analyzed and selected for commercially beneficial phenotypes, such as plant topography and chemical composition, including individual steviol glycoside content, overall steviol glycoside content, and rebaudioside A to stevioside ratio. The chosen plants were allowed to mature (about 90 days) before flowering and seeding. Beneficial features were also evaluated, selected, and identified as A01, A03, A05, A06, A07, and A08 from the progeny. The genetic variations among the selected progeny were evaluated by SNP markers. Tissue culture was used to clone and maintain selected commercially elite plants (Jackson 2021).

10.4.2.3 Recurrent Selection

After the population has been selected for the principal trait being transmitted, the recurrent selection is used to choose for performance within successive segregating progeny generations. The recurrent selection was employed by S.W Seed Company for the identification of new cultivars SW107, SW129, and SW207 from the natural open-pollinated populations of the USA. Cultivar SW107 was selected based on its leaf taste and steviol glycoside content. SW 129 was identified as a cultivar based on its high plant vigor, Reb-A content, and less bitterness (Shock and Parris 2017a). SW207 was selected based on its overwinterness, plant vigor, and high sweetness (Shock and Parris 2017b). All these selections were patented by S.W seed company, USA from 2012 to 2013, and a cross between the 'selection 395' and the 'SW 129' Stevia lines yielded the 'SW 227' Stevia line, which has considerable branching and high Reb-A and total glycosides (US PP28977 P3) (Shock and Parris 2018).

10.4.2.4 Polyploidy Breeding

It is the method of selection of superior plants after inducing polyploidy using chemicals like colchicine (Sattler et al. 2016). Polyploidization has been used as a potent plant breeding tool in current Stevia breeding in addition to the standard cross breeding method. Stevia autotetraploids have much larger leaves, thicker leaves, and higher chlorophyll content, as well as shorter internodes (Yadav et al. 2013). Apart from that, the manipulation of ploidy level altered the plant features like number, area and thickness of leaves as well as glycoside content (Hegde et al. 2015). The treatment of 0.20% colchicine for 12 h caused the production of autotetraploid plants in Stevia (Xiang et al. 2019). The polyploidy resulted in plants with larger leaves and enhanced production of Reb A, Reb B, and total glycosides with a low amount of stevioside with the overexpression of CPPS1 and KAH genes. Apart from that, polyploidy also altered the resistant genes (Markosyan et al. 2021b). Recently an aneuploid Stevia cultivar AP 1(2X + 1) with a high Reb M content by inducing polyploidy in Stevia variety' 814,011' using colchicine was developed (Markosyan et al. 2021a, b, c).

A combined method of anther culture and colchicine treatment was employed to develop new Stevia cultivar T60. To achieve this, the breeders used the Paraguayan variety, 'Criola' to produce haploid plants through anther culture. Then they were induced to diploid using colchicine. The resulted progeny was screened for Reb-A content and the selected plants were released as T60 cultivar (US PP22,593 P3) (Garnighian 2012).

10.4.2.5 Synthetic Breeding or Polycross Breeding

A synthetic breeding or cross between multiple parental varieties in all possible combinations is used in selective plant breeding to choose plants having one or more desirable features. The cross's offspring usually reproduce through vegetative reproduction. Synthetic cultivars are an efficient method of Stevia breeding (Kumar et al. 2011). 'AC Black Bird' and 'PTA-44' are two commonly used synthetic cultivars of Stevia (Al-Taweel et al. 2021).

Recently, a new cultivar of Stevia having high Reb D content via the combined method of polycrossing and mutation breeding was identified. To accomplish this, the breeders used a variety of germplasm from China and Argentina that was allowed to polycross with all possible combinations, and the seeds of F1 progeny were gathered, with half of them being treated with gamma irradiation for mutation and the other half being left untreated. The high Reb D content of the developing plants from the F1 seeds was chosen. The chosen mutants were subjected to a second dosage mutation, and the offspring were evaluated for increased Reb D content. Both the method and cultivar were recently patented with patent no. US 2021/0068360 A1 (Brower et al. 2021).

10.4.3 Traditional Breeding Limitations

The molecular variances of new breeds from their parents are not assessed because traditional breeding procedures entail selection based on phenotypic features. Because Stevia is a self-incompatible plant, cross-breeding is the only way to reproduce it naturally. This will result in plants that are genetically and phenotypically distinct. To circumvent this, the only way for large-scale reproduction of each cultivar is vegetative propagation. This will take a long time and will be expensive in terms of labor costs. Furthermore, the heritability of novel characters in cultivars generated using plant breeding procedures like selective breeding, polyploidy breeding, mutation breeding, and others must be tested over multiple generations before being released to the general public. It will take 5–10 years to complete this process. Because the key target of Stevia breeding is to increase Reb-A content, each selection is made after glycoside analysis using HPLC. This was also extremely expensive (Al-Taweel et al. 2019).

10.4.4 Biotechnological Approaches

Since the traditional breeding methods are facing various obstacles for large-scale production, biotechnology has broad application in research related to Stevia breeding. Plant tissue culture is the major biotechnology strategy that will boost the mass production of Stevia in both qualitative and quantitative aspects. This is most widely accepted in terms of the ability to employ cellular screening to retrieve valuable genetic variants (somaclones) and the ability to alter specific genes using DNA recombination technology. According to Khan et al. (2014), somaclonal variation can lead to the development of new desirable clones/variants with improved agronomic features including high yield, survival rate, disease, and pest resistance. In addition, a paradigm that blends the good aspects of transgenic technology with both organic and conventional farming is needed to handle the enormous agricultural load of today. Previous studies report that the steviol glycoside content in in-vitro cultures of Stevia is higher than that of in vivo cultured plants, confirming the efficiency of tissue culture (Javed and Yücesan 2022).

Biotechnological approaches also facilitate steviol glycoside production, through emerging strategies like metabolomics, transcriptomics, proteomics, and genomics (Basharat et al. 2021). Metabolic engineering gives promising results in the mass production of sweet compounds from Stevia through enzyme modification, bioreactor production, artificial synthesis, etc. (He et al. 2022).

10.5 Molecular Breeding

10.5.1 Molecular Marker-Assisted Breeding

Molecular markers are small repeated sequences unique to each organism. The analysis of molecular markers will be supportive for the study of phylogeny and genetic variability between various cultivars and species of Stevia (Dyduch-Siemińska et al. 2020). The identification of various molecular markers present in Stevia will help to boost the production of new varieties with desired qualities. With the knowledge of molecular markers specific to attributed characters, the selection process in plant breeding will become fast because it can be accessed at a young stage of the plant itself apart from waiting for the plant to mature. High accuracy and reduced labor cost also add advantages to molecular marker-assisted breeding (Dyduch-Siemińska et al. 2020). Recent chromosomal genome analysis conducted by Xu et al. (2021) reported that more than 4/5% of the Stevia genome contains repetitive sequences, the data opens up the possibilities of various molecular markers which can aid molecular breeding.

There are few studies that examined the genetic variability in Stevia genotypes using molecular markers like RAPD, ISSR, etc. in small populations. Chester et al. (2013) used RAPD markers to analyze genetic variability among Indian accessions of Stevia along with metabolic study to identify the superior one. Bhandawat et al. (2014) developed functionally relevant microsatellite markers from the EST sequence data available in NCBI for the characterization of Stevia germplasm diversity. The fidelity of in vitro cultures of Stevia was checked by Hassanen and Khalil (2013) using RAPD, ISSR, and AFLP markers. Chen et al. (2014) identified potential SSR and SNP markers from the cDNA library created via RNA sequencing and designed primers to facilitate Stevia breeding studies. Recently, Dyduch-Siemińska et al. (2020) used RAPD primers to investigate genetic variation among Stevia genotypes procured from various areas of the globe, and cluster analysis of the resulting data led to the categorization of the examined populations based on similarities and differences. The findings open up the possibility of cross-breeding germplasms having fewer genetic similarities to develop new varieties with desired qualities.

Since the less reproducibility of RAPD markers, recent research has focused on microsatellite (SSR) markers due to their extended polymorphism, reproducibility, and even distribution in the genome. The effectiveness of the combined use of RAPD and ISSR markers in the analysis of genetic variability within various Indian Stevia germplasms has been proven (Sharma et al. 2016a). However, Ribeiro et al. (2021) analyzed the genetic variability within 31 Brazilian genotypes of Stevia using microsatellite markers and identified the superior genotype. The genetic variability within the Paraguay population of Stevia and its artificial selected populations was studied by Luz et al. (2020) with ISSR markers. The findings of Hadia et al. (2008) also support the accuracy of ISSR markers in the genetic variability studies. The significance of SCoT markers in the analysis of genetic relationships was elucidated by Hashem et al. (2021).

It has been suggested that molecular methods are suitable and effective tools for determining genetic stability, stress tolerance, and somaclonal variation identification. Al-Taweel et al. (2019) identified the salt-tolerant plants with the aid of cDNA-SCoT marker. Few studies have reported on the efficiency of molecular markers for checking the fidelity of tissue cultures. Khan et al. (2014) employed ISSR markers for analyzing the genetic fidelity of somaclones produced from transformed Stevia leaves. The fidelity of in vitro cultures of Stevia was assessed by RAPD, ISSR, and AFLP markers (Hassanen and Khalil 2013). Iwaki et al. (2022) found that Stevia plants with high Reb C can be identified by SNP polymorphism. By utilizing molecular markers to analyze genetic diversity both within and between crop plant species, breeders can choose superior genotypes to be used as parents in hybridization programs or as direct varieties (Luz et al. 2020; Dyduch-Siemińska et al. 2020).

10.5.2 Functional Genomics

There are several molecular studies on *S. rebaudiana* that have been documented using various molecular markers. However, the creation of a linkage map of the Stevia genome by using RAPD markers lays the groundwork for Stevia molecular genetics studies (Yao et al. 1999). In 2013, Hodgins et al. (2014) analyzed the transcriptomes of various Asteraceae members including Stevia, and disclosed EST data through the Compositae genome project. With the understanding of the necessity of genome analysis in Stevia, some private consortiums like PureCircle/Coca-Cola/Keygen are nowadays giving more focus on Stevia genome mapping (Cosson et al. 2019). In addition, the whole genome sequence of Stevia from Behnke Nurseries, USA was released and the sequences are available at genebank with accession numbers, SRR6792730 and GCA_009936405 (O'Neill and Pirro 2020). A genome covering 1416 Mb based on chromosomal sequencing of the Chinese cultivar of Stevia was constructed to facilitate molecular research in Stevia (Xu et al. 2021). In a nutshell, functional identification of genes that control Stevia's key quality and agronomic traits is challenging because of the less accessible sequencing data. Due to this reason, molecular biology approaches have been employed sparingly for the breeding purposes of Stevia. Around 160 Stevia sequences are currently accessible in the National Center for Biotechnology Information (NCBI) database (Chen et al. 2014). The functional genomic analysis is mostly limited to the genes associated with SG biosynthesis in Stevia. Recent genomic study provides insights into the transcriptional regulation and biosynthesis of steviol glycosides (SVGlys) in *S. rebaudiana* under varied photoperiodic conditions by revealing distinct gene expression patterns associated with the biosynthesis pathways of StGlys and gibberellins in leaves and stems. This knowledge provides prospective molecular targets for metabolic engineering to maximize the production of SVGlys, meeting the increasing need in the commercial and industrial sectors for natural sweeteners (de Andrade et al. 2023). Similarly, the work of Zhang et al. (2023) clarified the function of SrMYB1 as a transcriptional repressor of SrUGT76G1, a crucial gene in Stevia rebaudiana's steviol glycoside biosynthesis, and offered insights into the regulatory network governing the metabolism of steviol glycosides. Likewise, the research by Xu et al. (2024) identified and analyzed a diverse array of R2R3-MYB genes in Stevia, revealing their potential roles in regulating the biosynthesis of steviol glycosides, with specific emphasis on key candidate genes exhibiting significant expression differences between Stevia varieties with distinct steviol glycoside content.

The understanding of the genetic mechanism of Reb A production is crucial for breeding research because the commercial industries prefer Stevia with more rebaudioside. As per the previous studies, the genes, UGT85C2, UGT74G1, and UGT76G1 are all involved in the production of rebaudioside A (Richman et al. 2005). Several studies have used various methodologies, including RNAi, base mutation, and enzyme tests, to examine the function and mechanism of UGT76G1. An amino acid substitution mutation in the UGT76G1 gene brought about a shift in the enzyme's secondary and tertiary structure, associated with decreased Reb-A content (Yang et al. 2014). In 2016, Olsson et al. found that this gene is responsible for eight distinct enzymatic activities related to glycoside synthesis (Olsson et al. 2016). The role of SrUGT76G1 in the Reb C production from dulcoside A via 1,3-glycosylation was also analyzed (Kim et al. 2019). The functional variations of five UGT76G1 mutants were also reported (Zhang et al. 2019). However, UGT76G1 has been evaluated for application as a biocatalyst in the industrial production of a wide range of glycosylated plant products owing to its extensive activity (Lee et al. 2019). All of these studies hint at the significance of this gene in Stevia breeding for steviol glycoside production with the desired composition. The study of Li et al. (2023) revealed five essential regulators (SrbHLH22, SrbHLH111, SrbHLH126, SrbHLH142, and SrbHLH152) that are important for rebaudioside among the 159 bHLH transcription factors in *S. rebaudiana*. Through subcellular localization analysis and dual luciferase reporter assays, a biosynthesis was validated, providing fresh perspectives on the control of steviol glycoside biosynthesis and possible uses in molecular breeding.

The production of mutants whose phenotype is compared to the control plants is one way of revealing the specific function of a plant gene (Petit et al. 2020). Recently, the implementation of miRNA in genetic studies has got attention because of its capacity to regulate various genes related to the biosynthetic pathways of important secondary metabolites and such studies could aid in genotype selection and alteration to achieve high-performing plant genotypes with higher secondary metabolite yields and biomass. Saifi et al. (2015) identified a total of 11 miRNA engaged in the steviol glycoside pathway in which two of them showed the same expression as their target mRNA while others are against target mRNA. Later on, Saifi et al. (2019) found the specific function of these two miRNAs (miR319g and miRStv11), via overexpressing them in Stevia leaves via *Agrobacterium*-mediated transformation and knockdown experiments and revealed that miRStv11 enhanced the KAH gene expression and miR319g have repressive effects on KO, KS, UGT85C2. So, the study suggests that these miRNAs can be used for breeding programs in Stevia for high rebaudioside fractions through the genetic engineering program.

10.5.3 Bioinformatics

Bioinformatic studies in Stevia have revolutionized our understanding of its genetic makeup and metabolic processes, primarily through transcriptomic analysis. High-throughput sequencing methods are employed in these investigations to examine patterns of gene expression and pinpoint critical genes implicated in significant metabolic processes, like the biosynthesis of steviol glycosides. For example, Chen et al. (2014), Singh et al. (2017) and Zhang et al. (2020), have discovered putative regulatory genes and clarified the molecular mechanisms behind steviol glycoside synthesis. Studying genetic variability within Stevia populations and comprehending physiological and transcriptional processes have also been accomplished through the use of transcriptome analysis (Xiang et al. 2019; Zhang et al. 2019; Yang et al. 2014).

Additionally, by comparing the genomes of Stevia with those of other related plant species, bioinformatics facilitates evolution studies by providing insights into the evolutionary history and specialized adaptations of Stevia (Xu et al. 2021). Utilizing bioinformatic techniques, Hashem et al. (2021) assessed the phylogeny of Egyptian Stevia cultivars through gene comparisons that are conserved. Moreover, bioinformatic analyses support the development of better cultivation techniques by optimizing in vitro conditions for the mass production of Stevia plants and secondary metabolites (Esmaeili et al. 2018; Kahrizi et al. 2018; Ghorbani et al. 2017; Hashempoor et al. 2018; Al-Taweel et al. 2019; Ghasemi-Omran et al. 2021; Thakur et al. 2021). Furthermore, a bioinformatic analysis of UGT genes was carried out by Mirzaei and Shakoory-Moghadam (2022), which disclosed structural quirks of the proteins these genes encode, like drought resistance.

10.6 Tissue Culture Applications

In last few decades, the popularity of Stevia as a sweet enhancer has boosted, necessitating an alternative to conventional propagation methods. Tissue culture-based studies, especially in vitro regeneration strategies, will help in large scale production of the desired plants (Pande and Gupta, 2013). Stevia seeds are usually sterile and there is always a slow mode of propagation through stem cuttings (Yang et al. 1981). So, in-vitro production will always help in mass production of the desired biomass.

Studies suggest that shoots of Stevia can be produced or regenerated on the MS medium supplemented with PGR or without PGR, but the MS medium supplemented with cytokinins like BAP and KN increased the multiplication, whereas for rooting, auxins like IBA and NAA play a major role (Sharma et al. 2023a, b). Along with PGRs, studies revealed that various parameters like nitrogen source, macro salts, micro salts play the major role in the in-vitro multiplication of the plantlets (Ibrahim et al. 2008a, b). MS medium formulated with 1650 mg/L NH_4NO_3 and 950 mg/L of KNO_3 showed the best response with respect to multiplication, shoot length and biomass (Akbari et al. 2017). Regeneration of shoots was seen when aerial explants (leaf, shoots) of Stevia inoculated onto the MS medium supplemented with 8.87 mM of BAP along with 5.71 mM of IAA and rooting was induced from the in-vitro shoots when sub-cultured onto the half- strength MS medium supplemented with 4.90 mM IBA (Sivaram and Mukundan, 2003). In-vitro shoots were developed from axillary position of nodal explants when inoculated on the MS medium supplemented with 1.5 mg/L BAP along with 0.5 mg/L KN, and rooting of the same shoots were developed when cultured on MS medium supplemented with 0.1 mg/L IAA (Ahmed et al. 2007). MS medium supplemented with 2.0 mg/L BAP along with 1.13 mg/L IAA helped in multiple shoot regeneration upon bud breakage with maximum number of shoots and highest multiplication efficiency. Best rooting efficiency was observed when 2.0 mg/L of IBA was supplemented onto the MS medium (Debnath, 2008). Internodal and apical shoot explants when cultured on MS medium supplemented with 0.3 mg/L KN helped in regeneration of maximum number of shoots and roots were induced on MS medium supplemented with 2.0 mg/L IBA (Taware et al. 2009). Nodal explants of Stevia when cultured on MS medium supplemented with 0.1 mg/L BAP induced shoots and subculturing them on 3.5 mg/L BAP supplemented medium helped in multiplication of shoots (Sairkar et al. 2009). Shoot tip explants when cultured on MS medium supplemented with 1.0 mg/L of BAP along with 0.5 mg/L of IAA showed best results in shoot induction when compared to leaf and nodal explants. Rooting was seen best on half-strength N6 medium (Chu medium) supplemented with 1.0 mg/L IAA (Anbazhagan et al. 2010). Nodal explants when cultured on MS medium supplemented with 0.5 mg/L BAP along with 0.25 mg/L KN induced multiple shoots and maximum roots were induced on MS medium supplemented with 1.0 mg/L IBA (Razak et al. 2014). When BAP and KN were used in different combinations to micropropagate Stevia from axenic nodal segments in MS medium, the results indicated that root development was preferable when plant growth regulators were not present, and that 10.0 mg/L BAP was the most effective at promoting shoot formation (Jadid et al. 2024).

Studies reveal that liquid medium helped in proliferation of multiple shoots. MS medium supplemented with 1.0 mg/L BAP along with 2.0 mg/L KN and 1% sucrose concentration induced the highest number of nodes from single shoot with highest length. It is also observed that plastic growth containers having liquid medium showed highest in-vitro multiplication rate (Modi et al. 2012). Modified MS medium formulated with higher concentration of copper ions (almost 5 times) showed increased shoot bud induction, broad leaves and improved overall biomass (Kalpana et al. 2010). Rooting efficiency for the in vitro shoots was improved when treated with growth retardants like alar and cycocel. It was also observed that 0.5 mg/L alar (B9; succinic acid 2-2-dimethyl hydrazide) along with 1.0 mg/L IBA showed best rooting and acclimatization efficiency for the shoots induced on 2.0 mg/L BAP enriched MS medium (Hassanen and Khalil 2013). Recent research shows that Stevia can be successfully micropropagated using meta-Topolin (mT), which leads to improved shoot initiation, multiplication, and proliferation when compared to other cytokinins. A high survival rate, genetic stability, and increased production of rebaudioside-A were attained in acclimatized plantlets through optimized culture conditions, which included mT supplementation and subsequent root induction. These results highlight the effectiveness of mT for large-scale in vitro propagation of Stevia with improved traits (Subrahmanyeswari et al. 2023a). Furthermore, Rodriguéz-Páez et al. (2024) successfully developed efficient in vitro micropropagation protocols for three elite *S. rebaudiana* genotypes (L020, L102, and Morita II) by optimizing the use of cytokinins and auxins, particularly BAP and NAA, achieving high rates of shoot multiplication and highlighting the importance of genotype-specific protocols.

Indirect organogenesis through callus was established from leaf explants when cultured on MS medium supplemented with 4.44 mM BAP along with 1.34 mM NAA, and callus was induced on MS medium supplemented with 11.31 mM 2,4-D along with 2.22 mM BAP (Janarthanam et al. 2009). Shoot apex and internode explants were cultured on MS medium supplemented with 0.1 mg/L 2,4-D to induce callus, and further subculturing on MS medium supplemented with BAP and 2,4-D helped in the induction of multiple shoots, and subsequent rooting was seen when shoots were transferred to MS medium supplemented with IBA (Taware et al. 2009). Induction of callus was seen when leaf explants cultured on MS medium supplemented with 2.0 mg/L 2,4-D along with 1.0 mg/L KN, and transferring the callus to shoot induction medium fortified with 5.0 mg/L BAP along with 1.0 mg/L NAA showed shoots with highest regeneration potential (Sairkar et al. 2009). In-vitro androgenesis of Stevia helped in production of embryos when cultured on MS medium supplied with various amount of BAP. Effect of temperature on androgenesis was studied, and it is observed that MS medium supplemented with 2.0 mg/L NAA along with 0.5 mg/L KN showed best embryo formation ratio (Uskutoğlu et al. 2019).

10.6.1 Synthetic Seeds

Synthetic seeds will help in regeneration and conservation of species which have issues with conventional breeding. In Stevia synthetic/artificial seeds help in regeneration and multiplication as they have very poor seed germination efficiency. Explants like shoot apex and axillary bud regions were encapsulated using 3% sodium alginate and 100 mM calcium chloride. Plant conversion rate efficiency decreased by 86–10% from 15 to 60 days as the storage duration of these synthetic seeds increased (Ali et al. 2012). Studies suggest that 4% calcium alginate can be used as a gelling agent for the shoot tip explants of Stevia and plant conversion rate is recorded maximum on medium supplemented with 0.05 M mannitol (Nower, 2014). Nodal explants were also taken as explants for synthetic seeds production using 2% sodium alginate and 1% sucrose supplemented MS medium. The germination efficiency was seen at around 80% if the seeds were stored at 0 °C (Chotikadachanarong, 2018). When later buds of Stevia were encapsulated using 2.5% sodium alginate and 1% calcium chloride showed best encapsulation efficiency and they were germinated well on MS medium supplemented with 1.0 mg/L BAP along with 1.5 mg/L KN (Shaafi et al. 2021). For the purpose of mass-propagating *S. rebaudiana*, Subrahmanyeswari et al. (2023b) developed an incredibly effective alginate-encapsulation method. The resultant synthetic seeds showed faster root initiation, greater shoot proliferation and elongation, and faster germination than non-encapsulated shoot tips. Plantlets regenerated from synthetic seeds showed clonal fidelity to the parent plant, providing a dependable technique for stevia's quick and genetically homogeneous in vitro propagation.

10.7 Genetic Engineering and Genome Editing

Understanding biosynthetic genes, their transcription process and regulation in metabolic pathways is critical for *S. rebaudiana* genetic engineering aimed at novel cultivars with improved agronomic and metabolomics features (Tamura et al. 1984; Kazmi et al. 2020). Moreover, a trustworthy technology for Stevia transformation is necessary to be built for future research involving genetic engineering and plant breeding programs for developing novel features like increased sweetness and adaptive to pests and diseases. In this regard, Zheng et al. (2019) successfully implemented a method for transforming Stevia plants using *Agrobacterium* with GFP gene and they also produced transgenic Stevia with overexpressing SrDXS1 and SrKAH which showed superior morphological and biochemical characteristics. This transformation strategy will be useful for subsequent overexpression or knockdown experiments of newly discovered genes from the Stevia RNA-seq database. Kim et al. (2019) also used *Agrobacterium* for the evaluation of transgenic Stevia overexpressing the SrUGT76G1 gene. Both the studies are evidence that the production of plants with more sweet components is possible through transgenic Stevia.

Gene silencing is commonly employed to identify the function of a specific gene by suppressing its action mostly using RNA interference (Ghag et al. 2023). The function of SrKA13H and SrUGT85C2 in the steviol glycoside biosynthesis pathway in transgenic Stevia was analyzed by Guleria and Yadav (2013) using *Agrobacterium*-mediated gene silencing through RNA interference. The *Agrobacterium*-mediated transformation of Stevia was also employed by Saifi et al. (2019) for the identification of the role of miRNA related to SVglycoside biosynthesis. Gene silencing was conducted by the transformation of Stevia leaves with a knockdown construct having anti miRNA sequences (miR319g and miRStv11) to revalidate their function. The co-expression analysis was done by the transformation of Stevia leaves using an equally mixed population of agrobacterium incorporated with miRStv 11 and anti-miR319g and found that the increased expression of KAH (Kaurenoic acid-13-hydroxylase) and KO (Kaurene oxidase), KS (Kaurene synthase), and UGT85C2 (UDP-glycosyltransferase 85C2) genes.

In order to successfully transform Stevia leaves using the direct regeneration method, Khan et al. (2014) allowed co-cultivation of *Agrobacterium* strains with leaves of nodal cultures of Stevia. Plants that had been genetically changed were permitted to create somaclones. Except for one somaclone, the ISSR study demonstrated that the somaclones are genetically closest to their parents. The study also shows that somaclones produced more stevioside and produced less Reb A than their parents, indicating that they could be used as a cultivar for high stevioside production. Taak et al. (2020) established an *Agrobacterium*-mediated transformation system with a bar gene for herbicide resistance. The type of explant, duration of pre-incubation, acetosyringone (As) concentration, cell density of *Agrobacterium*, duration of Agro-inoculation, and co-cultivation, selection regime, and the composition and concentration of plant growth regulators (PGRs) have all been competently optimized by Taak et al. (2020) for the Agrobacterium-mediated transformation of Stevia which could be useful for introducing other genes into the Stevia genome. Wu et al. (2020) also produced transgenic Stevia with overexpressing UGT76G1gene using *Agrobacterium*-mediated transformation.

Because extracting high-quality SGs from plants is currently more difficult, researchers have focused their efforts on elucidating the metabolic processes used by engineered microbes to produce SGs. The expression of different glycosyltransferases in microbes is crucial for modified microorganisms to synthesize distinct SGs (Zhou et al. 2021). Furthermore, with the help of proteomics, new efforts to design steviol glycoside production in yeast aim to supply important molecules to prevent variation caused by different *S. rebaudiana* cultivars and growing conditions (Feng et al. 2020). The 3D structure of UGT76G1 developed by Lee et al. (2019) could aid the large-scale industrial synthesis of customized steviol glycosides. Dewitte et al. (2016) also validated the potential of recombinant *E. coli* strains having UGT-76G1 isolated from Stevia for the production of steviol glycosides.

In addition to it, the transient transformation of *Nicotiana benthamiana* was carried out by recombinant Agrobacterium harbouring GFP-UGT76G1 gene to analyze the recombinant gene expression. They reported the production of Reb A in *N. benthamiana* leaves as a result of transformation. The study also proved that the incorporation of GFP will not affect the expression of UGT76G1 whether it is 5′ end or in 3′ end (Petit et al. 2020).

10.7.1 Methodologies

In the case of Stevia, most of the experiments employed SV glycoside biosynthetic genes for genetic transformation because of their industrial importance. In most of the reports, the gene was identified and isolated from cDNA libraries created using RNA sequences from leaves of Stevia and amplified by PCR. The primers were designed using the sequencing data publicly available in NCBI. Then these amplified sequences were cloned into expression vectors such as *E. coli*. Selectable markers and regulatory elements are two must features of vectors. The selectable marker helps to identify the cloned cells and regulatory elements help in the mRNA production, essential for gene expression. Commonly used selection marker genes in Stevia transformation are npt3, hyg resistance gene, GUS, GFP, etc., and the successfully cloned vector will be used for transformation either by a direct method like microprojectile, sonication, etc. or with the aid of Agrobacterium (Khan et al. 2014; Zheng et al. 2019).

The methodology for the transformation of Stevia using Agrobacterium is well established in earlier reports. The generalized protocol for Stevia transformation in the light of previous studies (Kim et al. 2019; Wu et al. 2020; Taak et al. 2020; Zheng et al. 2019) is given below.

(1) Pre-culture: nodal segments or callus will be pre-cultured on agar plate with MS Basal + BAP (2.0 mg/L) for 1–2 weeks.
(2) Agro inoculation: Agrobacterium strains which are transformed with vector construct at log phase will be pelleted and resuspended in liquid MS infiltrated to leaves or callus cultures or nodal cultures using syringe (OD600 = 0.4–0.6) for 20–30 min in liquid MS medium (pH-5.6).
(3) Co-cultivation: The agrobacterium-infected explants were transferred to a co-cultivation medium (MS Basal + BAP (2.0 mg/L) + Acetosyringone (100 M) for 3 days in the dark for 2 days.
(4) Transfer for callus induction: Explants will be vigorously rinsed twice in sterile deionized H_2O and once in MS media supplemented with 300 mg/L cefotaxime before soaking for another 20 min in MS media with cefotaxime. The washed explants were then placed on callus induction media (1 mg/L BA + 0.5 mg/L IAA + 125 mg/L cefotaxime + 50 mg/L kanamycin) or MS basal + BAP (2.0 mg/L) + cefotaxime (250 mg/L) for 3–4 weeks at 25 °C in the dark for 1–2 weeks.
(5) 1st selection: selection of the transformed plants was carried out by allowing the cultures to grow in selection media (MS Basal + BAP (1.0 mg/L) + NAA (0.5 mg/L) + cefotaxime (250 mg/L) + glufosinate ammonium or hygromycin (2 mg/L) for 20–30 days or by using fluorescent microscope for GFP signaling.
(6) 2nd selection: All the shoots from each explant were dissected into small pieces and allowed to grow in second selection media (MS basal + BAP (1.0 mg/L) + NAA (0.5 mg/L) + cefotaxime (250 mg/L) + glufosinate ammonium or hygromycin (4 mg/L) for 20–30 days.

(7) The shoots from the second selection were grown for one month in elongation media (MS basal + GA_3 (1.0 mg/L) + glufosinate ammonium or hygromycin (4 mg/L) before being placed in rooting media (Half- MS basal + glufosinate ammonium or hygromycin (4 mg/L).
(8) Hardening: the transformed plantlets were then allowed to grow in soilrite with liquid MS medium in indoor conditions for 2–3 weeks, then were shifted to soil in glass house conditions and were established for field conditions.
(9) The transformed plants were genetically analyzed for confirming the transformation.

Current studies open the door to whole plant genetic engineering, which can produce transgenic plants with higher yields through *Agrobacterium tumefaciens*-mediated genetic modifications through direct regeneration, somaclone development, or indirect regeneration. These methods can result in cultivars with improved traits like disease resistance, increased climatic adaptability, and optimal output of beneficial phytomolecules.

10.8 Mutation Breeding

Mutation is the cost and time-effective method to produce genetic variation in *S. rebaudiana* (Chiew et al. 2016). Previous studies reported the effectiveness of both physical (gamma irradiation) and chemical mutagens (EMS) in manipulating the Stevia genome (Kumar et al. 2011). Establishment of higher-yielding *S. rebaudiana* genotypes as an aspect of EMS and gamma radiation mutagen treatment using direct regeneration from the leaf explants was performed by Khan et al. (2015). The most successful treatment for selecting variations via direct shoot bud induction was dosages of 0.4% (v/v) EMS and 0.95 KR gamma radiation, which was shown to be the most effective of all the treatments. The EMS-treated plants exhibit high stevioside and rebaudioside content compared to control with the overexpression of UGT74G1 and UGT76G1. While the gamma irradiation plants showed comparatively high rebaudioside and low stevioside content with overexpression of UGT76G1 (Khan et al. 2015).

Singh et al. (2020) reported that 5KR gamma-irradiation causes mutant Stevia genotypes with a prolonged vegetative phase. They also showed enhanced stevioside content but a reduction in Reb-A content with a relatively lower expression of UGT 85C2 and 74G1. The effectiveness of gamma radiation in inducing mutation by germination studies of Malaysian varieties of Stevia was also reported (Abdullah et al. 2021). Recently, the seeds of wild-type Stevia seeds in Thailand were treated with 1% EMS and were found to have a special combination of glycosides, low stevioside, and Reb-A and a high Reb C content. The high Reb C content Stevia genotypes developed through mutation breeding are patented with a patent number, US11284576B2 (Iwaki et al. 2022). The molecular analysis of natural mutant Stevia, Z05, with low Reb-A content in *S. rebaudiana* revealed nonsense substitution mutation in UGT76G1 gene caused the structural changes and finally led to the non-functional enzyme (Yang et al. 2014). The study of Zhang et al. (2019) also supports that mutation in UGT76G1 can alter the SV glycoside contents. These reports suggest that site-specific mutation in the UGT76G1 gene can be used as a potential method for breeding novel Stevia cultivars having altered glycoside composition.

10.9 Hybridization

With the increase in metabolic diseases like type II diabetes and obesity, the need for Stevia leaves and various SVglys is also increasing. It is anticipated to continue to rise (Cosola et al. 2021) which demands the necessity of new improved cultivars to meet the global need. Since the plant Stevia is self-incompatible, heterozygous, and with less explored genomic data, cross-breeding is the possible method for making genetically improved varieties (Khan et al. 2021). The hybridization technique involves the selection of parents of superior quality and their crossing followed by an evaluation of the progeny for suitable features. Intraspecific, interspecific, and intergeneric hybridizations are all common conventional hybridization techniques. In the case of Stevia, the majority of hybridization research is done within the species to develop novel varieties.

The male and female components of the Stevia flower are present in the same blossom. Prior to the blossoming of the flower, the female parent's floral buds are emasculated by manually removing the male anthers. Pollen from the male flowers of the parent plants is manually applied on the stigma of the preceding emasculated bloom during flowering. First-generation (F1) hybrid seed is the term used to describe the seed that results from the cross. Plants bearing an F1 hybrid (half of their genetic composition originating from the male parent and half from the female parent) are created when this seed is sown. At meiosis, genes are segregated, resulting in a second-generation (F2) seed which has a distinct genetic combination unique from the parents.

10.9.1 Somatic Cell Hybridization

Somatic cell hybridization in Stevia is still in the primary stage. However, Lopez-Arellano et al. (2015) introduced a protocol for isolation and culture of protoplast from *Stevia rebaudiana* leaves using various enzymes and reported the successful regeneration of plants from somatic embryos derived from protoplast culture. This data will be helpful for future biotechnological studies such as somatic cell hybridization.

10.9.2 Hybrid Cultivars

- AKHL 1: The biparental cross between Seed parent 'Eirete' (unpatented) and pollen parent AKH/EM1 (unpatented) resulted in a progeny with high Reb-A content and low stevioside content which was patented as Stevia variety AKHL 1 (Patent No.US PP23164 P3) (Britos 2012).
- AKHL 4: Stevia variety AKHL 4 was developed by crossing between seed parent 'AKH/G.8.D' (with high stevioside content) and pollen parent 'Eirete' (high Reb-A content). AKHL4 was selected for the steviol glycoside composition of high Reb-A and low stevioside and patented (Patent No. US PP23728 P3) (Britos 2013).
- Stevia plants '12-05-005', '12-05-149' and '12-05-144'**:** new and unique Stevia plants '12-05-005', '12-05-149' and '12-05-144' developed through a bi-parental controlled cross of two selections from an open-pollinated Stevia population in East Lansing, Michigan. In East Lansing, Michigan, in 2012, female parent '10–34' (unpatented) was crossed with male parent '10–19' (unpatented). The plants that resulted from the controlled cross of '10–34' and '10–19' were chosen as Stevia '12-05-149'. (Patent No.: US PP22593 P3) (Warner et al. 2017a) which have a low stevioside content and comparatively high Reb M and Reb D contents, Stevia '12-05-144' (Patent No.: US PP22593 P3) (Warner et al. 2017b) with a high Reb C and Reb M concentration and Stevia '12-05-005'(USPP 27902P3) which have high Reb A, Reb C, Reb M and total steviol glycosides (Warner et al. 2016).
- Stevia 320032: Stevia cultivar, '320032' (Patent No.:WO2021061805A1) with high Reb A content developed as a result of a biparental cross between the female Stevia variety 'Eirete' (unpatented) and the proprietary male Stevia variety 'PC Star 2' (unpatented) which was conducted in Ganzhou, Jiangxi Province, China in 2013 (Markosyan et al. 2021a).
- Stevia cultivar 807086: Stevia cultivar 807086 (Patent No. US 9675015 B2) is the result of a tri-cross between the proprietary F1 female Stevia line 'YF003' (from the cross of the female parent 'AKH EM1' and the male parent 'AKH L4') and the male Stevia line 'PC Star 2' at Ganzhou, Jiangxi Province, People's Republic of China (Li et al. 2017a).
- Stevia cultivar 814011: The proprietary female Stevia variety 'YF001' (derived from a cross between the female parent, Stevia variety 'Eirete' and male parent Stevia variety 'AKHL 1') and the proprietary male variety 'PC Star 2' were tri-crossed in Ganzhou, Jiangxi Province, People's Republic of China to develop Stevia cultivar '814,011' (Patent No.US 9668451 B2) (Li et al. 2017b).
- Stevia cultivar 817096: The progeny of a cross between Stevia line 'YF002' (derived from a cross between female parent 'AKH/G.8.D' (unpatented) and male father 'AKH L1') as male parent and 'AKH/G.8.D' as a female parent was back-crossed to produce Stevia cultivar '817,096' (Patent No. US 9668450 B2) which has a high Reb D content (Li et al. 2017c).

- Stevia cultivar 16,265,046 (US 2021/0259184 A1) and 16,228,013 (US 11284578 B2): These varieties have high Reb D, Reb M, and high yield as the result of a biparental cross between the proprietary female Stevia variety '814011' and the proprietary male Stevia variety 44004 (unpatented elite high Rebaudioside M line) in Ganzhou, Jiangxi Province, the People's Republic of China in September 2015 (Markosyan et al. 2022a, 2022b, c).

10.10 Conclusions and Prospects

10.10.1 An Overview of the Current Status

The demand for Stevia is growing in tandem with the rise in metabolic illnesses like diabetes. Because zero-calorie natural sweeteners are becoming increasingly popular around the world, Stevia farming should be expanded. To meet worldwide demand, plant breeding researches are mainly focusing on the release of new cultivars with high leaf yield, superior agronomic qualities, and the ability to endure diverse biotic and abiotic challenges, as well as an enhanced ratio of sweet compounds, such as rebaudioside.

So far, Stevia breeding efforts have achieved significant development. The glycoside composition has boosted by 20% as a result of recent Stevia breeding work. Additionally, there seems to be improvement in the proportion of Reb A to stevioside, which was 0.36:1early but is now 0.96:1in certain varieties. The data presented in this chapter clearly demonstrates the rising interest in improving *S. rebaudiana* genetics using both traditional and novel biotechnological methodologies by scientists, researchers, and plant breeders. Stevia plant breeding will continue to rely heavily on traditional methods, but the integration of novel biotechnological approaches in tandem with traditional methods will help to enhance the effectiveness of breeding attempts in this species.

10.10.2 Current Research Initiatives to Combat Global Climate Change

Plant breeding research for drought-resistant varieties appears to be the major priority in plant science, due to the negative impacts of drought in agricultural sector. Understanding how plants respond to water scarcity is essential for managing plant water consumption effectively, as water resources are becoming more limited in various regions of the globe (Behroozi et al. 2017).

The molecular interruption of signaling pathways is one of the most severe impacts of drought. This has an impact on key metabolite biosynthesis and, as a result, plant growth is hindered. Alam (2021) conducted a genomic analysis based on EST sequences and functional annotation using bioinformatic tools and reported that, the main transcriptional factors, WD40-like protein, MYB-HB like, AP2-EREBP, C2H2, Hap3/NF-YB, bHLH, C2C2- CO-like, CW-Zn, FHA-SMAD, Nin like, SBP3, TIFY, Tc-PD, Znf-B, and bIP have an impact in the metabolic pathway of steviol glycosides. The results open the scope of making transgenic Stevia plants with increased tolerance to drought stress by overexpressing these genes with the aid of genetic engineering (Alam 2021). Recent studies on the identification and characterization of important genetic elements, such as the C2H2 zinc finger protein gene family, are made easier by the genome analysis of *S. rebaudiana*, which also clarifies the roles these genes play in a variety of physiological processes and drought stress responses (Nikraftar et al. 2024).

Soil salinization has become a serious environmental issue around the world, and it is likely to worsen as a result of projected climate change (Hassani et al. 2021). Earlier reports focuses the impact of salinity on Stevia (Aghighi Shahverdi et al. 2019; Aghighi Shaverdi, 2018). Gerami et al. (2020) studied the enhanced salt tolerance ability of Stevia plants using chitosan under in vitro salt stress conditions. Ghasemi-Omran et al. (2021) found that application of melatonin can improve Stevia cultivation under salt stress. In *S. rebaudiana* under drought, salinity, and waterlogging stresses, the study by Pal et al. (2023) finds a significant upregulation of genes including GST, SOD, APX (ROS detoxification), alpha-trehalose-phosphate synthase, S-adenosylmethionine decarboxylase (osmotic adjustment), CAX, NHX, CNGS, VPPase, VATPase (ion transporters), and PIP1, TIP (water channels). These genes offer potential targets for genetic improvement to promote crop sustainability in Stevia, along with important regulators such as HSFA2, DREB1A, and DREB2A. They also help to enhance plant resilience (Pal et al. 2023).

10.10.3 Future Perspectives

Since the fungal attack has been considered as a major challenge in Stevia cultivation, more focus on Stevia breeding for disease resistance with the involvement of modern biotechnological tools is needed. Apart from that, breeding programs should aim to improve the production of minor rebaudiosides Reb C, D and M due to its high sweetening potential, which adds market value to Stevia. Because of the substantial decrement of sweet compounds after flowering (Libik-Konieczny et al. 2021), development of Stevia cultivars with prolonged vegetative phases should also be considered in future breeding studies.

To meet the enhancing industrial demand, biotechnological approaches are widely applied in the commercial production of steviol glycoside. Since the rDNA technology for transgenesis involves the incorporation of the antibiotic-resistant gene along with the gene of interest, the chances of getting products with antibiotic resistance are more. Hence steviol glycosides have vast applications in the food industry, there should be proper systems for monitoring the safety of transgenic products

(Younes et al. 2022). So future research is needed for alleviating the drawbacks of recombinant steviol glycosides.

Numerous in vitro and in vivo investigations on Stevia preparations and steviol glycosides give mounting evidence that these chemicals could be employed in dietary therapy to assist certain physiological processes, including glucose metabolism in diabetes (Kurek and Krejpcio 2019). However, the molecular mechanisms behind the antidiabetic activity are unknown, and the dose–response relationships for human use have not been well studied, necessitating further research. Stevioside and rebaudioside A are the major steviol glycoside components in Stevia getting attracted by researchers. The extensive studies on other steviol glycosides like dulcoside, steviol, steviolbioside, etc. are essential for the evaluation of the potential use of Stevia. Several selective steviosides, such as dulcoside (very low concentration in Stevia), have been shown to have carcinogenic effects in the laboratory experiments (Khan et al. 2021). Hence the future breeding programs for developing novel varieties that are promptly propagable, high in rebaudioside content, and free of dulcoside are crucial. It is evident that 2–5% of Stevia leaves contain polyphenols which are identified as potential antioxidants and add therapeutic properties to Stevia. Nowadays, most steviol glycosides processors discard the polyphenols in solid Stevia waste generated while steviol glycosides production (Myint et al. 2020). It would be preferable if researchers concentrate their efforts on the marketing of polyphenols as medical or cosmetic products.

Appendix 1: Genetic Resources

Cultivar name	Place	Character	References
RSIT 94-1306	Canada	High Stevioside (17.25%)	Kumar et al. (2011)
RSIT 94-751	Canada	High Reb A (11.82%)	Kumar et al. (2011)
RSIT 95-166-13	Canada	High Rebaudioside-C (14.4%)	Kumar et al. (2011)
RSIT 94-1838	Canada	High Rebaudioside-C (10%)	Kumar et al. (2011)
RSIT 94-1829	Canada	High Rebaudioside-C (6.9%) and stevioside (7.2%)	Kumar et al. (2011), Brandle et al. (1998)
RSIT 94-1833	Canada	High Rebaudioside-C (7.4%)	Brandle et al. (1998)
RSIT 94-1560	Canada	High Rebaudioside-C (5.59%)	Brandle et al. (1998)
AKHL 1	USA	High yield and disease resistance	Al-Taweel et al. (2021)

(continued)

(continued)

Cultivar name	Place	Character	References
Shoutian No. 3	China	Stress Resistant	Sun et al. (2021)
Sugar high A3	Canada	Bushy plant with larger leaves	Xu et al. (2021), Al-Taweel et al. (2019)
Spanti	Spain	Small leaves	Al-Taweel et al. (2019)
Brazil Zairai	Japan	7.65% stevioside and 3% Reb A	Kumar et al. (2011) Brandle et al. (1998)
SF 6	China	High stevioside to rebaudioside ratio (1: more than 2.54)	Morita and Bu (2000)
SM4	China	High Yield	Kumar et al. (2011)
MDS-13 and MDS-14:	India	It is suitable for cultivation in the regions with high temperatures and low rainfall	Shweta (2020)
SRB-123	India	Harvesting 3–4 times a year. Suitable for Ecaan plain	Shweta (2020)
SRB-128	India	Suitable for Southern and Northern Indian climate	Shweta (2020)
SRB-512	India	Suitable for northern latitudes	Shweta (2020)
CIM-Mithi	India	High yield with a greater number of branches	Lal et al. (2011)
CIMAP MADHU	India	High stevioside (12.57%)	Lal et al. (2011)
ACC 1, ACC 2	India	High leaf yield	Lal et al. (2011)
Madhuguna	India	High yield and total glycoside content (5.56% stevioside and 2.27% Reb A)	Kumar et al. (2011)
Madhuguni	India	High yield and total glycoside content	Kumar et al. (2011)
HIM STEVIA	India	Delayed flowering	Lal et al. (2011)
Candy	Israel	High seed germination percentage	Parris et al. (2016)
Morita SF6	Japan	High stevioside to rebaudioside ratio (1: more than 2.54)	Al-Taweel et al. (2021) Morita and Bu (2000)
AC Black Bird	Japan	High level of total glycosides (14%) with less stevioside (0.03%) and enhanced Reb A (12.49%)	Kumar et al. (2011; Jim Brandle (1998)
ATCC Accession No. PTA-444	Japan	Reb A 2.56 times more than stevioside	Kumar et al. (2011)
Suweon 2	Korea	High yield (22% above) and steviosides (more than 12%)	Kumar et al. (2011)

(continued)

(continued)

Cultivar name	Place	Character	References
Suweon 11	Korea	Thick leaves	Kumar et al. (2011), Lee et al. (1982)
Eirete	Paraguay	Hybrid Of Peruguayan varieties	Cosson et al. (2019) Angelini et al. 2018)
Criolla	Paraguay	Wild variety	Angelini et al. (2018)
Morita II	Paraguay	High dietary fiber (87.79%), good for dietary supplements	Angelini et al. (2018), Campos et al. (2014)
Morita III	Paraguay	Drought resistant	Angelini et al. (2018)
Katupyry	Paraguay	High sweetness	Angelini et al. (2018)

Appendix 2: Research Institutes/industries working on Stevia

Institute name	Specialization research activities	Contact information and website
Purecircle Stevia Institute, Chicago	Releasing new varieties with high Reb contents	http://www.purecircle.com
Coca Cola Company	Application of Stevia in soft drinks and other foods	https://www.coca-colacompany.com/
Sweet Green Fields International Co., Limited	Stevia sweetener with improved solubility with a cyclodextrin	www.sweetgreenfields.com
Tate & Lyle Ingredients Americas Llc	Stevia blends containing Rebaudioside B	https://www.tateandlyle.com/
Morita Kagaku Kogyo Co., Ltd.	Manufacturing and selling of Stevia sweeteners	https://www.morita-kagaku-kogyo.co.jp/main/html
Chenguang Biotech Group Co., Ltd.	Industrial utilization method for *Stevia rebaudiana* and stevioside and chlorogenic acid of *Stevia rebaudiana*	http://www.cn-cg.com
Suntory Holdings Ltd	Application in beverage industry	https://www.suntory.com/
Council of Scientific And Industrial Research	Enzyme assisted extraction of steviol glycosides	https://www.csir.res.in/

(continued)

(continued)

Institute name	Specialization research activities	Contact information and website
Anhui Science And Technology University	Protoplast isolation from Stevia	http://www.ahut.edu.cn/
Temasek Life Sciences Laboratory Limited	Transgenic studies on Stevia	http://www.tll.org.sg/
S&W Seed Company	Stevia breeding	https://swseedco.com/
Ta Stevia Co Ltd.	Application in food industry	https://realStevia.com/
Sweet Green Fields USA Llc	Plant breeding and industrial applications	https://sweetgreenfields.com/
Csir Ihbt	Popularizing Stevia in India, processing of Eco friendly products and releasing new cultivar/s	https://www.ihbt.res.in
Keygene	Sequencing of Stevia	https://www.keygene.com/
EverStevia.Co Canada	Stevia planting material and commercial products manufacturing	http://www.everStevia.com/
Wostevia	Manufacturing value added products from stevia	http://www.wostevia.com
Stevia Technology	Provides technical assistance for stevia farming	https://www.steviashantanu.com/

References

Abdelsattar AM, Elsayed A, El-Esawi MA, Heikal YM (2023) Enhancing *Stevia rebaudiana* growth and yield through exploring beneficial plant-microbe interactions and their impact on the underlying mechanisms and crop sustainability. Plant Physiol Biochem 198:107673

Abdullah S, Fauzi NYM, Khalid AK et al (2021) Effect of gamma rays on seed germination, survival rate and morphology of *Stevia rebaudiana* Hybrid. Mal J Fund Appl Sci 17(5):543–549

Abdullateef R, Osman M (2011) Effects of stem cutting types, position and hormonal factors on rooting in *Stevia rebaudiana* Bertoni. J Agric Sci 4(1):49

Aghighi Shahverdi M, Omidi H, Tabatabaei SJ (2019) Stevia (*Stevia rebaudiana* Bertoni) responses to NaCl stress: growth, photosynthetic pigments, diterpene glycosides and ion content in root and shoot. J Saudi Soc Agric Sci 18(4):355–360

Aghighi Shaverdi M (2018) Morpho-physiological response of Stevia (*Stevia rebaudiana* Bertoni) to salinity under hydroponic culture condition (a case study in Iran). Appl Ecol Environ Res 16(1):17–28

Ahmed MB, Salahin M, Karim R, Razvy MA, Hannan MM, Sultana R et al. (2007) An efficient method for in vitro clonal propagation of a newly introduced sweetener plant (*Stevia rebaudiana Bertoni*) in Bangladesh. Am J Sci Res 2:121–125

Ajami M, Seyfi M, Abdollah Pouri Hosseini F et al (2020) Effects of Stevia on glycemic and lipid profile of type 2 diabetic patients: a randomized controlled trial. Avicenna J Phytomed 10(2):118–127

Akbari F, Arminian A, Kahrizi D, Fazeli A (2017) Effect of nitrogen sources on some morphological characteristics of in vitro *Stevia rebaudiana* Bertoni. Cell Mol Biol (Noisy-Le-Grand) 63(2):107–111

Alam P (2021) Functional annotations of ESTs of *Stevia rebaudiana* involved in abiotic stress signaling through computational approach. Saudi J Biol Sci 28(4):2602–2612

Ali A, Shahu R, Balyan P et al (2022) Antioxidation and antiglycation properties of a natural sweetener: *Stevia rebaudiana*. Sugar Tech 24(2):563–575

Ali A, Gull I, Majid A, Saleem A, Naz S, Naveed NH (2012) In vitro conservation and production of vigorous and desiccate tolerant synthetic seeds in *Stevia rebaudiana*. J Med Plants Res 6:1327–1333. https://doi.org/10.5897/jmpr11.1443

Al-Taweel SK, Abdel-Aziz RM, Rabea KM, Khaled KAM (2019) Studying cDNA SCoT in response to salinity stress in *Stevia rebaudiana* Bertoni. SABRAO J Breed Genet 51:281–294

Al-Taweel S, Azzam C, Khaled KAM, Abdel-Aziz R (2021) Improvement of Stevia (*Stevia rebaudiana* Bertoni) and steviol glycoside through traditional breeding and biotechnological approaches. SABRAO J Breed Genet 53: 88–111

Anbazhagan M, Kalpana M, Rajendran R, Natarajan V, Dhanavel D (2010) In vitro production of *Stevia rebaudiana Bertoni*. Emirates J Food Agric 22:216–222. https://doi.org/10.9755/ejfa.v22i3.4891

de Andrade MVS, Lucho SR, do Amaral MN, Braga EJB, Ribeiro PR, de Castro RD (2023) Long-day photoperiodic requirements for steviol glycosides and gibberellins biosynthesis and bio-sweetener levels optimization in *Stevia rebaudiana* Bertoni. Ind Crops Prod. 204:117363

Andújar I, González M, García-Ramos JC, Hajari E, Bogdanchikova N, Pestryakov A et al (2023) Are silver nanoparticles the "silver bullet" to promote diterpene production in *Stevia rebaudiana*? Plant Cell Tiss Org Cult 155(2):447–453

Angelini LG, Martini A, Passera B, Tavarini S (2018) Cultivation of *Stevia rebaudiana* Bertoni and associated challenges. Sweetners 35-55. https://doi.org/10.1007/978-3-319-27027-2_8

Arumugam B, Subramaniam A, Alagaraj P (2020) Stevia as a natural sweetener: a review. Cardiovasc Hematol Agents Med Chem 18(2):94–103

Badran AE, Abd Alhady MRA, Hassan WA (2015) In vitro evaluation of some traits in *Stevia rebaudiana* (Bertoni) under drought stress and their relationship on stevioside content. Am J Plant Sci 06(5):746–752

Basharat S, Huang Z, Gong M et al (2021) A review on current conventional and biotechnical approaches to enhance biosynthesis of steviol glycosides in *Stevia rebaudiana*. Chin J Chem Eng 30:92–104

Behroozi P, Baghizadeh A, Saei A, Kharazmi S (2017) Quantitative analysis of uridine diphosphate glycosyltransferase UGT85C2, UGT74G1 and UGT76G1 genes expression in *Stevia rebaudiana* under different irrigations. Russ J Plant Physiol 64(1):67–72

Benelli C, Carvalho LSO, El Merzougui S, Petruccelli R (2021) Two advanced cryogenic procedures for improving *Stevia rebaudiana* (Bertoni) cryopreservation. Plants 10(2):277. https://doi.org/10.3390/plants10020277

Bhandawat A, Sharma H, Nag A et al (2014) Functionally relevant novel microsatellite markers for efficient genotyping in *Stevia rebaudiana* Bertoni. J Genet 93:e75-81

Brandle J (1998) *Stevia rebaudiana* with altered steviol glycoside composition. US Patent: https://patents.google.com/patent/US6255557B1

Brandle J, Sys EA, Marsolais AA (1998) Stevia plant named "RSIT 95-166-13 US Patent: https://patents.google.com/patent/USPP10563P

Britos ERA (2012) Stevia plant named "AKH L1." US Patent: https://patents.google.com/patent/USPP23164P3

Britos ERA (2013) Stevia plant named "AKH L4." US Patent: https://patents.google.com/patent/USPP23728P3

Brower RJ III, Carlson TL, Dang B et al (2021) Stevia plants with an increased rebaudioside D content. US Patent https://patents.google.com/patent/WO2014146084A1

Campos M, Barbosa-Martín E, Matus-Basto et al (2014) Comparison of chemical and functional properties of *Stevia rebaudiana* (Bertoni) varieties cultivated in Mexican Southeast. Am J Plant Sci 05:286–293. https://doi.org/10.4236/ajps.2014.53039

Castañeda-Saucedo MC, Tapia-Campos E, Ramírez-Anaya JDP, Beltrán J (2020) Growth and development of Stevia cuttings during propagation with hormones in different months of the year. Plants 9(3):https://doi.org/10.3390/plants9030294

Chakma A, Afrin F, Rasul MG, Maeda H, Yuan C, Shah AKMA (2023) Effects of extraction techniques on antioxidant and antibacterial activity of stevia (*Stevia rebaudiana* Bertoni) leaf extracts. Food Chem Adv 3:100494

Chen J, Hou K, Qin P et al (2014) RNA-Seq for gene identification and transcript profiling of three *Stevia rebaudiana* genotypes. BMC Genomics 15:571

Chester K, Tamboli ET, Parveen R, Ahmad S (2013) Genetic and metabolic diversity in *Stevia rebaudiana* using RAPD and HPTLC analysis. Pharm Biol 51(5):771–777

Chiew M, Lai K, Hussein S, Abdullah J (2016) A review on induced mutagenesis of *Stevia rebaudiana* Bertoni. Pertanika J Scholarly Res Rev 2(3):77–85

Chotikadachanarong K (2018) Artificial seed production from nodal explant of *Stevia rebaudiana* Bertoni. Burapha Sci J 23(1):247–254

Chowdhury AI, Rahanur Alam M, Raihan MM et al (2022) Effect of Stevia leaves (*Stevia rebaudiana* Bertoni) on diabetes: a systematic review and meta-analysis of preclinical studies. Food Sci Nutr 00:1–11. https://doi.org/10.1002/fsn3.2904

Cosola C, Caggiano G, Palieri R, Gesualdo L (2021) Treating diseases associated with metabolic syndrome. In: Galanakis CM (ed) Steviol glycosides. Academic, pp 243–275

Cosson P, Hastoy C, Errazzu LE et al (2019) Genetic diversity and population structure of the sweet leaf herb, *Stevia rebaudiana* B., cultivated and landraces germplasm assessed by EST-SSRs genotyping and steviol glycosides phenotyping. BMC Plant Biol 19(1):436

de Oliveira VM, Forni-Martins ER, Magalhães PM, Alves MN (2004) Chromosomal and morphological studies of diploid and polyploid cytotypes of *Stevia rebaudiana* (Bertoni) Bertoni (Eupatorieae, Asteraceae). Genet Mol Biol 27(2):215–222

Debnath M (2008) Clonal propagation and antimicrobial activity of an endemic medicinal plant *Stevia rebaudiana*. J Med Plants Res 2:45–51

Dewitte G, Walmagh M, Diricks M et al (2016) Screening of recombinant glycosyltransferases reveals the broad acceptor specificity of Stevia UGT-76G1. J Biotechnol 233:49–55

Dyduch-Siemińska M, Najda A, Gawroński J et al (2020) *Stevia rebaudiana* Bertoni, a source of high-potency natural sweetener-biochemical and genetic characterization. Molecules 25(4). https://doi.org/10.3390/molecules25040767

Esmaeili F, Ghaheri M, Kahrizi D et al (2018) Effects of various glutamine concentrations on gene expression and steviol glycosides accumulation in *Stevia rebaudiana* Bertoni. Cell Mol Biol 64(2):1–5

Feng Y, Yao M, Wang Y et al (2020) Advances in engineering UDP-sugar supply for recombinant biosynthesis of glycosides in microbes. Biotechnol Adv 41:107538

Frederico AP, Ruas PM, Marin-Morales MA, et al (1996) Chromosome studies in some Stevia. Cav. (Compositae) species from Southern Brazil. Braz J Genet 19(4):605–609

Ghag, S.B., Alok, A., Rajam, M.V. P. Suprasanna. (2023). Designing climate-resilient crops for sustainable agriculture: a silent approach. J Plant Growth Regul 42:6503–6522

Gantait S, Das A, Banerjee J (2018) Geographical distribution, botanical description and self-incompatibility mechanism of genus stevia. Sugar Tech 20(1):1–10

Gantait S, Das A, Mandal N (2015) Stevia: A comprehensive review on ethnopharmacological properties and in vitro regeneration. Sugar Tech 17(1):95–106

Garnighian GV (2012) Stevia plant named "T60." US Patent: https://patents.google.com/patent/US20110271413P1

Gerami M, Majidian P, Ghorbanpour A, Alipour Z (2020) *Stevia rebaudiana* Bertoni responses to salt stress and chitosan elicitor. Physiol Mol Biol Plants 26(5):965–974

Ghasemi-Omran VO, Ghorbani A, Sajjadi-Otaghsara SA (2021) Melatonin alleviates NaCl-induced damage by regulating ionic homeostasis, antioxidant system, redox homeostasis, and expression of steviol glycosides-related biosynthetic genes in vitro cultured *Stevia rebaudiana* Bertoni. In Vitro Cell Development Biol Plant 57(2):319–331

Ghorbani T, Kahrizi D, Saeidi M, Arji I (2017) Effect of sucrose concentrations on *Stevia rebaudiana* Bertoni tissue culture and gene expression. Cell Mol Biol 63(8):33–37

Góngora Salazar VA, Vázquez Encalada S, Corona Cruz A, Segura Campos MR (2018) *Stevia rebaudiana*: a sweetener and potential bioactive ingredient in the development of functional cookies. J Funct Foods 44:183–190

Guleria P, Yadav SK (2013) Agrobacterium mediated transient gene silencing (AMTS) in *Stevia rebaudiana*: insights into steviol glycoside biosynthesis pathway. PLoS ONE 8(9):e74731

Hadia A, Badawy OM, Hafez AM (2008) Genetic relationships among some Stevia (*Stevia rebaudiana* Bertoni) accessions based on ISSR analysis. Res J Cell Mol Biol 2:1–5

Hajihashemi S (2021) Agronomic practices. In: Galanakis CM (ed) Steviol glycosides. Academic, pp 31–56

Hashem MM, AbdelHamid RI, AbuelMaaty S et al (2021) Differential UGT76G1 and start codon-based characterization of six Stevia germlines in Egypt. Biocatal Agric Biotechnol 33:101981

Hashempoor S, Ghaheri M, Kahrizi D et al (2018) Effects of different concentrations of mannitol on gene expression in *Stevia rebaudiana* Bertoni. Cell Mol Biol 64:28–31

Hassanen SA, Khalil RMA (2013) Biotechnological studies for improving of Stevia (*Stevia rebaudiana* Bertoni) in vitro plantlets. Middle East J Sci Res 14(2):93–106. https://doi.org/10.5829/idosi.mejsr.2013.14.1.1948

Hassani A, Azapagic A, Shokri N (2021) Global predictions of primary soil salinization under changing climate in the 21st century. Nat Commun 12(1):6663

He B, Bai X, Tan Y et al (2022) Glycosyltransferases: Mining, engineering and applications in biosynthesis of glycosylated plant natural products. Synth Syst Biotechnol 7(1):602–620

Hegde SN, Rameshsing CN, Vasundhara M (2015) Characterization of *Stevia rebaudiana* Bertoni polyploids for growth and quality. Med Plants 7(3):188

Hernández KV, Moreno-Romero J, Hernández de la Torre M et al (2022) Effect of light intensity on steviol glycosides production in leaves of *Stevia rebaudiana* plants. Phytochemistry 194:113027

Hodgins KA, Lai Z, Oliveira LO et al (2014) Genomics of Compositae crops: reference transcriptome assemblies and evidence of hybridization with wild relatives. Mol Ecol Resour 14(1):166–177

Hossain F, Islam MT et al (2017) Cultivation and uses of Stevia (*Stevia rebaudiana* Bertoni): a review. Afr J Food Agric Nutr Dev 17(4):12745–12757

Ibrahim IA, Nasr MI, Mohammed BR, El-Zefzafi MM (2008a) Plant growth regulators affecting in vitro cultivation of *Stevia rebaudiana*. Sugar Tech 10:254–259. https://doi.org/10.1007/s12355-008-0045-6

Ibrahim IA, Nasr MI, Mohammedm BR, El-Zefzafi MM (2008b) Nutrient factors affecting in vitro cultivation of *Stevia rebaudiana*. Sugar Tech 10:248–253. https://doi.org/10.1007/s12355-008-0044-7

Iwaki K, Ono E, Hirai T, Miyagawa K (2022) High rebaudioside C-content Stevia plant. US Patent: https://patents.google.com/patent/US11284576B2

Jackson MC (2021) Stevia cultivars a01, a03, a05, a06, a07, and a08. US Patent: https://patents.google.com/patent/US20190289806A1

Jadid N, Anggraeni S, Ramadani MRN, Arieny M, Mas'ud F (2024) In vitro propagation of Indonesian stevia (*Stevia rebaudiana*) genotype using axenic nodal segments. BMC Res Notes 17(1):45

Jan SA, Habib N, Shinwari ZK et al (2021) The anti-diabetic activities of natural sweetener plant Stevia: an updated review. SN Appl Sci 3(4):517

Javed R, Yücesan B (2022) Impact of *Stevia rebaudiana* culturing in liquid medium: elevation of yield and biomass, mitigation of steviol glycosides. PPASB 59(1):69–75

Judickaitė A, Lyushkevich V, Filatova I et al (2022) The potential of cold plasma and electromagnetic field as stimulators of natural sweeteners biosynthesis in *Stevia rebaudiana* Bertoni. Plants 11(5). https://doi.org/10.3390/plants11050611

Kahrizi D, Ghaheri M, Yari Z et al (2018) Investigation of different concentrations of MS media effects on gene expression and steviol glycosides accumulation in *Stevia rebaudiana* Bertoni. Cell Mol Biol 64(2):23–27

Kaplan B, Turgut K (2019) Improvement of rebaudioside A diterpene glycoside content in *Stevia rebaudiana* Bertoni using clone selection. Turk J Agric for 43(2):232–240

Kazmi A, Khan MA, Mohammad S et al (2020) Biotechnological production of natural calorie free steviol glycosides in *Stevia rebaudiana*: an update on current scenario. Curr Biotechnol 8(2):70–84

Khalil SA, Zamir R, Ahmad N (2014) Selection of suitable propagation method for consistent plantlets production in *Stevia rebaudiana* (Bertoni). Saudi J Biol Sci 21(6):566–573

Khan SA, Rahman L ur, Verma R, Shanker K (2015) Physical and chemical mutagenesis in *Stevia rebaudiana*: variant generation with higher UGT expression and glycosidic profile but with low photosynthetic capabilities. Acta Physiol Plant 38(1):4

Khan SA, Ur Rahman L, Shanker K, Singh M (2014) *Agrobacterium tumefaciens*-mediated transgenic plant and somaclone production through direct and indirect regeneration from leaves in *Stevia rebaudiana* with their glycoside profile. Protoplasma 251(3):661–670

Khan SA, Verma P, Ur Rahman L, Parasharami VA (2021) Exploration of biotechnological studies in low-calorie sweetener *Stevia rebaudiana*: present and future prospects. In: Aftab T, Hakeem KR (eds) Medicinal and aromatic plants. Academic, pp 289–324

Janarthanam B, Gopalakrishnan M, Lakshmi Sai G, Sekar T (2009) Plant regeneration from leaf derived callus of *Stevia rebaudiana* Bertoni. Plant Tissue Cult Biotechnol 19:133–141. https://doi.org/10.3329/ptcb.v19i2.5430

Kalpana M, Anbazhagan M, Natarajan V, Dhanavel D (2010) Improved micropropagation method for the enhancement of biomass in *Stevia rebaudiana* Bertoni. Recent Res Sci Technol 2(1):008–013

Kim MJ, Zheng J, Liao MH, Jang I-C (2019) Overexpression of SrUGT76G1 in Stevia alters major steviol glycosides composition towards improved quality. Plant Biotechnol J 17(6):1037–1047

Kumar Y, SinghS DhyaniD, AhujaP S (2011) A review on the improvement of Stevia [*Stevia rebaudiana* (Bertoni)]. Can J Plant Sci 91:1–27. https://doi.org/10.4141/cjps10086

Kurek JM, Krejpcio Z (2019) The functional and health-promoting properties of *Stevia rebaudiana* Bertoni and its glycosides with special focus on the antidiabetic potential—a review. J Funct Foods 61:103465

Laha S, Subrahmanyeswari T, Verma SK, Kamble SN, Singh S, Bhattacharyya S, et al. (2023) Biogenic synthesis, characterization and application of silver nanoparticles as biostimulator for growth and rebaudioside-A production in genetically stable stevia (*Stevia rebaudiana* Bert.) under in vitro conditions. Ind Crops Prod 197:116520

Lal RK, Chandra R, Gupta MM, Singh AK, Panday R. (2011) Registration of a high yielding variety CIMAP Madhu of Stevia (*Stevia rebaudiana*). J Med Aromatic Plant Sciences 33(1):77–80

Lata H, Chandra S, Wang YH et al (2014) in vitro germplasm conservation of elite *Stevia rebaudiana* Bertoni. Acta Hortic 303–308

Lee SG, Salomon E, Yu O, Jez JM (2019) Molecular basis for branched steviol glucoside biosynthesis. Proc Natl Acad Sci USA 116(26):13131–13136

Lee JI, Kang KH, Park HW (1982). New high rebaudioside-A Stevia variety "Suweon 11". The Research Reports of the Office of Rural Development (Korea R.)

Li Y, Qiu Y, Xu X, Luo M (2023) Genome-wide identification of SrbHLH transcription factors highlights its potential role in rebaudioside A (RA) biosynthesis in *Stevia rebaudiana*. BMC Plant Biol 23(1):352

Libik-Konieczny M, Capecka E, Tuleja M et al (2021) Synthesis and production of steviol glycosides: recent research trends and perspectives. Appl Microbiol Biotechnol 105:3883–3900. https://doi.org/10.1007/s00253-021-11306-x

Li SW, Bu YC, Markosyan A (2017a) Stevia cultivar "807086." US Patent: https://patents.google.com/patent/US9675015B2

Li SW, Bu YC, Markosyan A (2017b) Stevia Cultivar "814011." US Patent: https://patents.google.com/patent/US9668451B2

Li SW, Bu YC, Markosyan A (2017c) Stevia cultivar "817096." US Patent: https://patents.google.com/patent/US9668450B2

Lopez-Arellano M, Dhir S, Albino NC et al (2015) Somatic embryogenesis and plantlet regeneration from protoplast culture of *Stevia rebaudiana*. Biotechnol J Int 1–12

Luz GC, Strioto DK, Mangolin CA, de Machado MFPS (2020) ISSR markers to assess genetic diversity of cultivated populations from artificial selection of *Stevia rebaudiana* (Bert.) Bertoni. Breed Sci 70(4):508–514

Mahajan M, Thakur BK, Pal PK (2021) Moisture stress and nitrogen availability modulate the secondary metabolite profiles, enzymatic activity, and physiological and agronomic traits of *Stevia rebaudiana*. Plant Physiol Biochem 162:56–68

Markosyan A, Jing R (2018) Self-compatible Stevia varieties, breeding methods therefrom and methods of preparing novel compositions using said varieties. US Patent: https://patents.google.com/patent/US20180070553A1

Markosyan A, Jing R, Ong SS et al (2021a) Stevia cultivar "320032" with super high rebaudioside a content. World Patent: https://patents.google.com/patent/WO2021061805A1

Markosyan A, Ong SS, Jing R et al (2021b) Aneuploid Stevia cultivar "ap-1." US Patent: https://patents.google.com/patent/US20210298259A1

Markosyan A, Ong SS, Jing R et al (2022a) High rebaudioside M Stevia plant cultivars and methods of producing the same. US Patent: https://patents.google.com/patent/US11284577B2

Markosyan A, Ong SS, Jing R et al (2021c) Stevia cultivar "16265046." US Patent: https://patents.google.com/patent/US20210259184A1

Markosyan A, Ong SS, Jing R et al (2022b) Stevia cultivar "16228013." US Patent: https://patents.google.com/patent/US11284578B2

Martini A, Tavarini S, Macchia M et al (2017) Influence of insect pollinators and harvesting time on the quality of *Stevia rebaudiana* (Bert.) Bertoni seeds. Plant Biosyst Int J Dealing Aspects Plant Biol 151(2):341–351

Mirzaei S-M (2022) Bioinformatics analysis and pharmacological effect of *Stevia rebaudiana* in the prevention of type-2 diabetes. Cell Mol Biomed Rep 2(2):64–73. https://doi.org/10.55705/cmbr.2022.336232.1035

Mlambo R, Wang J, Chen C (2022) *Stevia rebaudiana*, a versatile food ingredient: the chemical composition and medicinal properties. J Nanomater 2022. https://doi.org/10.1155/2022/3573005

Modi AR, Patil G, Kumar N, Singh AS, Subhash N (2012) A simple and efficient in vitro mass multiplication procedure for *Stevia rebaudiana* Bertoni and analysis of genetic fidelity of in vitro raised plants through RAPD. Sugar Tech 14:391–397. https://doi.org/10.1007/s12355-012-0169-6

Morita T, Bu Y (2000) Variety of *Stevia rebaudiana* Bertoni. US Patent; https://patents.google.com/patent/US6031157A

Myint KZ, Wu K, Xia Y et al (2020) Polyphenols from *Stevia rebaudiana* (Bertoni) leaves and their functional properties. J Food Sci 85(2):240–248

Nikraftar S, Ebrahimzadegan R, Majdi M, Mirzaghaderi G (2024) Genome-wide analysis of the C2H2-ZFP gene family in *Stevia rebaudiana* reveals involvement in abiotic stress response. Sci Rep 14(1):6164

Nower AA (2014) In vitro propagation and synthetic seeds production: an efficient methods for *Stevia rebaudiana* Bertoni. Sugar Tech 16:100–108. https://doi.org/10.1007/s12355-013-0228-7

Olsson K, Carlsen S, Semmler A et al (2016) Microbial production of next-generation Stevia sweeteners. Microb Cell Fact 15(1):207

O'Neill K, Pirro S (2020) The complete genome sequence of *Stevia rebaudiana*, the Sweetleaf. F1000Res 9:751

Özyiğit Y, Uçar E, Turgut K (2015) The effect of different pollination methods on seed yield and germination features in *Stevia rebaudiana* Bertoni. Turk J Agric Res (TUTAD) 2(2):114

Pal P, Masand M, Sharma S, Seth R, Singh G, Singh S et al (2023) Genome-wide transcriptional profiling and physiological investigation elucidating the molecular mechanism of multiple abiotic stress response in *Stevia rebaudiana* Bertoni. Sci Rep 13(1):19853

Pande SS, Gupta P (2013) Plant tissue culture of *Stevia rebaudiana* (Bertoni): a review. J Pharmacog Phytotherapy 5:26–33. https://doi.org/10.5897/JPP13

Parris CA, Shock CC, Qian M (2016) Dry leaf and steviol glycoside productivity of *Stevia rebaudiana* in the Western United States. HortScience 51:1220–1227

Peteliuk V, Rybchuk L, Bayliak M et al (2021) Natural sweetener *Stevia rebaudiana*: functionalities, health benefits and potential risks. EXCLI J 20:1412–1430

Petit E, Berger M, Camborde L et al (2020) Development of screening methods for functional characterization of UGTs from *Stevia rebaudiana*. Sci Rep 10(1):15137

Pigatto GB, Gomes EN, de Cássia TJ et al (2018) Efectos de ácido indolbutírico, posiciones de estacas de tallo y sustratos sobre la propagación vegetativa de *Stevia rebaudiana* Bertoni. Rev Colomb Cienc Hortic 12(1):202–211

Putnik P, Bezuk I, Barba FJ et al (2020) - Sugar reduction: *Stevia rebaudiana* Bertoni as a natural sweetener. In: Agri-food industry strategies for healthy diets and sustainability. Academic, Boston, pp 123–152

Ramesh K, Singh V, Megeji NW (2006) Cultivation of Stevia [*Stevia rebaudiana* (Bert.) Bertoni]: a comprehensive review. In: Advances in agronomy. Academic Press, Boston, pp 137–177

Ramos-Tovar E, Muriel P (2017) Stevia as a putative hepatoprotector. In: Muriel P (ed) Liver pathophysiology. Academic, Boston, pp 715–727

Rashid Z, Ahngar TA, Nazir A et al (2021) Research technology of Stevia. Cutting-Edge Res Agric Sci 6:1–14

Razak UNAA, Ong CB, Yu TS, Lau LK (2014) In vitro micropropagation of *Stevia rebaudiana* Bertoni in Malaysia. Brazilian Arch Biol Technol 57:23–28. https://doi.org/10.1590/S1516-89132014000100004

Ribeiro MM, Diamantino T, Domingues J et al (2021) *Stevia rebaudiana* germplasm characterization using microsatellite markers and steviol glycosides quantification by HPLC. Mol Biol Rep 48(3):2573–2582

Richman A, Swanson A, Humphrey T et al (2005) Functional genomics uncovers three glucosyltransferases involved in the synthesis of the major sweet glucosides of *Stevia rebaudiana*. Plant J 41(1):56–67

Rodriguéz-Páez LA, Pineda-Rodriguez YY, Pompelli MF, Jimenez-Ramirez AM, Genes-Avilez OJ, Jaraba-Navas J de D et al (2024) Micropropagation protocols for three elite genotypes of *Stevia rebaudiana* Bertoni. Horticulturae 10(4):404

Saifi M, Nasrullah N, Ahmad MM et al (2015) In silico analysis and expression profiling of miRNAs targeting genes of steviol glycosides biosynthetic pathway and their relationship with steviol glycosides content in different tissues of *Stevia rebaudiana*. Plant Physiol Biochem 94:57–64

Saifi M, Yogindran S, Nasrullah N et al (2019) Co-expression of anti-miR319g and miRStv_11 lead to enhanced steviol glycosides content in *Stevia rebaudiana*. BMC Plant Biol 19(1):274

Sakai A, Engelmann F (2007) Vitrification, encapsulation-vitrification and droplet-vitrification: a review. Cryo Lett 28(3):151–172

Salehi B, López MD, Martínez-López S et al (2019) *Stevia rebaudiana* Bertoni bioactive effects: from in vivo to clinical trials towards future therapeutic approaches. Phytother Res 33(11):2904–2917

Sairkar P, Chandravanshi MK, Shukla NP, Mehrotra NN (2009) Mass production of an economically important medicinal plant *Stevia rebaudiana* using in vitro propagation techniques. J Med Plants Res 3:266–270

Sattler MC, Carvalho CR, Clarindo WR (2016) The polyploidy and its key role in plant breeding. Planta 243(2):281–296

Schiatti-Sisó IP, Quintana SE, García-Zapateiro LA (2022) Stevia (*Stevia rebaudiana*) as a common sugar substitute and its application in food matrices: an updated review. J Food Sci Technol. https://doi.org/10.1007/s13197-022-05396-2

Shaafi B, Saeed Moosavi S, Reza Abdollahi M, Sarikhani H (2021) The optimized protocols for production, adaptation and keeping of the produced artificial seeds from encapsulated lateral buds in *Stevia rebaudiana* (Bertoni). Agrotechniques Ind Crop 24–35. https://doi.org/10.22126/ETIC.2021.6246.1004.

Sharma N, Kaur R, Era V (2016a) Potential of RAPD and ISSR markers for assessing genetic diversity among *Stevia rebaudiana* Bertoni accessions. Indian J Biotech 15(1):95–100

Sharma S, Walia S, Singh B, Kumar R (2016b) Comprehensive review on agro technologies of low-calorie natural sweetener Stevia (*Stevia rebaudiana* Bertoni): a boon to diabetic patients. J Sci Food Agric 96(6):1867–1879

Sharma S, Gupta S, Jain R, Kothari SL, Kachhwaha S (2023a) SiO2 nanoparticles as elicitor for increased rebaudioside-A in *Stevia rebaudiana* micropropagated in solid and liquid cultures: a comparative study. Plant Cell Tiss Org Cult 155(2):541–552

Sharma S, Gupta S, Kumari D, Kothari SL, Jain R, Kachhwaha S (2023b) Exploring plant tissue culture and steviol glycosides production in *Stevia rebaudiana* (Bert.) Bertoni: a review. Collect FAO Agric 13(2):475

Shatnawi, Shibli, Abu-Romman, et al (2011) Clonal propagation and cryogenic storage of the medicinal plant *Stevia rebaudiana*. Span J Agric Res 9(1):213–220

Shock CC, Parris CA (2017a) Stevia plant named "SW 107." US Patent: https://patents.google.com/patent/USPP27937P3

Shock CC, Parris CA (2017b) Stevia plant named "SW 129." US Patent: https://patents.google.com/patent/USPP28373P3

Shock CC, Parris CA (2018) Stevia plant named "SW 227." US Patent: https://patents.google.com/patent/USPP28977P3

Shweta (2020) How to grow Stevia; get full information of Stevia cultivation in India. In: Jagran K (ed). https://krishijagran.com/agripedia/how-to-grow-Stevia-get-full-information-of-Stevia-cultivation-in-india/

Singh DP, Kumari M, Prakash HG et al (2019) Phytochemical and pharmacological importance of Stevia: a calorie-free natural sweetener. Sugar Tech 21(2):227–234

Singh G, Pal P, Masand M et al (2020) Comparative transcriptome analysis revealed gamma-irradiation mediated disruption of floral integrator gene(s) leading to prolonged vegetative phase in *Stevia rebaudiana* Bertoni. Plant Physiol Biochem 148:90–102

Singh G, Singh G, Singh P et al (2017) Molecular dissection of transcriptional reprogramming of steviol glycosides synthesis in leaf tissue during developmental phase transitions in *Stevia rebaudiana* Bert. Sci Rep 7:11835

Sivaram L, Mukundan U (2003) In vitro culture studies on *Stevia rebaudiana*. In Vitro Cell Develpment Biol Plant 39:520–523. https://doi.org/10.1079/IVP2003438

Smitha GR, Umesha K (2012) Vegetative propagation of Stevia [*Stevia rebaudiana* (Bertoni) Hemsl.] through stem cuttings. J Trop Agric Food Sci 50:72–75

Subrahmanyeswari T, Gantait S, Kamble SN, Singh S, Bhattacharyya S (2023a) Meta-Topolin-induced regeneration and ameliorated rebaudioside-A production in genetically uniform candy-leaf plantlets (*Stevia rebaudiana* Bert.). S Afr J Bot 159:405–418

Subrahmanyeswari T, Laha S, Kamble SN, Singh S, Bhattacharyya S, Gantait S (2023b) Alginate encapsulation of shoot tips and their regeneration for enhanced mass propagation and germplasm exchange of genetically stable *Stevia rebaudiana* Bert. Sugar Tech. 25(3):542–551

Sun Y, Xu X, Zhang T et al (2021) Comparative transcriptome analysis provides insights into steviol glycoside synthesis in Stevia (*Stevia rebaudiana* Bertoni) leaves under nitrogen deficiency. Plant Cell Rep 40(9):1709–1722

Taak P, Tiwari S, Koul B (2020) Optimization of regeneration and Agrobacterium-mediated transformation of Stevia (*Stevia rebaudiana* Bertoni): a commercially important natural sweetener plant. Sci Rep 10(1):16224

Talevi A (2022) Potential medicinal effects and applications of Stevia constituents. Phytochem Rev 21(1):161–178

Tamura Y, Nakamura S, Fukui H, Tabata M (1984) Comparison of Stevia plants grown from seeds, cuttings and stem-tip cultures for growth and sweet diterpene glucosides. Plant Cell Rep 3(5):180–182

Taware S, Mukadam D, Chavan A, Taware SD (2009) Comparative studies of in vitro and in vivo grown plants and callus of Stevia rebaudiana (Bertoni). Int J Integr Biol 9(1);10–15

Thakur K, Ashrita SA et al (2021) Steviol glycoside accumulation and expression profiling of biosynthetic pathway genes in elicited in vitro cultures of *Stevia rebaudiana*. In Vitro Cell Development Biol Plant 57(2):214–224

Tuleja M, Santocki M, Dziurka M et al (2021) The search towards cyto-embryological and physiological obstacles in sexual reproduction of *Stevia rebaudiana* Bertoni. Sci Hortic 288:110342

Uskutoğlu T, Uskutoğlu D, Turgut K (2019) Effects on Pre-treatment and different tissue culture media for androgenesis in *Stevia rebaudiana* Bertoni. Sugar Tech 21:1016–1023. https://doi.org/10.1007/s12355-019-00722-z

Wang J, Zhao H, Wang Y et al (2020) A review of Stevia as a potential healthcare product: Up-to-date functional characteristics, administrative standards and engineering techniques. Trends Food Sci Technol 103:264–281

Wang Y, Xu P, Wang W, Jia X, Zhu L, Yin H (2023) Oligosaccharides increased both leaf biomass and steviol glycosides content of *Stevia rebaudiana*. Plant Physiol Biochem 202:107937

Warner R, Beaudry R, Hancock J (2016) Stevia plant named "12–05–005." US Patent: https://patents.google.com/patent/USPP27902P3

Warner RM, Beaudry R, Hancock J, Vallejo VA (2017a) Stevia plant named "12–05–149." US Patent: https://patents.google.com/patent/USPP27939P3

Warner RM, Beaudry R, Hancock J, Vallejo VA (2017b) Stevia plant named "12–05–144." US Patent: https://patents.google.com/patent/USPP27938P3

Wu Q, La Hovary C, Chen H-Y et al (2020) An efficient *Stevia rebaudiana* transformation system and in vitro enzyme assays reveal novel insights into UGT76G1 function. Sci Rep 10(1):3773

Xiang Z-X, Tang X-L, Liu W-H, Song C-N (2019) A comparative morphological and transcriptomic study on autotetraploid *Stevia rebaudiana* (Bertoni) and its diploid. Plant Physiol Biochem 143:154–164

Xu X, Yuan H, Yu X et al (2021) The chromosome-level Stevia genome provides insights into steviol glycoside biosynthesis. Hortic Res 8(1):129

Xu X, Yang Y, Zhang T, Zhang Y, Tong H, Yuan H (2024) Systematic analysis of the R2R3-MYB transcription factor gene family in *Stevia rebaudiana*. Ind Crops Prod 210:118123

Yadav AK, Singh S, Rajeev R (2014) Self-incompatibility evidenced through scanning electron microscopy and pollination behavior studies in *Stevia rebaudiana*. Indian J Agric Sci 84(1):93–100

Yadav AK, Singh S, Yadav SC, Dhyani D, Bhardwaj G, Sharma A, Singh B (2013) Induction and morpho-chemical characterization of *Stevia rebaudiana* colchiploids. Indian J Agric Sci 83(2):159–165

Yang Y-H, Huang S-Z, Han Y-L et al (2014) Base substitution mutations in uridinediphosphate-dependent glycosyltransferase 76G1 gene of *Stevia rebaudiana* causes the low levels of rebaudioside A: mutations in UGT76G1, a key gene of steviol glycosides synthesis. Plant Physiol Biochem 80:220–225

Yang YW, Hsing YI, Chang WC (1981) Clonal propagation of *Stevia rebaudiana* Bertoni through axillary shoot proliferation in vitro. Bot Bull Acad Sin 22:57–62

Yao Y, Ban M, Brandle J (1999) A genetic linkage map for *Stevia rebaudiana*. Genome 42(4):657–661

Younes M, Aquilina G et al (2022) Safety of the proposed amendment of the specifications for enzymatically produced steviol glycosides (E 960c): Rebaudioside D produced via enzymatic bioconversion of purified Stevia leaf extract. EFSA J 20(5):e07291

Zaidan UH, Mohamad Zen NI, Amran NA et al (2019) Biochemical evaluation of phenolic compounds and steviol glycoside from *Stevia rebaudiana* extracts associated with in vitro antidiabetic potential. Biocatal Agric Biotechnol 18:101049

Zhang S, Liu Q, Lyu C et al (2020) Characterizing glycosyltransferases by a combination of sequencing platforms applied to the leaf tissues of *Stevia rebaudiana*. BMC Genomics 21(1):794

Zhang S-S, Chen H, Xiao J-Y et al (2019) Mutations in the uridine diphosphate glucosyltransferase 76G1 gene result in different contents of the major steviol glycosides in *Stevia rebaudiana*. Phytochemistry 162:141–147

Zhang T, Zhang YX, Sun YM, Xu XY, Wang YJ, Chong XR et al (2023) Isolation and functional analysis of SrMYB1, a direct transcriptional repressor of SrUGT76G1 in *Stevia rebaudiana*. J Integr Agric 22(4):1058–1067

Zheng J, Zhuang Y, Mao H-Z, Jang I-C (2019) Overexpression of SrDXS1 and SrKAH enhances steviol glycosides content in transgenic Stevia plants. BMC Plant Biol 19(1):1

Zhou X, Gong M, Lv X et al (2021) Metabolic engineering for the synthesis of steviol glycosides: current status and future prospects. Appl Microbiol Biotechnol 105(13):5367–5381

Chapter 11
Phytoconstituents, Medicinal Uses and Conservation of Upastree (*Antiaris toxicaria* Lesch)

Suraj Umdale, Amit Mirgal, and Nikhil Gaikwad

Abstract The Upastree (*Antiaris toxicaria* Lesch.) is a "Poisonous Plant" belongs to the family Moraceae (Fig family) and is distributed throughout the world, predominantly in the tropical areas of Africa, Asia, and Australia. The genus *Antiaris* consists of only single known species, *A. toxicaria*. The milky latex of the upastree contains one of the world's most deadly phytoconstituent known as a cardiac glycoside "Antiarin". Due to poisonous or toxic nature of latex, upastree is commonly known as the poison arrow tree. Also, the stem of the upastree is used for making plywood and veneer. The upastree latex contains a variety of pharmacologically important phytoconstituents of medicinal applications such as antimicrobial, anti-arrhythmic, anti-osteoporosis, anti-tumor, anticonvulsant, antidepressant-like activity, treating mental illness, and epilepsy. The present chapter summarizes the potential economic and health benefits, Agroforestry uses, cultivation practices, germplasm diversity, and conservation of upastree. The efforts should be taken for germplasm collection, evaluation, cultivation, and conservation, by application of tissue culture, cryopreservation, and breeding techniques. Moreover, the upastree will be promoted as potential plantation species and further encouraged for sustainable utilization as a source of commercial timber.

Keywords Antiaris · Antimicrobial · Pharmacology · Poison · Upastree

---

S. Umdale
Department of Botany, Jaysingpur College Jaysingpur, (Affiliated to Shivaji University), Kolhapur, Maharashtra, India

A. Mirgal · N. Gaikwad (✉)
Laboratory of Cytogenetics and Plant Breeding, Department of Botany, Shivaji University, Kolhapur, Maharashtra, India
e-mail: nbgaikwadsuk@gmail.com

J. M. Al-Khayri et al. (eds.), *Biodiversity and Genetic Improvement of Medicinal and Aromatic Plants II*, Advances in Plant Breeding Strategies 10,
https://doi.org/10.1007/978-3-031-81857-8_11

11.1 Introduction

The *Antiaris toxicaria* Lesch. commonly known as upastree is a poisonous, deciduous, forest tree species of the fig family Moraceae. Upastree is widely distributed in the tropical forest regions of Asia, tropical Africa, Australia, Indonesia, China, Philippines, and various other tropical islands (Ravikumar and Sankar 2009; Upadhyay et al. 2010; Umdale et al. 2020). In India, the upastree is sporadically distributed along the Western Ghats (Kerala, Karnataka, Goa, and Maharashtra) and the Western Ghats (Andhra Pradesh and Tamil Nadu) (Singh et al. 2001; Ravikumar and Vijaya Sankar 2009). The upastree is a large forest tree that grows up 30–50 m tall. The stem is a small-scale source of lightweight hardwood timber with a density of 250–540 kg/m^3 and is often used for making wooden drums in central Uganda (Omeja et al. 2005). The bark is grayish in color and consists of toxic milky latex. The bark is used for making cloth dyes and paints. The *Muthuvan* tribe in Kerala, India traditionally uses upastree bark for the preparation of Maravuri cloth (Manithottam and Fancis 2008). The major poisonous phytoconstituents present in upastree latex is cardiac glycoside known as Antiarin (Kopp et al. 1992). In ancient times, latex is used as a poison for arrows, darts, and blowdarts in Southeast Asian countries (Darmadi 2018; Umdale et al. 2020).

Botanical Classification (APG-IV 2016):

Kingdom: Plantae (Photosynthetic Eukaryotes)
Clade: Tracheophytes (Vascular Plants)
Clade: Angiosperms (Flowering plants, Seeds within fruits)
Clade: Eudicots (Two cotyledons)
Clade: Rosids (Bitegmic, crassinucellate ovules)
Order: Rosales (Usually perfect flowers, several stamens, and an inferior ovary)
Family: Moraceae (Presence of milky or watery latex)
Genus: *Antiaris* Lesch.
Species: *A. toxicaria* Lesch.

Chromosome number-$2n = 24, 48$

11.1.1 Local Names

Upastree is know with different local names in the world which are represented as follows.

English—Upastree, Antiaris, Blow dart, False mvule, False iroko, Bark cloth tree, Poison arrow tree

Akan—Foto, Kyenkyen
Angola—Sansama; Kirundu
Burmese—Hmyasceik, aseik

Chinese—Poison arrow tree
Filipino—Upas
French—Ako
Ghana—Kyenkyen, Chenchen
Indonesian—Bemoe
Javanese—Ancar, Upas
Malay—Antjar, Antiaris, Ipoh, Terap
Malayalam—Maravuri,
Mandinka—Jafo
Marathi—Chandkuda, Chand—phal
Nigerian—Oro, Ogiovu
Swahili—Mkunde
Tamil—Aranthelli, Mara—uri, Nettavil
Tanzania—Mlulu, Mkuzu
Thai—Yang nong
Uganda—Mumaka, Kirundu
Wolof—Man, Kan
Zaire—Bonkonko

11.1.2 Taxonomy

The genus *Antiaris* consists of only single known species, *A. toxicaria.* However, the genus *Antiaris* shows wide diversity in plant morphology. At present, the *A. toxicaria* comprises four subspecies and two varieties shown in Table 11.1. Ugwoke et al. (2017) and Porchselvi and Muthulakshmi (2018) reported the correct identification of *Antiaris toxicaria* with the help of pharmacognostic and phytochemical and anatomical studies respectively.

11.1.3 Botanical Description

Upastree is deciduous, monoecious forest tree of 30–50 m tall with dome canopy; bark is 10–20 mm thick greyish colour, with lenticels, the exudates is milky or watery. Leaves are simple, alternate, oval in shape and 5.5–16 cm × 4–11 cm; stipule small, cauducous; petiole 2–7 mm long, velvety; leaf apex acuminate or obtusely acuminate, margin entire, glabrous; lateral nerves 5–10 pairs, parallel, prominent, reticulate. Flowers are unisexual, greenish-yellow; male receptacle axillary, with 2–7 tepals; peduncle velvety; receptacle surrounded by imbricating bracts; stamens 2–6; filaments erect; anthers yellow; female flowers solitary, present in disk shaped heads; tepals absent; ovary superior, unilocular with single ovule; style two. Fruits fleshy, single seeded bright red coloured drupes with 2–3 cm in diameter. The seed are edible

Table 11.1 Diversity of *A. toxicaria* species

S. No.	Name of subspecies/variety	Nomenclature status	Link
1	*Antiaris toxicaria* var. *Africana* Scott-Elliot ex A. Chev	Accepted	http://www.theplantlist.org/tpl1.1/record/kew-2642231
2	*Antiaris toxicaria* subsp. *humbertii* (Leandri) C. C. Berg	Accepted	http://www.theplantlist.org/tpl1.1/record/kew-2642233
3	*Antiaris toxicaria* subsp. *Macrophylla* (R. Br.) C. C. Berg	Accepted	http://www.theplantlist.org/tpl1.1/record/kew-2642235
4	*Antiaris toxicaria* subsp. *Madagascariensis* (H. Perrier) C. C. Berg	Accepted	http://www.theplantlist.org/tpl1.1/record/kew-2642236
5	*Antiaris toxicaria* var. *usambarensis* Engl.) C. C. Berg	Accepted	http://www.theplantlist.org/tpl1.1/record/kew-2642237
6	*Antiaris toxicaria* subsp. *Welwitschii* (Engl.) C. C. Berg	Accepted	http://www.theplantlist.org/tpl1.1/record/kew-2642239

and 10–15 mm long. The birds, bats, antelopes, monkeys and humans eat and are responsible for seed dispersal (Fig. 11.1).

11.1.4 Species Distribution

Upastree is distributed in tropical areas mostly in the semi-arid, rainforest, even in swamp forests, and found throughout the Old World tropics. It is native to Australia, Cameroon, China, Democratic Republic of Congo, Fiji, India and Indian subcontinents, Indonesia, Java, Kenya, Peninsular Thailand, Malaysia, Vietnam, Nigeria, Papua New Guinea, Philippines, Sierra Leone, Sudan, Sumatra (also Enggano, Riau-Lingga Island), Singapore, Solomon Islands, Sulawesi, Tonga, and Uganda. In India, it is sporadically distributed in the Western Ghats, Eastern Ghats region and Andaman Islands (Orwa et al. 2009; Ravikumar and Sankar 2009; Upadhyay et al. 2010; Mirgal et al. 2013; Umdale et al. 2020) (Fig. 11.2).

Phytoconstituents in Upastree

The latex of the upastree consists of several cardiac glycosides, antiarins (α-antiarin and β-antiarin) (Fig. 11.3), angiogenin, antioside, antiarosides, antiogoside, antiarojavoside, antialloside, γ-lactone, β-sitosterol, betulinic acid, chalcones, cymarol, D-javose, flavonoids, flavanones, malayoside, oleanolic acid, periplogenin, peripalloside, strophalloside, strophanthidol, toxicariosides (A, B, C, D, N, O, J, K, L), strophanthidin, strophalloside, 7 phenylpropanoid and lignin derivatives, ursolic acid, uposide (Muehlradt et al. 1964; Hano et al. 1991; Que et al. 2009; Jiang et al. 2009; Dong et al. 2011; Zuo et al. 2013; Li et al. 2015).

Fig. 11.1 *Antiaris toxicaria* Lesch. John Lindley (1866) (CC 0)

Potential Economic and Health Benefits

The upastree bark latex is used as medicine since ancient times. Though it consists of poisonous phytoconstituents, it is advocated as a cardiac stimulant in a small dose. Lin et al. (2003) reported the use of leaves and bark of *Antiaris toxicaria* as medicine by Dai ethnic group from China. The bark, leaves, and seeds are used as an anodyne, astringent, febrifuge, and vermifuge, and to treat mental illnesses, hepatitis, and dysentery (Lin et al.2003; Dai 2010). The bark decoction is given in the treatment of leukemia, and splenomegaly and also consists of anticancer activities (Bosu and Krampah 2005). The latex consist of Antiarins are said to act on the heart more powerfully than digitalin. Beta-antiarin is more potent than alpha-antiarin. Latex, in small quantities, is a mild cardiac and circulatory stimulant, whereas in large quantities it acts as a myocardial poison. It stimulates intestinal and uterine contractions (Khare 2007). In African countries, a minute quantity of latex is applied to wounds, cuts, and

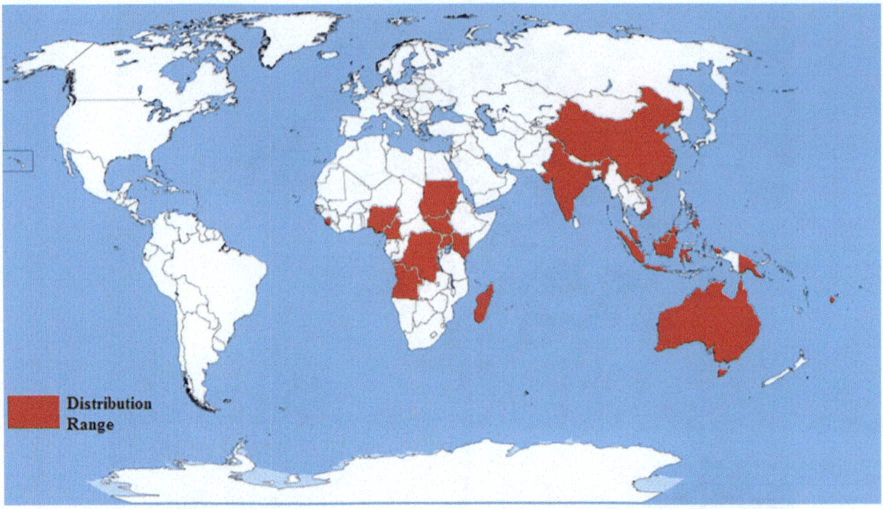

Fig. 11.2 The world map shows species distribution of the Upastree (*A. toxicaria*)

skin diseases viz. eczema, and leprosy (Mei et al. 2007). The roasted seeds are eaten as a treatment for small growths on the body (Darmadi 2018). Also, the bark is the source of tannins and is used for dyeing and making paints. The mature, well-grown upastree gives about 100–500 g of latex in 2 days. The bark latex is traditionally used on darts, blowdarts, and arrows as a toxin during the war, as a fish poison, and for hunting in South-East Asian countries. Gupta and Sharma (2017) reported that specific plant based principle/material can be used during autopsy material or on the crime spot by scientists and the forensic investigators, where from *Antiaris toxicaria* leaves and bark are used as investigating material. Methanolic extracts of leaves, bark and roots of *Antiaris toxicaria* reported the presence of galic acid, catechins, chlorogenic acid, caffeic acid, ellagic acid, epigallo catechin, routine, isoquercitrin, quercitrin, quercetin and kaempferol. Secondary metabolites present in the plant that can be used in medicinal preparations and will be candidate species for developing a natural insecticide (Subiono et al. 2022). Numpulsuksant et al. (2021) reported from their study that *Antiaris toxicaria* has highest use value which indicates that the species has great importance for daily life to the local people from Maha Sarakham province of Thailand. Different biological activities of *A. toxicaria and A. toxicaria* var *Africana* are described in Table 11.3 and 11.4.

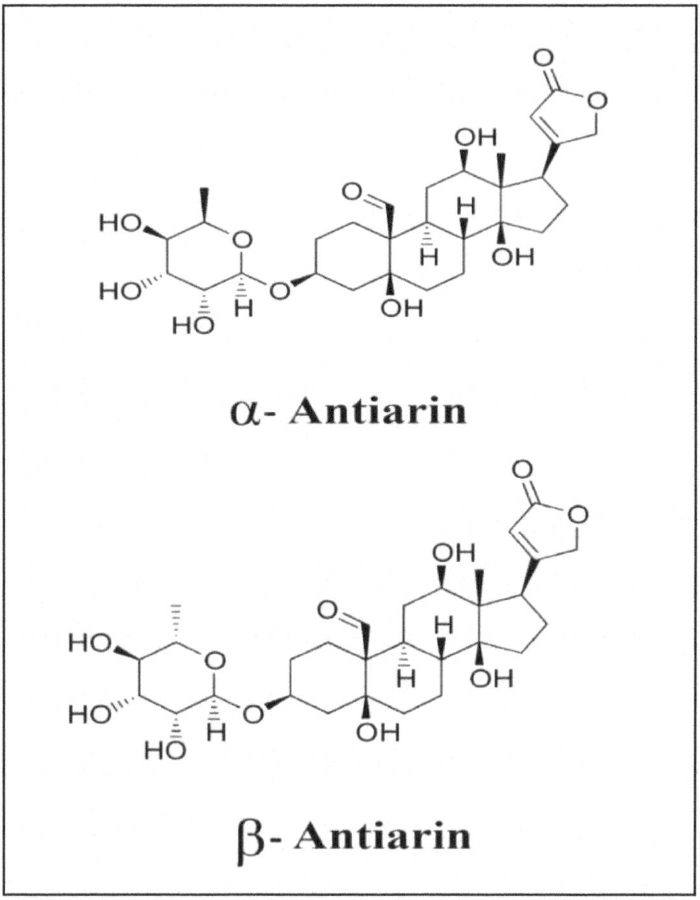

Fig. 11.3 The structural representation of major phytoconstituent, antiarin in upastree

11.2 Agroforestry Uses

The upastree is an important forest tree in the international market for its veneer and plywood. It is a reasonable source of lightweight hardwood with a density of 250–540 kg/m^3. It is an excellent, fast-growing shade tree and also wood is used as a source of marginal fuel. It is often traded in mixed consignments of lightweight hardwood. The wood is also useful in making wooden drums (Omeja et al. 2005). In Ghana, the total stock volume of upastree above 30 cm diameter was 825 m^3/km^2 in 2001 (Bosu and Krampah 2005). The tree bark is being used for the preparation of Maravuri cloth by the Muthuvan tribe in Kerala, India (Manithottam and Fancis 2008). Upastree has a wide distribution range and adaptability to diverse eco-geographical regions of the world; this shows it has an excellent forest tree for restoration of native forests and establishment of woodland gardens (Ken 1997). The wood is also used for making

Table 11.3 Various biological activities reported in different parts of the upastree (*A. toxicaria*)

S. No.	Effect	Plant part	Extract	References
1	Anticonvulsant, antidepressant and neuroprotective effect	Leaf, bark latex	Aqueous extract	Mante et al. (2012, 2013, 2017) Agbaje et al. (2014) Antwi et al. (2019) Yeboah et al. (2019) Ilesanmi et al. (2022)
2	Antioxidant activity	Stem latex	Methanolic	Gan et al. (2008)
		Leaves	Methanolic	Kemila and Krishnaveni (2019)
3	Anti-microbial activity (antibacterial)	Stem bark Leaves Seeds	Aqueous Methanolic and aqueous ethanolic	Gopalakrishnan et al. (2015) Kemila and Krishnaveni (2016) Dong et al. (2011)
	Anti-microbial activity (Antifungal)	Leaves	Aqueous Methanolic	Kemila and Krishnaveni (2016)
4	Anti-cancer/ anti-tumour	Stem bark	Methanolic	Kuete et al. (2009)
		Stem bark	Ethanolic	Shi et al. (2014)
		Latex	Aqueous	Jiang et al. (2008) Dai et al. (2009) Gan et al. (2009) Liu et al. (2013)
		Seeds	Ethanolic	Dong et al. (2011) Zuo et al. (2013) Wu et al. (2014) Huang et al. (2017) Zhao et al. (2018)
5	Anti-proliferative activity	Latex	Aqueous	Zheng et al. (2021)
6	Cyto-toxic activity	Latex	Aqueous	Hu et al. (2021)
7	Cardiovascular activity	Latex	Acetone Ethanol	Fujimoto et al. (1983) Dai et al. (2009)
		Stem bark Seeds	Ethanolic Methanolic	Shi et al. (2010) Agarwal et al. (2014)
8	Antimalarial	Leaf Stem bark	Ethyl acetate Ethanolic	Lacroix et al. (2011) Nondo et al. (2015)
9	Pesticidal	Leaves Bark Root	Methanolic	Subiono et al.(2022)

Table 11.4 Various biological activities reported in different parts of the upastree (*A. toxicaria* var *africana*)

S. No.	Effect	Plant part	Extract	Reference
1	Antioxidant activity	Stem bark Leaf	Methanolic Ethanolic	Kuete et al. (2009) Thiam et al. (2022)
2	Anti-cancer/anti-tumour	Stem bark Leaf	Methanolic Ethanolic	Kuete et al. (2009), Vouffo et al. (2010) Thiam et al. (2022)
3	Anti inflammatory	Leaf	Ethanolic	Thiam et al. (2022)
4	Cardiovascular activity	Leaf	Ethanolic	Thiam et al. (2022)
5	Relieve pain of chest, rheumatic, respiratory and stomach pain	Stem bark	Ethanolic	Banso and Mann (2008)
6	Antibacterial	Leaf	Methanolic	Kuete et al. (2009)
7	Against leprosy and sore throat	Leaf	Methanolic	Kuete et al. (2009)

furniture, lightweight boxes, crates, toys, particle boards, blackboards beer canoes sometimes used as substitutes for making cork. The bark is used for manufacturing paper, mats, hull walls, and rough clothing (Bosu and Krampah 2005). The bark has a high concentration of tannins that are used in traditional clothes dyeing and paints. In Africa, due to presence of a high concentration of tannins in the bark of the upastree is traditionally used in dyeing clothes and paints. The upastree is susceptible to termites and marine borers therefore, is used as a source of timber after the proper wood treatment (Orwa et al. 2009). In Ghana, farmers give preference to the upastree as a shade tree in cocoa farms due to its higher adaptability (Graefe et al. 2017).

11.3 Growth and Development

Upastree has a wide distribution range and adaptability to diverse eco-geographical regions. The tree grows well-drained soils and requires sunlight. It is found in forests above sea level to 1800 m altitude. It is fast growing tree and will reach 30–50 m in height (Full mature size) in 20 years. In Africa, the seeds are grown in pots, and saplings are transplanted in the desired location. The saplings will reach 40 cm with low intensity of light but require full light for further growth. In forests and farmland, the upastree grows 50 cm tall and 1 cm of diameter per year in height in appropriate environmental conditions. The upastree is recommended for the roadside tree plantation (Bosu and Krampah 2005). In West Africa, flowering occurs between November and February and fruiting occurs from February to March. However, In India, it blooms between October and January and mature fruit fall occurs during April to June (Mirgal et al. 2013). The fruits and seeds are edible therefore seed dispersal occurs via animals such as birds, monkeys, chimpanzees, and antelopes. Frugivores like *Cercopithecus campbelli* and *Cercopithecus petaurista,* hornbills, turacos, plantain eaters and fruit bats are the effective contributors to the distribution of *A. toxicaira* seeds in Africa (Kankam and Oduro 2009). Reduction in population of dispersers can cause seedling recruitment and possibly the genetic structure of tropical trees, as a result conservation of important frugivore seed dispersers should be part of sustainable management of *A. toxicaria* (Cordeiro and Howe 2001) Upastree produces a large number of seeds, but seeds are recalcitrant and lose viability (Mirgal et al. 2016a, b). The rate of seed germination rate is higher, however, a high mortality rate is noticed in the first year (Bosu and Krampah 2005; Mirgal et al. 2013). Viability loss of excised recalcitrant *A. toxicaria* axes seemed to be activated by mechanical or physical damage, instead of metabolic damage under rapid drying with silica gel Xin et al. (2010). Desiccation tolerance of recalcitrant *Antiaris toxicaria* seeds was enhanced by nitric oxide treatment which acts via protein S-nitrosylation and carbonylation reported by Bai et al. (2011). The seed germination rate is 70–95% within the period of three months. The rate of seed germination is improved using the application of Gibberellic acid and a combination of various soil types (Mirgal et al. 2016a, b).

11.4 Cultivation Practices and Challenges

The upastree grows in well-drained soils and requires full sunlight for further growth and development. The fresh seeds show high germination percentage (70–95%). The seeds are grown in pots with soil and sawdust in a ratio of 70:30, which is helpful for the enhancement of seed germination percentage (Mirgal et al. 2016a, b). The saplings of an average height of 60 cm are transplanted in the desired location. In Northern Côte d'Ivoire, the late transplantation or sapling height of about 2 m tall is recommended for plantation due to the higher grazing pressure of cattle and wild animals. The upastree has a good self-pruning ability and further fire protection will be offered for growth and development. The major challenge of upastree cultivation is generating awareness about the hazardous properties of tree latex. The development of silviculture practices for effective cultivation of upastree in social forests, Agroforestry plantations, social forest, and natural forests. As per Mligo (2019), it was observed that good post fire regeneration of *Antiaris toxicaria* seedling from seeds in rigde top and valley bottom areas compared to other plants from Pugu forest reserve of Tanzania.

11.5 Germplasm Biodiversity and Conservation

The genus *Antiaris* shows wide diversity in plant morphology. Upastree is noted as endangered species in China and got protection from the Chinese Government (Fu 1989). The upastree is distributed in fragmented populations due to the continued conversion of rainforests to rubber plantations (Zhu et al. 2006). In Africa, the upastree is more common in forests, but over-exploitation has rapidly reduced the wild populations (Orwa et al. 2009). The study of genetic erosion of natural upastree populations is difficult due to its distribution in low densities in forests (Mirgal et al. 2013). The upastree is naturally observed in patches of deciduous to semi-evergreen forests in the Western Ghats of India and is almost sporadically found in sacred groves. Due to the low regeneration of upastrees, even in a protected forest, small populations are recorded (Mirgal et al. 2013; Schmidt et al. 2017). Despite a wide distribution, the species needs to be protected; over-exploitation and environmental deterioration have caused populations to decline annually and even become extinct in some areas (Mirgal et al. 2013; Umdale et al. 2020). Lan et al. (2014) recommended the need for germplasm conservation and restoration projects. Upastree produces a large number of seeds and is easily collected from the soil. The seeds are recalcitrant and lose viability very soon (Mirgal et al. 2016a, b). The efforts should be taken for the cryopreservation of upastree seeds. The upastree shows wide diversity in plant morphology. Therefore, efforts should be taken for the collection, evaluation, and conservation of diverse germplasm.

11.6 Tissue Culture Studies

Tissue culture studies was carried out by Lizhen et al. (2007), where multiplication of upastree was reported by using micropropagation culture on MS medium supplemented with 6- BA 2.5 (mg/l) combined with NAA (0.5%). Rooting was observed on half MS medium supplemented with 1.0 mg/l NAA.

11.7 Genetic Engineering and Gene Editing

Zhang et al. (2018) has carried out the complete chloroplast genome of *Antiaris toxicaria* plant and it was revealed that upastree was closely related to the genus Ficus within the family Moraceae. This valuable genomic information of *Antiaris toxicaria* could help us to identify and utilize medicinal potential of it.

11.8 Mutation Breeding

In *Antiaris toxicaria* induced mutation studies are not reported, however natural population developed from seeds reported albino plants with 5% frequency Mirgal et al. (2016a, b). Out of these albino plants only 4–5% plants survived after 2 years of growth. Those mutants are producing albino leaves at tip (young leaf), whereas same leaves producing chlorophyll after maturity at the base.

11.9 Conclusions and Future Prospects

Upastree is majestic forest tree species having tremendous benefits for humankind. It is a potential source of wood for Agroforestry and has pharmacological importance. However, due to over-exploitation and environmental deterioration, the upastree populations have been declining faster in forest areas. The efforts should be taken for the collection, evaluation, and conservation of the elite germplasm of the upastree. The application of tissue culture technique, cryopreservation, breeding program, and will be encouraged for further evaluation of the upastree. The attempts should be taken for cultivation, germplasm conservation, and sustainable utilization of the upastree. Also, initiatives should be taken for the multiplication of germplasm, and habitat restoration for the conservation of upastree. The cultivation of the upastree is currently being promoted with providing proper Silvicultural practices to the native communities. Upastree is fast growing forest tree, it must get endorsement as a potential plantation species. Also, efforts should be taken to promote the upastree as a source of commercial timber.

References

Agarwal P, Akhade M, Laddha K, Narkhede S, Mirgal A, Salunke C (2014) Quantification of convallatoxin in *Antiaris toxicaria* leusch seeds by RP-HPLC. Anal Chem Lett 4:172–177. https://doi.org/10.1080/22297928.2014.925821

Agbaje EO, Ishola Ismail O, Oniyire, Joel A (2014) Antidepressant, anxiolytic, and anticataleptic effects of aqueous leaf extract of *Antiaris toxicaria* Lesch. (Moraceae) in mice: possible mechanisms of actions. J Basic Clin Physiol Pharmacol 25(4):429–438. https://doi.org/10.1515/jbcpp-2013-0054

Angiosperm Phylogeny Group (2016) An update of the angiosperm phylogeny group classification for the orders and families of flowering plants: APG IV. Bot J Linn Soc 181(1):1–20. https://doi.org/10.1111/boj.12385

Antwi S, Asiedu-Larbi J, Martey NK (2019) Safety and effectiveness of mist antiaris, a herbal preparation for treatment of peripheral neuropathy. Biomed Res Int 2019:2607872. https://doi.org/10.1155/2019/2607872

Bai X, Yang L, Tian M, Chen J, Shi J, Yang Y, Hu X (2011) Nitric oxide enhances desiccation tolerance of recalcitrant *Antiaris toxicaria* seeds via protein S-nitrosylation and carbonylation. PLos ONE 6(6):e20714. https://doi.org/10.1371/journal.pone.0020714

Banso A, Mann A (2008) Evaluation of antibacterial properties of flavonoid fraction from *Antiaris africana* (Engl). J Appl Biosci 12:665–670

Bosu PP, Krampah E (2005) *Antiaris toxicaria* Lesch. [Internet] Record from PROTA4U. Louppe D, Oteng-Amoako AA and Brink M (eds) PROTA (Plant resources of tropical Africa/Resources végétales de l'Afrique tropicale), Wageningen, Netherlands. http://www.prota4u.org/search.asp

Cordeiro NJ, Howe HF (2001) Low recruitment of trees dispersed by animals in African forest fragments. Conserv Biol 15:1733–1741

Dai HF, Gan YJ, Que DM, Wu J, Wen ZC, Mei WL (2009) Two new cytotoxic cardenolides from the latex of *Antiaris toxicaria*. J Asian Nat Prod Res 11(9):832–837. https://doi.org/10.1080/10286020903164285

Dai HF (2010) Records of Li Folk Medicine, vol. 2 China Science and Technology Press, Beijing, p 16

Darmadi H (2018) Sumpit (Blowgun) as traditional weapons with Dayak High Protection. J Educ Teach Learn 3(1):113. https://doi.org/10.26737/jetl.v3i1.601

Dong WH, Mei WL, Zhao YX, Zeng YB, Wang H, Dai HF (2011) A new drimanese squiterpenoid glycoside from the seeds of *Antiaris toxicaria*. J Asian Nat Prod Res 13(6):561–565. https://doi.org/10.1080/10286020.2011.573479

Fu LG (1989) The rare and endangered plants in China. Shanghai Education Press, Shanghai [In Chinese.]

Fujimoto Y, Suzuki Y, Kanaiwa T et al (1983) Studies on the Indonesian *Antiaris toxicaria* sap. J Pharmacobiodyn 6(2):128–135. https://doi.org/10.1248/bpb1978.6.128

Gan Y, Mei W, Zeng Y, Han Z, Dai H (2008) Liposoluble components and their antioxidant activities from *Antiaris toxicaria* latex. Redai Yaredai Zhiwu Xuebao 16(2):144–147

Gan YJ, Mei WL, Zhao YX, Dai HF (2009) A new cytotoxic cardenolide from the latex of *Antiaris toxicaria*. Chin Chem Lett 20(4):450–452. https://doi.org/10.1016/j.cclet.2008.12.043

Gopalakrishnan S, Philip Kaupa S, Lakshmi YS, Banu F (2015) Antimicrobial activity of synthesized silver nanoparticles and phytochemical screening of the aqueous extract of *Antiaris toxicaria*. Indian J Drugs Dis 4(1):1–4

Graefe S, meyer-Samed LF, Chauvette K et al (2017) Evaluating farmers' knowledge of shade trees in different cocoa agro-ecological zones in Ghana. Hum Ecol 45:321–332. https://doi.org/10.1007/s10745-017-9899-0

Gupta VK, Sharma B (2017) Forensic applications of Indian traditional toxic plants and their constituents. Forensic Res Criminol Int J 4(1):27–32

Hano Y, Mitsui P, Nomura T, Kawai T, Yoshida Y (1991) Two new dihydrochalcone derivatives, Antiarones J and K, from the root bark of *Antiaris toxicaria*. J Nat Prod 54:1049–1055

Hu QY, Zhang XK, Wanga JN (2021) Malayoside, a cardenolide glycoside extracted from *Antiaris toxicaria* Lesch, induces apoptosis in human non-small lung cancer cells via MAPK-Nur77 signaling pathway. Biochem Pharmacol 190:114622. https://doi.org/10.1016/j.bcp.2021.114622

Huang YH, Sun Y, Huang FY, Li YN, Wang CC, Mei WL, Dai HF, Tan GH, Huang C (2017) Toxicarioside O induces protective autophagy in a sirtuin-1-dependent manner in colorectal cancer cells. Oncotarget 8(32):52783–52791. https://doi.org/10.18632/oncotarget.17189

Ilesanmi OB, Akinmoladun AC, Elusiyan CA, Ogungbe IV, Olugbade TA, Olaleyeb MT (2022) Neuroprotective flavonoids of the leaf of *Antiaris africana* Englea against cyanide toxicity. J Ethnopharmacol 282:114592. https://doi.org/10.1016/j.jep.2021.114592

Jiang MM, Dai Y, Gao H, Zhang X, Wang GH et al (2008) Cardenolides from *Antiaris toxicaria* as potent selective Nur77 modulators. Chem Pharm Bull 56(7):1005–1100. https://doi.org/10.1248/cpb.56.1005

Jiang MM, Gao H, Dai Y, Zhang X, Wang NL, Yao XS (2009) Phenylpropanoid and lignan derivatives from *Antiaris toxicaria* and their effects on proliferation and differentiation of an osteoblast-like cell line. Planta Med 75:340–345

John Lindley (1866) Image of Antiaris freely available on Google with License (CC 0)

Kankam BO, Oduro W (2009) Frugivores and fruit removal of *Antiaris toxicaria* (Moraceae) at Bia Biosphere Reserve, Ghana. J Trop Ecol 25:204–210

Kemila P, Krishnaveni C (2016) Phytochemical and antimicrobial evaluation of the leaves of *Antiaris toxicaria* Lesch. Int J Appl Adv Sci Res (IJAASR) 1(1):160

Kemila P, Krishnaveni C (2019) Pharmacognostical and antioxidant properties of the leaves of *Antiaris toxicaria* Lesch. Int J Pharm Biol Sci-IJPBSTM 9(1):369–373. https://doi.org/10.21276/ijpbs.2019.9.1.48

Ken F (1997) Plants for a future: edible and useful plants for a healthier world. pp 320

Khare CP (2007) Indian medicinal plants an illustrated dictionary. Springer Science + business media, LLC

Kopp B, Bauer WP, Bernkop-Schnurch A (1992) Analysis of some Malaysian dart poisons. J Ethnopharmacol 36(1):57–62. https://doi.org/10.1016/0378-8741(92)90061-u

Kuete V, Vouffo B, Mbaveng AT, Vouffo EY, Siagat RM, Dongo E (2009) Evaluation of *Antiaris africana* methanol extract and compounds for antioxidant and antitumor activities. Pharm Biol 47(11):1042–1049. https://doi.org/10.3109/13880200902988595

Lacroix D, Prado S, Kamoga D, Kasenene J, Namukobe J, Krief S, Dumontet V, Mouray E, Bodo B, Brunois F (2011) Antiplasmodial and cytotoxic activities of medicinal plants traditionally used in the village of Kiohima, Uganda. J Ethanopharmacology 133:850–855

Lan QY, Luo YL, Qiu YP, Jin BQ, He HY, Yin SH (2014) Storage of recalcitrant seeds of *Antiaris toxicaria* (Moraceae). Seed Sci Technol 42:97–100. https://doi.org/10.15258/sst.2014.42.1.11

Li XS, Zhu JJ, Zhao H, Li SL, Hao XJ, Yao XS, Tang JS (2015) Three new compounds from the bark of *Antiaris toxicaria*. Phytochem Lett 13:182–186

Lin Y, Yi Z, Zhao Y (2003) Chinese Dai medicine colourful illustrations. Yunnan Nationality Press, Kunming

Liu Q, Tang JS, Hu MJ, Liu J, Chen HF et al (2013) Antiproliferative cardiac glycosides from the Latex of *Antiaris toxicaria*. J Nat Prod 76(9):1771–1780. https://doi.org/10.1021/np4005147

Lizhen J, Shuzhen Z, Benpeng Y, Xue Y (2007) Tissue culture of Jianxue Sealing Throat. Commun Plant Physiol 43(3):512

Manithottam J, Francis MS (2008) Preparation of marvuri from *Antiaris toxicaria* (Pers.) Lesch. by muthuvans of Kerala. Indian J Trad Know 7:74–76

Mante PK, Adongo DW, Wood E (2017) Anticonvulsant effects of *Antiaris toxicaria* aqueous extract: investigation using animal models of temporal lobe epilepsy. BMC Res Notes 10:167. https://doi.org/10.1186/s13104-017-2488-x

Mante PK, Adongo DW, Kukuia KKE, Ameyaw EO, Woode E (2012) Neuropharmacological assessment of an aqueous bark extract of *Antiaris toxicaria* (Pers.) Lesch. (Moraceae) in rodents. Am J Pharmacol Toxicol 7(4):123–134. https://doi.org/10.3844/ajptsp.2012.123.134

Mante PK, Adongo DW, Woode E, Kukuia KKE, Ameyaw EO (2013) Anticonvulsant effect of *Antiaris toxicaria* (Pers.) Lesch. (Moraceae) aqueous extract in rodents. Int Sch Res Not 2013:519208. https://doi.org/10.1155/2013/519208

Mei WL, Gan YJ, Dai HF (2007) Advances in studies on chemical constituents of *Antiaris toxicaria* and their pharmacological activities. Chin Herb Drugs 39:151–154

Mirgal AB, Rane AD, Gunaga RP, Narkhede SS, Bhave SG (2013) A Note on stand dynamics of *Antiaris toxicaria* (Pers.) Lesch., a rare plant of konkan region of Western Ghats. Indian Forester 139(12):1161–1162

Mirgal AB, Gunaga RP, Salunkhe CB (2016) Seed traits, germination pattern and seedling vigour in *Antiaris toxicaria* (Pers.) Lesch., a rare plant species of Western Ghats, J Appl Nat Sci 8(3):1710–1713. https://doi.org/10.31018/jans.v8i3.1027

Mirgal AB, Salunkhe CB, Gunaga RP, Narkhede SS (2016) A note on albinism in *Antiaris toxicaria* (Pers.) Lesch., a rare plant of konkan region of Western Ghats. Life Sci Leafl 78:52–54

Mligo C (2019) Post fire regeneration of indigenous plant species in the Pugu forest reserve Tanzania. Global Ecol Conserv 18:e00611. https://doi.org/10.1016/j.gecco.2019.e00611

Muehlradt P, Weiss EK, Reichstein T (1964) Glycosides and Aglycons, CCLVIII, Cardenolides of *Antiaris toxicaria* seeds. 1. Separation and identification. Helvetica Chimica Acta 47(8):2164–2186

Nondo RSO, Zofou D, Moshi MJ, Erasto P, Wanji S, Ngemenya MN, Titanji VPK, Kidukuli AW and Masimba PJ (2015) Ethnobotanical survey and in vitro antiplasmodial activity of medicinal plants used to treat malaria in Kagera and Lindi regions, Tanzania. J Med Plants Res 9(6):179–192

Numpulsuksant W, Saensouk S, Saensouk P (2021) Diversity and ethnobotanical study of medicinal plants in Ban Hua Kua,Kae Dam District, Thailand. Biodiversitas 22(10):4349–4357. https://doi.org/10.13057/biodiv/d221027

Omeja P, Obua J, Cunningham AB (2005) Demand and supply of wood for drum making in central Uganda. Int For Rev 7(1):21–26. http://www.jstor.org/stable/43739488

Orwa C, Mutua A, Kindt R, Jamnadass R, Anthony S (2009) Agroforestree database: a tree reference and selection guide version 4.0. http://www.worldagroforestry.org/sites/treedbs/treedatabases.asp

Porchselvi C, Muthulakshmi S (2018) Pharmacognostic and phytochemical evaluation of *Antiaris toxicaria* (Pers). Lesch. J Med Plants Stud 6(6):217–221

Que DM, Mei WL, Wu J, Han Z, Dai HF (2009) Structure elucidation of flavonoids from *Antiaris toxicaria* roots. Chin J Org Chem 29:1371–1375

Ravikumar K, Vijay Sankar R (2009) *Antiaris toxicaria* (Moraceae)—a new distributional record to the Eastern Ghats. J Threat Taxa 1:58–59

Schmidt M, Zizka A, Traore S, Ataholo M, Chatelain C, Daget P et al (2017) Diversity, distribution and preliminary conservation status of the flora of Burkina Faso. Phytotaxa 304(1):001–215

Shi LS, Liao YR, Su MJ, Lee AS, Kuo PC et al (2010) Cardiac glycosides from *Antiaris toxicaria* with potent cardiotonic activity. J Nat Prod 73(7):1214–1222. https://doi.org/10.1021/np9005212

Shi LS, Kuo SC, Sun HD, Morris-Natschke SL, Lee KH, Wu TS (2014) Cytotoxic cardiac glycosides and coumarins from *Antiaris toxicaria*. Bioorg Med Chem 22(6):1889–1898. https://doi.org/10.1016/j.bmc.2014.01.052

Singh NP, P Lakshminarsimhan, S Karthikeyan, PV Prasann (2001) Flora of Maharashtra—Dicotyledons, vol 2, Botanical Survey of India, Culcutta, p 928

Subiono T, Sadarudin, Tavip MA (2022) Qualitative and quantitative phytochemicals of leaves, bark and roots of *Antiaris toxicaria* Lesch., a promising natural medicinal plant and source of pesticides. Plant Sci Today. https://doi.org/10.14719/pst.1896

Thiam K, Zhao M, Marchioni E, Muller CD, Diop YM, Julien-David D, Emhemmed F (2022) Cytotoxic and pro-apoptotic effects of leaves extract of *Antiaris africana* Engler (*Moraceae*). Molecules 27:7723. https://doi.org/10.3390/molecules27227723

Ugwoke CEC, Anze SPG, Nweze AE, Obisike CV (2017) Pharmacognostic evaluation of the leaf of *Antiaris toxicaria* (family:Moraceae). Int J Pharm Sci Res 8(6):2696–2700

Umdale SD, Mirgal AB, Shinde BN, Sawant RS, Salunkhe CB, Gaikwad NB (2020) Evaluation of genetic diversity in *Antiaris toxicaria* Lesch from sacred groves of the Western Ghats, India. Natl Acad Sci Lett 43:383–388. https://doi.org/10.1007/s40009-020-00878-3

Upadhyay GK, Srivastavaa SK, Ansari AA (2010) A new locality of *Antiaris toxicaria* subsp. *Macrophylla* (*Moraceae*) in the Andaman Islands, India, with a note on its conservation. J Jpn Bot 85:350–357

Vouffo B, Dongo E, Facey P, Thorn A, Sheldrick G, Maier A, Fiebig H, Laatsch H (2010) Antiarol Cinnamate and Africanoside, Cinnamoyl Triterpene and a Hydroperoxy-Cardenolide from the Stem Bark of *Antiaris africana*. Planta Med 76:1717–1723

Wu XL, Wu YL, Li HG et al (2014) A new periplogenin cardenolide from the seeds of *Antiaris toxicaria*. J Asian Nat Prod Res 16(4):418–421. https://doi.org/10.1080/10286020.2014.885506

Xin X, Jing XM, Liu Y, Song SQ (2010) Viability loss pattern under rapid dehydration of *Antiaris toxicaria* axes and its relation to oxidative damage. J Integr Plant Biol 52(5):434–441

Yeboah R, Thomford KP, Mensah R, Appiah AA, Thomford AK, Ocloo A (2019) A pilot study on the efficacy of an *Antiaris toxicaria* subsp. *africana* (Engl.) C.C. Berg based Ghanaian herbal product in the management of peripheral neuropathy. J Pharm Pharmacogn Res 7(1):59–66

Zhang L, Yang X, Mao X, Wang Z (2018)The complete chloroplast genome of *Antiaris toxicaria*, a medicinal and extremely toxic species. Mitochondrial DNA Part B 3(1):1100–1101

Zhao HG, Zhou SL, Lin YY et al (2018) Toxicarioside N induces apoptosis in human gastric cancer SGC-7901 cell by activating the p38MAPK pathway. Arch Pharm Res 41:71–78. https://doi.org/10.1007/s12272-017-0956-4

Zheng WP, Huang FY, Dai SZ, Wang JY, Lin YY, Sun Y, Tan GH, Huang YH (2021) Toxicarioside O inhibits cell proliferation and Epithelial-Mesenchymal transition by downregulation of Trop2 in lung cancer cells. Front Oncol 10:609275. https://doi.org/10.3389/fonc.2020.609275

Zhu H (2006) Forest vegetation of Xishuangbanna, south China. Forestry Studies in China 18:1–58. https://doi.org/10.1007/s11632-006-0014-7

Zuo WJ, Donga WH, Chen J, Zhao YX, Chen HQ, Mei WL, Dai HF (2013) Two new strophanthidol cardenolides from the seeds of *Antiaris toxicaria*. Phytochem Lett 6(1):1–4. https://doi.org/10.1016/j.phytol.2012.10.001

Index

J. M. Al-Khayri et al. (eds.), *Biodiversity and Genetic Improvement of Medicinal and Aromatic Plants II*, Advances in Plant Breeding Strategies 10,
https://doi.org/10.1007/978-3-031-81857-8

The manufacturer's authorised representative in the EU is Springer Nature Customer Service Centre GmbH, Europaplatz 3, 69115 Heidelberg, Germany. If you have any concerns regarding our products, please contact ProductSafety@springernature.com

Printed and bound by CPI Group (UK) Ltd, Croydon, CR0 4YY
30/09/2026
02225568-0007